Interpreting space:
GIS and archaeology

Applications of Geographic Information Systems

Series Editors:
Duane F. Marble, *The Ohio State University,* **and**
Donna J. Peuquet, *The Pennsylvania State University*

Interpreting space:
GIS and archaeology

Edited by

Kathleen M.S. Allen
Tulane University

Stanton W. Green
*The University of
South Carolina*

Ezra B.W. Zubrow
*State University of
New York at Buffalo*

Taylor & Francis
London • New York • Philadelphia
1990

USA	Taylor & Francis Inc., 1900 Frost Road, Suite 101, Bristol, PA 19007
UK	Taylor & Francis Ltd., 4 John Street, London WC1N 2ET

British Library Cataloguing in Publication Data

Interpreting space: GIS and archaeology.
1. Geography. Information systems. Applications of computer systems
I. Allen, Kathleen M. II. Green, Stanton W. III. Zubrow, Ezra B.W.
910.285

ISBN 0-85066-824-7

Library of Congress Cataloging-in-Publication Data

Interpreting space: GIS and archaeology / edited by Kathleen M. Allen, Stanton W. Green, Ezra B.W. Zubrow.
    p.   cm.
ISBN 0-85066-824-7
1. Archaeology—Research 2. Geography—Data processing
I. Allen, Kathleen M., 1949– II. Green, Stanton W. III. Zubrow, Ezra B.W.
CC83.I58 1990
930.1′028′5—dc20

Cover illustration: *Tender Terrain Was* by Walter Prochownik
Typeset by Electronic Village Ltd., Richmond, Surrey

Printed in Great Britain by Burgess Science Press, Basingstoke

Contents

Dedication

To Walter and Helen, Nancy, Beth and Wally
To Claudia, Harrison and Devin
To Nikita, Alanya, Alexis and Marcia

Preface

This is a book about some of the more exciting changes occurring in the ways archaeologists are interpreting space. Our longstanding and continuing interest in the application of spatial methods to archaeology has been rejuvenated by the revolution in geographic analysis made possible by geographic information systems (GIS). There has been a veritable explosion in symposia, presented papers, archaeological reports and publications on the use of GIS in archaeology. As is typical of times of rapid change, there has been little time and ability to discuss the things we are doing and, perhaps most significantly, to evaluate their methodological and theoretical implications. GIS applications are literally changing the way archaeologists are thinking about space—and this has profound ramifications in all sectors of the discipline.

This volume is a basic sourcebook on GIS for archaeologists, and social scientists in related disciplines (History, Historical Geography). It defines and critically examines GIS through general discussion and case studies. The attempt is to provide archaeologists with the information to be able to evaluate, purchase and apply GIS to their work. It includes work by prehistoric and historic archaeologists working in academic and applied contexts, as well as geographers who are contributing both generally to the application of GIS to the social sciences (Marble) and toward particular aspects of archaeological research (Stine, Decker, Lanter). Critical to the effective use of GIS by archaeologists is a solid understanding of the theory and method that underlies it. GIS is growing and changing remarkably fast—some estimates report an annual growth rate of 50 per cent in GIS use—projecting total hardware and software purchases in 1993 to 3.5 billion dollars (*Photogrammetric Engineering & Remote Sensing*, November 1989). It is essential that archaeologists do not let the GIS tail wag the anthropological dog—and for this a basic understanding of spatial process and the place of GIS in archaeology is a must for both research and practical purposes.

Plans for this book began in January 1989 at the Archaeological Congress in Baltimore, Maryland. Writing and revising has been pushed beyond the speed of anthropological light so as to make this book as timely as possible. To safeguard the possibility of producing an obsolete volume in a quickly changing GIS-world we have been updating chapters and developing our introductions through January 1990. The authors in the volume are to be thanked and congratulated for meeting truly remarkable deadlines and turnaround times—especially by academic standards. Consistent with the electronic nature of GIS technology, we have relied on fax and E-mail communications with the support of telephones and overnight express.

The compilation of this book began by soliciting updated versions of papers from recent archaeological conferences on GIS. These include the 1988 Society for American Archaeology Meetings, (Kvamme, Chapter 10; Farley *et al.*, Chapter 13; Marozas and Zack, Chapter 14; Warren, Chapter 18; Carmichael, Chapter 19; Altschul, Chapter 20; Zubrow, Chapter 24; Allen, Chapter 25), and the Archaeological Congress (Crumley and Marquardt, Chapter 7; Stine and Lanter, Chapter 8; Savage, Chapter 26; Green, Chapter 27; Madry and Crumley, Chapter 28). We have complemented these papers with chapters explicitly written for this book. These include a paper on the impact of GIS on the social sciences by Duane Marble (Chapter 2), a distinguished geographer in the field, a review of GIS applications in archaeology (Savage, Chapter 3), a survey of GIS in the United Kingdom (Harris and Lock, Chapter 4), a paper on the legal and practical issues of data and GIS access and freedom of information (Stine and Stine, Chapter 5), chapters on GIS hardware (Madry, Chapter 15) and software (Zubrow, Chapter 16) and a variety of applications (Williams *et al.*, Chapter 21; Jackson, Chapter 22; and Hasenstab and Resnick, Chapter 23).

Substantial commentaries introduce each section of the volume. These chapters guide the reader by critically discussing the pertinent issues. As such, they provide synthetic statements on GIS and its place in archaeology (Part I), GIS theory as it applies to archaeology and spatial processes in general (Part II), data sources, hardware and software, their integration and use (Part III), the ways in which GIS can be applied in archaeology (Part IV), and a summary critique of GIS in archaeology (Part V).

The book has been a truly collaborative effort on the part of the editors. Allen and Zubrow participated in the 1988 Society for American Archaeology symposium, while Green organized the 1989 Archaeological Congress GIS symposium that included the other editors. All introductions were outlined jointly by the editors and mutually critiqued. Each of the editors has read all the papers. The structure and substance of the book is the result of our editorial consensus allowing us to share equally the credit for the book's worth and the responsibility for any of its errors and misstatements.

Although computer technology is both an essential and creative part of GIS, we believe that much of the long term significance of the volume will rest on its treatment of the theoretical and methodological issues involved in interpreting spatial processes. If we understand the whys and wherefores of GIS, then the ever changing hardware, software and algorithms become tools toward the querying of important archaeological questions. The final value of this volume, therefore, rests on its success in clearly illustrating and explaining the use of GIS in archaeology and the way that geographic information systems can help us formulate better questions about spatial processes so that we can enhance the quality of our research.

Technology is not the solution to all problems. Ultimately, it is people who affect the environment and who have insights into their relationship with the landscape. The editors are very pleased that Walter Prochownik, an artist famous for his abstract landscapes, agreed to provide his conception of interpreting space for the cover. The spatial illusions of *Tender Terrain Was* reflect the relativity of landscape interpretation.

Kathleen Allen, Stanton Green and Ezra Zubrow
Barcelona, Spain (we wish)
January 1990

Acknowledgements

This volume is being mailed to the publisher one year to the day after its conception; a feat that would not be possible without the logistical and personal support of many people and the facilities of our respective institutions. We begin by thanking Robin Mellors of Taylor & Francis for his vision to include archaeological research in their *Applications of GIS* series. The project greatly benefited from the support, comments and ideas of Duane Marble and Donna Peuquet, the editors of this series. We thank the Department of Anthropology of Tulane University for providing assistance in the form of computers, personnel and postage; Charlene Patterson for the know-how to make sure that manuscripts were punctually and properly posted, and Jennifer Briggs for her superb editorial and word-processing assistance.

The last stages of this project were completed during two intensive, four day, sessions at the University of South Carolina (USC) and SUNY Buffalo. We express gratitude to these institutions for providing the necessary phone, fax, BITNET and photocopying support. From USC, we especially thank Dot O'Dell and Deannie Stevens (Department of Anthropology) for their southern hospitality (not only did they meet our endless clerical requests, but they did so with a smile); Professor Lester Lefton (Psychology) for the use of his fax; Homer Steadly (Humanities and Social Sciences Computer Lab) for once again providing an on-the-spot computer rescue by converting SPELLBINDER to WORDPERFECT; and Roy Stine (Department of Geography) for being on call for our GIS queries.

From SUNY Buffalo, we express our gratitude to: Mary Cannon and Jean Grela (Department of Anthropology) for answering all our secretarial needs; Jay Leavitt for assisting us on the journey from WORDSTAR to WORDPERFECT; Mark Winer for providing us with a PS2/3O with a 3.5 inch drive; Jim Stamos (Biology) for making his wonderful illustration talents available at such short notice; and Eleazer Hunt, Kim Bartolotta, Paul Burger and Eric Hansen for help with last minute editing and proofreading. The National Centers for Earthquake Research and for Geographic Information and Analysis provided technical assistance. A special handshake to Carol Huber for not wincing when we asked her to reformat all 29 chapters in three days, and then doing such a great job. Thanks to Sarah Jerauld for advice on the cover format. The project was made very special by Walter Prochownik: we only hope that this book can live up to its cover.

We also wish to thank Diana Christensen and David Carmichael for organizing the GIS symposium at the 1988 Society for American Archaeology meetings, and H. Dennison Parker (*GIS World*) for allowing us to reprint the results of his GIS software survey (Table 16.1).

Ultimate thanks to Claudia Green and Marcia Zubrow for their unquestioning support during the last hectic days of this project.

Contributors

Kathleen M.S. Allen
Department of Anthropology, Tulane University, New Orleans, LA 70118, USA,
AN08AAF @ TCSVM

Jeffrey H. Altschul
Statistical Research, P.O. Box 31865, Tucson, AZ 85751, USA

Frederick L. Briuer
Waterways Experiment Station, US Army Corps of Engineers

David L. Carmichael
Tetra Tech, Inc., 348 W. Hospitality Lane, Suite 300, San Bernardino, CA 924087,
USA

Carole L. Crumley
Department of Anthropology, University of North Carolina, Chapel Hill, NC 27514,
USA

T. Drew Decker
Department of Geography, University of South Carolina, SC 29208, USA

James A. Farley
Coordinator for Computer Resources, Arkansas Archeological Survey, Fayetteville,
AR 72702, USA

Stanton W. Green
Department of Anthropology, University of South Carolina, Columbia, SC 29208,
USA, N050032 @ UNIVSCVM

Trevor M. Harris
Department of Geology and Geography, West Virginia University, Morgantown,
WV 26506, USA

Robert J. Hasenstab
Bagdon Environmental Associates, Inc., 3 Normanskill Blvd., Delmar, NY 12054,
USA

Jack M. Jackson
P.O. Box 398, Hutto, TX 78634, USA

Kenneth L. Kvamme
Laboratory of GIS and Spatial Analysis, Arizona State Museum, University of
Arizona, Tucson, AZ 85721, USA

David P. Lanter
Department of Geography, University of California, Santa Barbara, CA, USA

W. Fredrick Limp
Arkansas Archeological Survey, P.O. Box 1249, Fayetteville, AR 72702, USA

Gary R. Lock
Institute of Archaeology, University of Oxford, 36 Beaumont Street, Oxford OX1 2PG, United Kingdom

Jami Lockhart
Arkansas Archeological Survey, P.O. Box 1249, Fayetteville, AR 72702, USA

Scott L.H. Madry
Associate Director, Cook College Remote Sensing Center, Rutgers University, New Brunswick, NJ 08903, USA

Duane F. Marble
Department of Geography, The Ohio State University, Columbus, OH 43210, USA

William H. Marquardt
Florida Museum of Natural History, University of Florida, Gainesville, FL 32601, USA

Bryan A. Marozas
c/o Ebert and Associates, 5800 McLeod NE, Suite C, Albuquerque, NM 87109, USA

Benjamin Resnick
GAI Consultants, Inc., 570 Beatty Road, Monroeville, PA 15146, USA

Stephen H. Savage
Department of Anthropology, University of South Carolina, Columbia, SC 29208, USA

Linda F. Stine
Department of Anthropology, University of South Carolina, Columbia, SC 29208, USA

Roy S. Stine
Department of Geography, University of South Carolina, Columbia, SC 29208, USA

Robert E. Warren
Illinois State Museum, Corner Spring & Edwards, Springfield, IL 62706, USA

Ishmael Williams
Arkansas Archeological Survey, P.O. Box 1249, Fayetteville, AR 72702, USA

James A. Zack
Environmental Systems Research Institute, Inc., 380 New York Street, Redlands, CA 92373, USA

Ezra B.W. Zubrow
Department of Anthropology, National Center for Geographic Information and Analysis, State University of New York at Buffalo, Buffalo, NY 14261, USA, APYEZRA @ UBVMS

PART I

Introduction

1

Approaching archaeological space: an introduction to the volume

Stanton W. Green

Introduction

This volume examines a most fundamental aspect of archaeological thinking and practice: the interpretation of behaviour and material culture over space. The spatial dimension is central to archaeology because it involves all levels of archaeological research—theory, method and practice. The *collection, analysis, interpretation and presentation* of archaeological data must actively and creatively take into account the spatial dimension. To a certain extent, archaeology can be viewed as a discipline involved in sampling space in order to understand human behaviour.

Despite the centrality of space to archaeological theory and practice there have been numerous frustrations and limitations with how archaeologists have been able to collect data about spatial behaviour and the ways in which they can analyze, interpret and present their conclusions. These obstacles derive from theoretical problems in partitioning behavioural and material patterns into arbitrary spatial units, and methodological issues relating to the systematic and concurrent consideration of space, time and form.

Geographic information systems are essentially spatially referenced databases that allow one to control for the distribution of form over space and through time. They are more than computerized cartography because they provide for the storage, mathematical manipulation, quick retrieval and flexible display of *spatially referenced data*. The many uses of this kind of database are both critically examined and demonstrated throughout this volume. Before introducing this discussion, let us first provide a bit of background on the need for such a methodology.

Questions and frustrations

We can introduce these theoretical and methodological limitations by listing some of the more typical spatial questions asked by archaeologists.

(1) How does one define activity areas?
(2) How, in fact, does one define a site?
(3) How does one compare and define types of behaviour over space?
(4) How does one spatially correlate social activities with subsistence activities?
(5) How does one spatially correlate the perceived and used cultural environment with the natural and physical environment?

(6) How does one compare artifact distributions over space?
(7) How can one determine the relative effects of different aspects of the natural environment on aspects of cultural environment?

These questions all emphasize spatial process, and draw upon archaeological theory and method. In some cases, such as that of site definition, there are questions as to the definition or even validity of spatial units of behaviour. The archaeological site itself, perhaps the seminal analytical unit of archaeology, is currently under challenge as a useful concept because of differing opinions as to its spatial validity (Foley, 1981; Dunnell and Dancey, 1983; Ebert, 1986). Can one partition human activities into spatial packages? The same holds true for the definition of activity areas. Here we see debate at the theoretical level and perhaps most especially at the methodological level where archaeologists strive to develop methods for statistically deriving artifact clusters (Whallon, 1973, 1974; Paynter *et al.*, 1974).

Other problems have derived from the need to compare different classes of artifacts, or categories of behaviour as they vary over space. Although manual comparison of lithic and ceramic distributions, for example, can provide a qualitative basis for interpretation, more systematically rigorous and quantitative analyses are frustrated either by a lack of method or the use of methods that are difficult to interpret (for example spatial autocorrelation).

Many of the problems noted above derive from the fact that classic statistics are not applicable to spatial data. For example, statistical comparisons can be made between areas of sites by comparing grids with regard to various artifact classes.This can be applied statistically using spatial autocorrelation where the fit of one or more variables to a repetitive pattern is evaluated. The problem with this approach is the lack of analytical or visual description. An autocorrelation value does not describe broadly a distributional pattern, nor does it visually present it.

Another way to look at spatial patterning is through interpolation. Here one can, in fact, describe the pattern. Mapping offers visual descriptions, while spatial regression such as trend surface analysis can be used to examine analytically and describe spatial trends. Here, however, the problem runs into three dimensional comparison. How can one compare the second order trend surface maps of one variable with another? Although there have been methods developed to do this, they run into the same problem as autocorrelation—by reducing space to a statistic it loses its descriptive force.

In the end, traditional approaches to the interpretation of spatial process and pattern all face the same problems:

(1) That statistics are difficult to use for describing and analyzing continuous data;
(2) That spatial data often have no boundaries so that classic set theory does not apply;
(3) That there are no inherent internal partitions to enable one to set up meaningful nested spatial units;
(4) That traditional statistics are not equipped to deal with the simultaneous description and correlation of multiple forms over space.

Laying these mathematical abstractions aside, the field archaeologist realizes all of these problems as s/he decides on the appropriate sizes for grid, sampling and recording units. We believe that GIS offers both methodological solutions as well as theoretical feedback for these problems.

The methodological strength of GIS

Geographic information systems provide a specific methodology for dealing with the dimensional and multi-variable problems outlined above. Although GIS have not conquered all of the problems, as we shall discuss, they provide a means for spatially referencing observations in ways that allow one to describe, analyze, compare and even mathematically manipulate multiple spatial distributions. Moreover, as computerized databases, they allow tremendous flexibility and speed in data entry, revision, analysis and, very significantly, data display.

GIS consist of two parts; a standard relational database which allows cross tabular searching, and a graphic or mapping database which is attached through one of numerous programming hooks. This allows one to look at space, time and form simultaneously, thus conquering many (but not all) of the methodological problems discussed above.

Spatial theory and landscape archaeology

Although GIS can be used to describe and analyze multiple variables, this does not itself mitigate the theoretical problems of defining appropriate spatial units of analysis as the basis for the interpretation of human behaviour over space. Here we run into problems of archaeological theory. Can one partition human activities into activity areas and sites?

Landscape theory provides the conceptual power necessary to take full advantage of GIS methodology. Some of these papers in this volume make explicit use of this combination of landscape archaeology and GIS (Crumley and Marquardt, Chapter 7; Savage, Chapter 26; Green, Chapter 27); while most others derive their importance from the implicit use of landscape archaeology for interpreting space.

In either case, we believe that the combination of landscape archaeology and GIS is one of the most profound and stimulating combinations in archaeological theory and method in the 20th century.

GIS and archaeology

The above arguments hold for all aspects of the increasingly diverse field of archaeology. Despite their common concern with people and material culture, archaeologists do many different things and indeed have many different research and practical concerns and objectives. Although this book emphasizes anthropological archaeology in academic and applied contexts, we can see applications in scientific archaeology, classical archaeology and archaeology and art history.

Anthropology's concern with behaviour and material culture is naturally tied to the spatial analysis of things and people over natural and cultural landscapes. The chapters in this book succeed as they do because they can consider natural and cultural variables in integrative ways. Archaeological observations can be cross-referenced spatially with features in the natural environment. Essentially, cultural landscapes can be created through the connection of natural and cultural factors.

Cultural resource management is increasingly taking advantage of GIS in developing policy and planning. Here research is done within an applied framework so that the development and maintenance of archaeological databases provide powerfully effective managing tools. Very significantly, many government agencies are using GIS to integrate archaeology into their general planning and management structure (Stine and Stine, Chapter 5; Zubrow, Chapter 16). The National Park Service programme, for example, has implemented GIS for developing scenic views and campsites. Multiple variables (such as slope, view, distance to roads and streams) are used to locate scenic views. Conversely, a particular location within a park can be examined in terms of its scenic value (what can be seen from it, and from where it can be seen). What is of importance in this type of approach is that it takes into account the visual structure of the landscape (Higuchi, 1988).

Although the intent of the National Park Service is not directed toward archaeological research, the geographical database developed can be of obvious benefit to the archaeologist. There is a clear correspondence between modern and prehistoric campsites and data collected for management purposes can also be used for research objectives. Madry and Crumley (Chapter 28) provide an elegant archaeological example. Using GIS, they view the historic Burgundy landscape from the perspective of hillforts. From this, they find that roads and paths are visible from the hillforts—a logical expectation but one that requires a GIS approach to be tested systematically.

As we continue to 'humanize' archaeology with actor-directed and decision-making approaches, we are in need of means to ask questions about how prehistoric and historic people 'saw' their world. Although anthropologists have long argued that people create their cultural environments, we have been hard pressed as archaeologists to find ways to study formally perceived past (or for that matter present day) environments. GIS, we would argue, provides such a means. And of course, the flexibility of GIS databases and analyses allows one to combine and manipulate wide varieties of variables as they might affect view and location. We are not limited to using the same locational arguments for 20th century campers and 5th millenium plant collectors—although the resemblances may in fact be greater than we often think.

As important in this case are the applied or management benefits of GIS. Archaeological concerns such as site sensitivity can be easily taken into account as an integrated part of the database and planning process using a GIS. Regional site inventories (all the rage in the 1970s and 1980s) are possible in flexible user friendly formats.

But we want the reader to know that this book provides the fundamentals for all areas in archaeology. Most obvious, perhaps, is its applicability to those areas where archaeology and the physical sciences join. GIS is already revolutionizing such areas as geomorphological and stratigraphic studies (Raper, 1989). Here the integration of multiple sources of geological data is used to develop three dimensional analysis and description. Landforms can be perceived from a variety of angles and at varying scales. The mathematical algorithms of GIS allow one, for example, to use topographic data to examine culturally relevant landform factors. Savage (Chapter 26), for example, uses GIS calculus to model the terrain roughness of a portion of the Savannah River Valley in order to model hunter-gatherer mobility patterns. In a similar manner Madry and Crumley (Chapter 28) recreate the landscape in Burgundy in order to examine the historic road system. This type of 'map-a-matics' (Cowen, 1988) exemplifies the power of GIS for mathematically manipulating accessible and typical data (in this case elevation) to produce both the calculation and visual description of variables that would be difficult to obtain without the use of a spatially referenced database. These derived variables can, in turn, be added to the original database. For example, a site inventory can then include calculated slope and change in slope for site locations.

Still, GIS can be applied beyond these more anthropological types of archaeological research. Although we have no examples in our book, we would argue that GIS could be applied in classical archaeology as it is excellent for mapping large areas and sites. The use of GIS in urban geography, for example with census mapping, could serve as an analogue for the work of classical archaeologists. Even in archaeology and art history, GIS would be useful for keeping track of the distribution of art objects over space and time and even as a mechanism for maintaining records, relating museum exhibits and the original location of their artifacts.

Contents of the book

As a sourcebook the book discusses what GIS are, how they work, how they can be applied, what kinds of GIS exist, the strengths and weaknesses of these systems, and how they can be acquired. The book contains chapters that:

(1) Describe GIS in theoretical and operational terms by persons who have been developing GIS theory and method (Crumley and Marquardt, Chapter 7; Stine and Lanter, Chapter 8; Warren, Chapter 9; Kvamme, Chapter 10);

(2) Discuss the impact of GIS on the social sciences and in particular archaeology (Marble, Chapter 2; Savage, Chapter 3);

(3) Discuss the variety of types of GIS that exist (Zubrow, Chapter 16);

(4) Discuss GIS data sources (remote sensing data, digital line graphs, etc.) and their integration (Stine and Decker, Chapter 12; Farley *et al.*, Chapter 13; Marozas and Zack, Chapter 14);

(5) Describe GIS hardware (Madry, Chapter 15) and software (Zubrow, Chapter 16) including cost and vendor;

(6) Discuss legal and practical issues of data access and use (Stine and Stine, Chapter 5)

(7) Apply GIS to prehistoric and historic archaeological projects in research and management settings (Warren, Chapter 18; Carmichael, Chapter 19; Altschul, Chapter 20; Williams *et al.*, Chapter 21; Jackson, Chapter 22; Hasenstab and Resnick, Chapter 23; Zubrow, Chapter 24; Allen, Chapter 25; Savage, Chapter 26; Green, Chapter 27; Madry and Crumley, Chapter 28);

(8) Finally, the book contains a list of GIS acronyms and a list of federal agencies who use and maintain GIS, the level of their use, and agency contacts.

We have encouraged authors to provide high quality figures to illustrate the visual capabilities of GIS. Visual display, always an essential aspect of map-making, takes on additional methodological and theoretical dimensions in GIS. We view GIS output as much more than descriptive display of data. Rather, we see it as a means for manipulating and analyzing data, as well as creatively feeding back to the ways in which we are thinking about a particular landscape or landscapes in general. It is essential to consider GIS output as the result of landscape modelling or recreation not as rote reconstruction. For this reason we emphasize GIS visual output as a substantial methodological aspect. Toward this end, we include several colour plates as illustrations of the types of output possible from GIS. The variety of outputs from GIS is great, and becoming greater as we write, because of the increasing sophistication of GIS software and printer and display screen hardware. We have attempted here to provide examples of the more typical GIS output displays available today to most practitioners. In the case of different authors using the same or similar systems, we include only a few illustrations of the colour output of this particular GIS, and have asked the authors to utilize half-tone or black and white line drawing representations of this output for their other illustrations.

Interpreting Space is offered as a theoretical, methodological and practical introduction to the use of geographic information systems in archaeology. Despite the rapid changes occurring in this field, we believe that the case studies and general essays offered will hold up as contributions to archaeological method and theory as they change our ways of thinking about the spatial aspect of past cultural systems.

References

Cowen, D. J., 1988, GIS vs. CAD vs. DBMS: what are the differences? *Photogrammetric Engineering and Remote Sensing*, **54**, 1551–1558

Dunnell, R. and Dancey, W., 1983, The siteless survey: a regional scale data collection strategy. In *Advances in Archaeological Theory and Method*, edited by M. Schiffer (New York: Academic Press) Volume 5, pp. 267–287

Ebert, J., 1986, *Distributional Archaeology: nonsite recording and analytical methods for application to the surface archaeological record*, Ph.D dissertation, Department of Anthropology, University of New Mexico

Foley, R., 1981, Off-site archaeology: an alternative approach for the short-sited. In *Patterns of the Past: Essays in Honour of David L. Clarke*, edited by I. Hodder, G. Isaac and N. Hammond, (Cambridge: Cambridge University Press), pp 157–183.

Higuchi, T., 1988, *The Visual and Spatial Structure of Landscapes* (translated by Charles Terry), (Cambridge, Massachusetts: MIT Press)

Paynter, R. W., Green, S. W., and Wobst, H. M., 1974, Spatial clustering: techniques of discrimination. Paper presented at the *Annual Meeting of the Society for American Archaeology, Washington, D.C.*

Raper, J., (Ed.), 1989, *Three Dimensional Applications in GIS*, (London: Taylor and Francis)

Whallon, R., 1973, Spatial analysis of occupation floors I: The application of dimensional analysis of variance. *American Antiquity* **38**, 320–328.

Whallon, R., 1974, Spatial analysis of occupation floors II: The application of nearest neighbour analysis. *American Antiquity* **39**, 16–34.

2

The potential methodological impact of geographic information systems on the social sciences

Duane F. Marble

The day-to-day necessity of dealing with space and spatial relationships represents one of the basic facets of any human society, and our accumulated knowledge about the spatial form of the world in which we live has been traditionally stored in the form of maps. My concern in this essay is with the potential impact of a critical new tool for the storage, analysis, and visualization of spatial data: the *geographic information system* or *GIS*. Most tool innovations in science are of a minor nature and, although the cumulative impact of these changes may be impressive, it is only rarely that a change occurs which revolutionizes a field to the point where many of the things that we do must be looked at from a completely different viewpoint. It is my contention that GIS is beginning to do this for geography, and that it will also do so for those portions of the social sciences which are concerned with, or at least should be concerned with, the spatial aspects of human society.

The present volume contains a number of essays examining the way in which these geographic information systems are beginning to impact research activities in one of the social sciences—archaeology. In the following discussion I hope to address not only the ways in which this new tool is impacting our handling of spatial data, but also what I strongly feel is its significant potential for the creation of changes in the way in which we approach spatially structured questions, not only in geography and archaeology, but in many of the other social sciences as well.

Traditional approaches to the storage and analysis of spatial data

Maps, in one form or another, have been with us for a very long time and their importance to society has often been noted. For example, the well known travel writer Paul Theroux in his essay *'Mapping the World'* (reprinted in Theroux, 1985) quotes Sir Alexander Hosie as saying: 'It would seem as though cartography were an instinct implanted in every nation with any claim to civilization.' Also, in the opening volume of their massive history of cartography, Harley and Woodward (1987) state that:

'Mapping—like painting—precedes both written language and systems involving number, and...there have been relatively few mapless societies in the world at large.'

Human societies are spatially structured and our pervasive and early concern with mechanisms for the storage of spatial data certainly derives from our desire to communicate with others about the nature of objects which we know exist at other locations. However early in human history it may have appeared, the very concept of the map as a spatial data store clearly represents a major intellectual development. Arthur Robinson, a noted cartographer, sums this up quite well:

'The use of a reduced, substitute space for that of reality, even when both can be seen, is an impressive act in itself; but the really awesome event was the similar representation of distant, out of sight, features. The combination of the reduction of reality and the construction of an analogical space is an attainment in abstract thinking of a very high order indeed, for it enables one to discover structures that would remain unknown if not mapped.'
(Robinson, 1982, p.1)

The representation of the world upon a sheet of paper involves the creator of the map with the complex and interlinked cartographic concepts of scale, spatial generalization and symbolization; critical topics in effective cartographic communication and ones which continue to be the focus of much modern cartographic research (Brassel, 1978; Marble, 1987).

In contemporary terms, the map constitutes an analogue database which is capable of storing data not only about the location of entities but also about the highly complex spatial relationships which exist between them. Strangely enough, the formal study of these very important spatial relationships has received attention from only a few geographic researchers (e.g., Peuquet, 1987) and what can be classed, at best, as only passing attention from researchers in the other social science disciplines (e.g., Piaget and Inhelder, 1956). As we will see, the lack of formal theory in this area has constituted a major roadblock to the subsequent development of effective spatial data handling systems.

Retrieval of information from the map database has been restricted, with minor exceptions based upon the use of simple manual tools such as the ruler, compass and planimeter, to what the human eye–brain combination is capable of extracting. This is especially true for the identification and retrieval of the complex spatial relationships contained in the map. Thus the traditional dilemma of the cartographic designer is clearly defined: how to insert the maximum amount of information into a given map without either making it unintelligible to the reader or increasing its physical size to an unmanageable level. This problem has been widely discussed. It has even reached the attention of Lewis Carroll. In *Sylvie and Bruno Concluded* there is the following conversation:

"'What a useful thing a pocket map is!' I remarked.
'That's another thing we've learned from your Nation,' said Mein Herr, 'map making! But we've carried it much further than you. What do you consider the largest map that would be really useful?'
'About six inches to the mile.'
'Only six inches!' exclaimed Mein Herr. 'We very soon got to six yards to the mile. Then we tried a hundred yards to the mile. And then came the

grandest idea of all! We actually made a map of the country, on the scale of a mile to a mile.'

'Have you used it much?' I enquired.

'It has never been spread out, yet,' said Mein Herr: 'the farmers objected: they said it would cover the whole country, and shut out the sunlight! So we now use the country itself, as its own map, and I assure you it does nearly as well.'"
(Carroll, 1894, pp. 616–617)

Dr. Michael Dobson, a well-known cartographer, once remarked that the 'it does nearly as well' at the end clearly established Mein Herr's identity as a cartographer.

A tremendous amount of cartographic investigation, based upon both research and pragmatic trial and error solutions, has been directed toward the development of viable solutions to this problem of map design. The results have been successful, with some notable exceptions, despite the fact that it often takes years of manual effort to create the final product (e.g., the compilation and production of a standard topographic map sheet). However, the final map—despite the best efforts of the cartographer—must always represent an uneasy compromise between the conflicting goals of maximization of information content and ease of visual extraction of that information.

Although maps appear at first sight as relatively simple iconic devices (so simple in fact that most societies do not even provide formal instruction in their use for many children), a major difficulty has been that the resultant documents have not lent themselves to easy use within the framework of more than the simplest attempts at data extraction. Complex queries, especially those which contain a quantitative component (such as multiple measurements of distance, direction or area), simply cannot be processed easily and quickly by the user even with the assistance of the simple tools available. This situation is greatly complicated by the fact that spatial data occur, or are capable of occurring, in vast quantities such as the more than 50,000 1:24,000 map sheets which form the basic coverage for the United States.

This may become a bit clearer if we consider a specific case, for example, the problem of a map user (for instance a recreation planner) who desires to obtain data pertaining to all the lakes in the State of Minnesota which are larger than two acres in extent, located within one-half mile of a paved highway, and which are at least partially in public ownership. There is little question that the requisite spatial data exist (somewhere!—but that is another problem), but even if it were freely available in map form finding the desired answer may require days or even weeks of effort on the part of the user. Why so long? Think of the steps involved: (1) finding the area of all lakes; (2) identifying those lakes of a proper size that fall, at least in part, within a one-mile buffer zone centred on the paved roads; (3) checking the pattern of land ownership for each final candidate; and, of course, (4) preparing the resultant report which must, obviously, contain a map!

I believe that this example makes it clear that after centuries of effort we have developed methods to store very substantial amounts of spatial data in traditional map formats (although, again somewhat strangely, there is as yet no clearly defined, quantitative measure of the data content of a map). Unfortunately it is also clear that the retrieval of subsets of these data is most often time consuming and expensive. It has been my experience that, as a result, a substantial amount of the spatial data which are stored in map form is heavily under-utilized and that many spatially-oriented activities take on sub-optimal forms.

I would also put forward the point of view, as will be developed later in this essay, that this inadequate access to spatial data has also had a significant impact upon the nature of scientific investigations of human spatial behaviour.

The development of geographic information systems

The post-World War II introduction of digital computer technology led to revolutionary developments in many fields, but it was not until the early 1960s that the first effective steps were taken to apply computer technology to the handling of spatial data. The first operational geographic information system (the Canada Geographic Information System) was designed by Dr. Roger Tomlinson to service the spatial data handling needs of the Canada Land Inventory (Tomlinson *et al.*, 1976). CGIS was the not only the first GIS, but it was to remain as one of the most technically advanced for nearly two decades since—aside from Tomlinson's unpublished dissertation, which did not appear until some years later—information on its development was never widely disseminated.

CGIS was created in response to a governmental need to handle spatial data pertaining to the land resources of all of southern Canada. Other systems were created shortly thereafter in the United States (notably those in the states of New York—a nearly immediate failure—and Minnesota—an ultimate, long-term success) but the overall failure rate for GIS remained uncomfortably high for many years. The widespread pattern of system failure in the late 1960s and early and mid-1970s is now generally held to have arisen out of a concentration on technical questions (many of which proved too complex for local programmers) and a substantial neglect of institutional factors. The process of successful GIS design today is a complex one, but it does substantially increase the overall chances of success (see, for example, Marble and Wilcox, in preparation).

After twenty-five years of development, the GIS is generally viewed today as a computer-based system for handling spatial data which is composed of four major sub-systems (Marble, 1984):

(1) A data entry subsystem which handles all problems connected with the translation of raw or partially processed spatial data (either in analogue or digital form) into an input stream of known and carefully controlled characteristics (the most common form of input today involves the manual digitizing of maps);

(2) A data storage and retrieval subsystem which accepts the input stream of spatial data and structures the database for efficient retrieval by the users of the GIS;

(3) A data manipulation and analysis subsystem which takes care of all data transformations initiated by the user and either carries out spatial analysis functions internally (e.g., the NETWORK module in ARC/INFO) or provides a two-way interface between the GIS and specialized spatial modelling systems; and

(4) The data visualization and reporting subsystem which returns the results of queries and analyses to the user in the form of maps and other graphics as well as in textual form. Regretfully, most GIS designers have yet to realize the visualization potential opened to them by the separation of data storage and data visualization in a computer environment and an amusing goal of many GIS operations is to create map output that cannot be distinguished from traditional manual forms (Marble, 1987).

This GIS software must, of course, operate within the constraints imposed by a complex computer environment (platform, operating system, etc.) and it requires, as well, a properly organized institutional structure in order to perform effectively (Marble and Wilcox, in preparation; Marble, in preparation).

One of the most significant innovations in GIS technology came about with the introduction, in the early 1980s, by the Environmental Systems Research Institute (ESRI) of the commercial ARC/INFO GIS. This GIS, which dominates the markets of the late 1980s, was the product of the vision of Jack Dangermond who founded ESRI a decade before. ARC/INFO marked the final and critical transition of GIS technology from a base consisting of non-standardized, 'home-brew' systems to the widespread use of a standardized industrial technology which could be adapted to the solution of a variety of spatial problems.

The consequences of the introduction of ARC/INFO are somewhat analogous to Henry Ford's introduction of the famous 'Model T' in the 1920s with its subsequent impressive impact upon individual mobility. This significant increase in individual mobility, together with governmental decisions to provide the necessary infrastructure (e.g., good rural roads) to support it, changed the spatial pattern of America as many small urban centres disappeared from the rural landscape and the larger ones grew even further. I suggest that the growing provision of standardized spatial databases by the government (e.g., the digital elevation models—DEMs— of the U. S. Geological Survey) and the private sector, represents an infrastructure investment comparable to the 'get the farmer out of the mud' road programme of the late 1920s and early 1930s. This infrastructure development, together with the introduction of ARC/INFO and its competitors, is leading to changes of similar magnitude.

The analogy to Ford's introduction of the Model T may be extended a bit further since the Model T, although a relatively cheap and basically dependable vehicle (which as Ford said, was available in any colour the customer desired as long as it was black!), lacked many of the features which are considered necessary and desirable in today's automotive technology. GIS technology, as it enters the 1990s, lacks many critical components which are crucial in future spatial data handling applications such as:

- The ability to handle spatial data which are known to be in error by specific amounts;
- The ability to store and manipulate three-dimensional as well as two-dimensional spatial data;
- The ability to deal with databases which contain an explicit temporal aspect; and
- The ability to work efficiently with large space-time databases (e.g., those on a planetary scale).

The Model T put critical transportation technology into the hands of a large segment of the relevant population and ESRI has clearly done the same for GIS technology with the thousands of ARC/INFO installations in place today.

Of the GIS subsystems noted above, the data manipulation and analysis subsystem is perhaps the most clumsy at present. For years GIS applications focused only upon the creation of new maps from old and little attention was given to any questions of spatial analysis which extended beyond the concepts of simple overlay. Only in the last few years has any significant attention been directed toward integration of the growing capabilities of the GIS and our moderately rich set of spatial analysis models. This must change if GIS technology is to reach its full potential.

Today we stand on the brink of a major revolution in spatial data handling. This revolution will ultimately be comparable to the original introduction of the map as an analogue database. It will impact our scientific views of the world as well as the way in which we, as individuals, deal with spatial structures and problems.

On the interaction of tools and problems

How is this revolution in spatial data handling going to impact the social sciences? We are now clearly in possession of a major new tool for handling large volumes of spatial data. Let us now examine the general question of tools and their impacts, and then turn to the specific case of the geographic information system.

Discussions on the impact of tools generally take a secondary place to discussions of changing theories in social science methodology. However a number of years ago I came upon a significant exception. Nearly thirty years ago T. C. Koopmans, a noted mathematical economist, took a year off to reflect upon the current status of economics. His thoughts were summarized in a short book of essays (Koopmans, 1957), one of which was entitled 'The Interaction of Tools and Problems in Economics'. In this latter essay he comments that:

> 'If we look with a historian's interest at the development of a science, however, we find that tools also have a life of their own. They may even come to dominate an entire period or school of thought. The solution of important problems may be delayed because the requisite tools are not perceived. Or the availability of certain tools may lead to an awareness of problems, important or not, that can be solved with their help. Our servants may thus become our guides, for better or for worse, depending on the accidents of the case. But in any case changes in tools and changes in emphasis on various problems go together and interact.' (p.170)

Koopmans gives us, I believe, a key to understanding what is likely to happen with respect to the GIS. The advent of GIS technology is, I feel, already demonstrating the start of a major impact upon the spatially-oriented activities of society. We are starting to change the ways in which we use spatial data in forestry, land management, marketing, and many other operational situations. How is this going to impact the social sciences and their view of human society?

The development of input-output analysis

One case that Koopmans believes clearly demonstrates the interaction between tools and problems in the social sciences arises out of the widespread use of input-output or inter-industry analysis models. A standard tool in economics, regional science and geography, input-output analysis basically assumes that the output (national or regional) of any productive sector of the economy (and we are fairly free to define these sectors as needed) can be directly related in a quantitative, linear fashion to the inputs which are used to create that output.

If we assume that the sole commodity produced by a representative productive sector is iron, then the assumptions of the model would tell us that for each ton of iron produced we can identify the quantity of each input (such as iron ore, coke, etc.) that is needed to create that output. The ratios of inputs to outputs (e.g., tons of ore required per ton of iron produced) are known as the *technological coefficients* of the linear production function for that sector and these are known to vary as production technology changes over time as well as from region to region. They are usually empirically determined.

The input-output model further assumes that such a linear production function can be defined for each of the sectors which make up the economy. Given the appropriate technological coefficients and a stated 'bill of goods' (that is a list of the desired net outputs from each of the sectors), it is then possible to view the economy as a system of linear equations where the unknowns are the flows between sectors. The solution of the system of linear equations permits the estimation of both the direct and indirect impacts arising from a specific bill of goods. For example, in the example given above, the direct impact arises since it requires iron ore in order to make iron. Secondary or indirect impacts arise since the extraction of iron ore requires tools made of iron. Making these in turn requires that more iron ore be processed into iron, and so forth. While it might appear that these 'rounds of effect' would go on indefinitely, we find in practice that they soon drop to a level which is small enough to ignore. In a similar manner we can make use of input-output analysis to determine, quickly and efficiently, the impacts of a decision to, say, build a new airport upon the overall economy of a metropolitan area assuming that the relevant technological coefficients are available.

The theoretical foundations of this very useful analysis tool may be traced back through the seminal theoretical and empirical work of Leontief (1941) to the theoretical constructs of Walras (1874) and Quesnay (early 1800s—see Kuczynski and Meek, 1972). Among Leontief's contributions were the reduction of the basic Walrasian model to a more restrictive linear form which permitted some empirical testing for an economy composed of a small number of highly aggregated sectors (he later won the Nobel Prize in Economic Science for this work).

Leontief's pioneering input-output study on the structure of the American economy appeared on the eve of the Second World War. The point which must be made here is that while his theoretical extensions of the earlier work of Quesnay and Walras opened up a number of interesting economic questions, there would have been no viable extensions of his empirical analysis without the means of solving, quickly and accurately, the large number of simultaneous linear equations (hundreds if not thousands of them) which are present in the Leontief model when the economy is disaggregated to interesting economic and spatial levels.

The ability of investigators to obtain solutions to such very large systems of linear equations was completely dependent upon the introduction of the digital computer following the end of the Second World War. As a result of subsequent decades of development of computer technology we can now solve input-output models of substantial size on home computers costing less than three thousand dollars. Without this revolution in computing, the widespread, practical application of input-output analysis would never have occurred. The current, widespread use of this tool to seek solutions for real world problems has also resulted, as Koopmans suggested, in the identification of significant new theoretical questions and in the development of new views of older ones (Pasinetti, 1977; Stone, 1979; Miller and Blair, 1985).

As an aside, an interesting parallel to the developments in input-output analysis can also be found in the area of numerical weather forecasting. Today we think nothing of making use of weather forecasts which are based upon massive space-time computer models of the atmosphere, and much of the current debate on global warming and its consequences rests upon the long-run forecasts derived from these models. However, when Richardson (1922) first developed the theory of numerical weather prediction, there was no technology available which was capable of implementing, on a useful space-time scale, the system of non-linear equations that he proposed. Again, we were forced to wait for the digital computer and, in this case as well as in input-output analysis, the tool that permitted the implementation of critical, pre-existing theory also gave rise to significant new theoretical investigations.

Simulation of the spatial diffusion of innovation

Some years before Koopmans was reflecting upon the status of economics, a Swedish geographer turned his attention to the creation of models for the spatial diffusion of innovation (Hägerstrand, 1953). Innovation diffusion has been, and still is, of substantial interest to researchers in nearly all social science fields. As a geographer, Hägerstrand's primary interest was in the spatial patterns of innovation adoption that occurred over time.

While the spatial patterns of innovation adoption have intrigued many social science researchers (including some in archaeology), Hägerstrand attempted to model this spatial process on an individual adopter basis. He utilized a Monte Carlo simulation technique which had been originally developed to assist in neutron flight path prediction in atomic physics (Hägerstrand, 1967). In this approach, a map of the area is defined which represents the spatial pattern of adoptions as two distinct point distributions (e.g., envisage a map of a small area with black dots representing those individuals who have already adopted the innovation and white dots those who represent potential adopters). He then put forth a set of models, of increasing behavioural complexity, to explain the mechanism through which white dots change to black dots over time. All of the models were based upon the notion of a series of discrete time periods ('generations' in his terminology). The behavioural components involved the spatial structure of communications between individuals as well as differences in the level of individual resistance to the adoption of the innovation.

Basically, Hägerstrand defined a series of probability distributions which set forth the way in which a current adopter would communicate with non-adopters (at other locations) during a specific time period. This involved explicit distance biases and, in some of his models, additional attenuation of the spatial contact fields by intervening physical features such as lakes. The model was originally implemented (by a team of graduate assistants utilizing paper and pencil methods!) for a small area in southern Sweden. For ease of analysis the area was broken into a limited number of grid cells not unlike the spatial disaggregation approach used in the Minnesota and other early GIS. The operational procedure involved the use of tab of random numbers to determine, for example, which specific non-adopter wou be contacted by an adopter within a given time period, as well as the resistance that specific non-adopter to acceptance of the innovation in question.

Because of the probabilistic nature of the analysis tool, it was clear that eac of the simulations would produce a different resultant spatial pattern. These pattern reflect, on the average, the very complex interactions between the underlyin probability distributions (e.g., the distance decay field for interpersonal communica tions) *and* the spatial distribution of adopters and non-adopters, as well as the location of the physical features in the landscape which act to bias the contact fields.

When run over and over again, the output of the models could be analyzed to determine the average spatial pattern as well as to examine the explicit variation which would occur around this mean.

When it became available in English, Hägerstrand's work had a substantial methodological impact upon American geography and his concepts were adapted to other space-time diffusion structures (Pitts, 1963; Morrill, 1965). Manual computation of the Monte Carlo models was impossible in any case containing an interesting amount of spatial detail and the models were rapidly adapted to the large mainframe computers which were becoming available to university researchers at that time (the work of Pitts, 1965 and 1967, was especially important in this effort; see also Marble and Bowlby, 1968). The Hägerstrand models as originally implemented by Pitts and Marble at Northwestern University in the 1960s were forced to operate upon a small spatial grid of only a few hundred cells. Today, OSU MAP-for-the-PC (a PC-based teaching GIS which is in use by a large number of universities) easily deals with systems of up to 50,000 grid cells in a desktop environment. Limitations in the 1960s software, led Marble (1972) to conclude that substantial difficulties in the application of the model prevented geographers from using it in a viable fashion.

Why did this occur? In the first instance, investigations of the models raised a substantial number of questions relating to statistical tests of the similarity of complex spatial patterns which could not be answered in a viable fashion. Secondly, the models called for highly disaggregate spatial data which could be obtained only at the cost of extensive field work, as well as data dealing with topics (e.g., the attenuation of individual contact fields by physical elements in the landscape) which cut across highly divergent portions of the discipline (e.g., physical and human geography). In the third instance, the multiple computer runs of the Hägerstrand model proved time consuming and expensive on the machines of the day. So expensive in fact were they that the costs were considered by funding agencies to be completely unacceptable for social science research. They were, however, well within the costs normally associated with contemporary physical science computations.

The first two of these factors are indeed representative of what Koopmans suggested: the use of the tool resulted in significant questions being asked of existing tools and concepts in the discipline. However, unlike input-output analysis, investigations of these new topics were never really incorporated into the research agenda of the discipline since the model development ground to a halt for lack of adequate computational resources. Why has it not been taken up again in light of the massive developments in computer technology which have taken place in the last twenty years? An interesting question for those concerned with geographic methodology to contemplate!

It seems clear to me from these two examples, and there are many more that come to mind, that Koopmans was indeed correct with respect to the strong interactions which may exist between tools and problems in the social sciences. Given this, we can now turn to the specific instance of geographic information systems. My thesis is, quite simply, that not only will we see interactions but that they will be major ones that will change the whole way in which we, as scientists, view human spatial behaviour.

GIS and the future of spatial analysis

Have any of these critical interactions begun to occur? In geography, they are clearly present—at least as noted by some insightful researchers. Peuquet (1987), for instance, clearly points some of them out:

> 'In the development of computerized geographic databases and information systems over the past twenty years, there have been many systems constructed for specific uses that proved to be extremely difficult or impossible to extend to new application contexts. Such difficulties are hard to predict and are often not discovered until after a substantial investment of time and money has been made....The topic of a conceptually high-level, unifying representational scheme for geographic phenomena has appeared recently within the context of geographic information systems with an added practical urgency.' (p. 376)

She also goes on to note, at the conclusion of her article, that: 'Spatial relationships are unique to locational information. These relationships are extremely important but not well understood in any formal sense' (p. 391). This, I believe, is a clear example of the growing interaction between the GIS and theoretical problems within geography. As Peuquet points out, interest in this question is not a new one (e.g., Nystuen, 1963) but current interest in this theoretical topic is clearly being driven by attempts effectively to extend the sphere of application of GIS technology. Of course, the question that Peuquet is addressing, the representation of geographic space, is also a crucial one for any of the social sciences concerned with the development of formal theory related to the spatial aspects of human behaviour.

Potential changes in the way social scientists view the world

Significant interactions do occur between tools and problems. The GIS, as an important new tool, is certainly starting to generate—at least within geography—some of the expected tool/theory interactions. But are we seeing only the larger changes that are altering the nature of scientific investigations of human spatial behaviour that were suggested earlier?

I would submit that social scientists concerned with human spatial behaviour, including my fellow geographers, have adopted a limited and myopic view of the subject, and that this myopia has been in large part the result of our inability to visualize, let alone model, the full scope of human spatial behaviour. When examined, most contemporary theories of spatial behaviour rest upon a foundation which has been simplified to the point of unreality. Why has this been the case? I would submit that a major reason has been the fact that we have lacked the tools which would permit us to organize and comprehend the data defining the *real* and extremely complex spatial environment in which human behaviour actually takes place.

The geographic information system is such a tool and the result of its continued use can only be an immediate and heightened awareness on the part of social science researchers of the spatial complexity which surrounds us and conditions much of our behaviour. I look forward to this shift in viewpoint since we have clearly been ignoring the really interesting problems for far too long.

Potential changes in human spatial behaviour

In closing, I should also point out that while I feel that GIS technology holds the potential to change our whole approach to the study of human spatial behaviour, we cannot as social scientists afford to ignore its potential for changing the structure of human spatial activities as well.

It was remarked earlier in this essay that a tremendous amount of spatial data stored in traditional map form was underutilized. Ultimately GIS will change this, not only on a business or governmental level, but on a personal one as well. Already major changes in the methods of distribution of spatial information are in the offing (see, for instance, McGranaghan *et al.*, 1987). If we as individuals and households currently operate in an atmosphere of limited spatial information which requires extensive and expensive search activities (which are not that different from primitive man's search for a water hole even if our searches involve a shop in downtown Chicago), we can only speculate on the changes which may occur in the spatial structure of society as individual access to spatial information is substantially increased.

Acknowledgments

I wish to acknowledge my appreciation to Professor Randall Jackson, a colleague in the Department of Geography at OSU, for providing me with much of the historical material pertaining to Input-Output Analysis.

References

Brassel, K., 1978, Manipulation Processes in Computer Cartography, *Proceedings, IEEE Computer Society Conference on Pattern Recognition and Image Processing*, pp. 214–219

Carroll, L., 1894, *Sylvie and Bruno Concluded*. In *The Complete Works of Lewis Carroll*, Introduction by Alexander Wollcot, (New York: Modern Library) 1936

Hägerstrand, T., 1953, Translated by A. Pred as *Innovation Diffusion As a Spatial Process*, (Chicago: University of Chicago Press)

Hägerstrand, T., 1967, On Monte Carlo Diffusion. In *Quantitative Geography, Part I: Economics and Cultural Topics*, edited by W. L. Garrison and D. F. Marble, Northwest University Studies in Geography, No. 13.

Harley, J. B. and Woodward, D., 1987, *The History of Cartography, Volume I: Cartography in Prehistoric, Ancient, and Medieval Europe and the Mediterranean*, (Chicago: The University of Chicago Press)

Koopmans, T. C., 1957, *Three Essays on the State of Economic Science*, (New York: McGraw-Hill Book Company)

Kuczynski, M. and Meek, R., 1972, *Quesnay's Tableau Economique*, (London: Macmillan)

Leontief, W. W., 1941, *The Structure of American Economy 1919–1929*. (Cambridge, Mass.: Harvard University Press)

McGranaghan, M., Mark, D. M., and Gould, M. D., 1987, Automated Provision of Navigation Assistance to Drivers, *The American Cartographer*, **14**, (2), 121–138

Marble, D. F., 1967, On Monte Carlo Simulation of Diffusion. In *Quantitative Geography, Part I:* Economic and Cultural Topics, edited by W. L. Garrison and D. F. Marble, Northwestern University Studies in Geography, No. 13

Marble, D. F. and Bowlby, S., 1968, Computer Programs for the Operational Analysis of Hägerstrand-type Spatial Diffusion Models. Technical Report No. 9, Task 389–140, No. 1228(33), Department of Geography, Northwestern University

Marble, D. F., 1972, Human Geography Simulations. In *Simulation in Social and*

Administrative Science, edited by H. Guetzkow, P. Kotler and R. L. Schultz. (Englewood Cliffs, NJ: Prentice-Hall, Inc.)

Marble, D. F. and Peuquet, D., 1983, The Computer and Geography: Some Methodological Comments. *The Professional Geographer*, **35**, (3)

Marble, D. F., 1984, Geographic Information Systems: An Overview. *Proceedings, Pecora Symposium*, IEEE Computer Society

Marble, D.F., 1987, The Computer and Cartography. *The American Cartographer*, **14**, (2), 101–103

Marble, D. F., 1988, Representation of Geographic Space: Toward a Conceptual Synthesis. *Annals of the Association of American Geographers*, **78**, (3), 375–394

Marble, D. F. and Peuquet, D., (Eds) 1990, *Introductory Readings in Geographic Information Systems*. (London: Taylor & Francis)

Marble, D. F. and Wilcox, D., (in preparation), *A New Model for the Design and Implementation of Geographic Information Systems*

Marble, D. F., (in preparation), *The Design and Implementation of Geographic Information Systems*

Miller, R. E. and Blair, P. D., 1985, *Input-Output Analysis: Foundations and Extensions*. (Englewood Cliffs, NJ: Prentice-Hall, Inc)

Morrill, R. L., 1965, The Negro Ghetto: Problems and Alternatives. *Geographical Review*, **55**, 339–362

Nystuen, J. D., 1963, Identification of Some Fundamental Spatial Concepts. *Papers of the Michigan Academy of Science, Arts, and Letters*, **48**, 373–384

Pasinetti, L. L., 1977, *Lectures on the Theory of Production*. (New York: Columbia University Press)

Peuquet, D. J., 1987, An Algorithm to Determine the Directional Relationship Between Arbitrarily-Shaped Polygons in the Plane. *Pattern Recognition*, **20**, 65–74

Piaget, J. and Inhelder, B., 1956, *The Child's Conception of Space*. Translated by F. J. Langdon and J. L. Lunzer. (London: Routledge and Kegan Paul)

Pitts, F. R., 1963, Problems in the Computer Simulation of Diffusion. *Papers, Regional Science Association*, **11**, 111–122

Pitts, F.R., 1965, Hager III and Hager IV: Two Monte Carlo Computer Programs for the Study of Spatial Diffusion Problems, Technical Report No. 4, Task 389–140, No. 1228(33), Department of Geography, Northwestern University

Pitts, F.R., 1967, MIFCAL and NONCEL: Two Computer Programs for the Generalization of the Hägerstrand Models to an Irregular Lattice, Technical Report No. 7, Task 389–140, No. 1228(33), Department of Geography, Northwestern University

Richardson, L. F., 1922, *Weather Prediction by Numerical Process*, (London: Cambridge University Press)

Robinson, A. H., 1982, *Early Thematic Mapping in the History of Cartography*, (Chicago: University of Chicago Press)

Stone, R., 1979, Where Are We Now? A Short Account of the Development of Input-Output Studies and Their Present Trends. In *Proceedings, Seventh International Conference on Input-Output Techniques*, Innsbruck, Austria, pp 1–39. (Reprinted in *Readings in Input-Output Analysis: Theory and Application*, 1986 edited by Ira John (New York: Oxford University Press)

Theroux, P., 1985, *Sunrise with Seamonsters*. (Boston: Houghton Mifflin Company)

Tomlinson, R. F., Marble, D. F. and Calkins, H. W., 1976, *Computer Handling of Geographic Data*, UNESCO Natural Resource Research Series, No. 13, (Paris: The UNESCO Press)

Walras, L., 1954, *Elements of Pure Economics*, edited by W. Jaffe, (London: George Allen and Unwin) (First published in French in 1874–1877)

3

GIS in archaeological research

Stephen H. Savage

Introduction

The development of geographic information systems (GIS) has been a recent phenomenon, and its use in archaeological research is more recent still—limited, in fact, to the last five or six years. Even among geographers the definition of a true GIS is still a matter of debate (Berry, 1987; Clarke, 1986; Cowen, 1988), so it is not surprising that some confusion should exist among archaeologists as to exactly what a GIS is, and what it can be used for in archaeology.

In spite of the general unfamiliarity of the subject with most archaeologists, several researchers have begun to explore different archaeological problems with GIS methods. Although the list of titles is still short, three main lines of research appear to be emerging: (1) site location models developed primarily for cultural resource management purposes; (2) GIS procedure related studies; and (3) studies that address larger theoretical concerns related to landscape archaeology through GIS methods. In this chapter I review these uses of GIS in archaeology. I place particular emphasis on the third line of research (even though the largest number of papers written to date deal with the first), since I feel that GIS can greatly facilitate the study of general questions related to the settlement, environment, and sociology of archaeologically studied populations.

Since there is still some debate about what constitutes a GIS, I first offer a basic definition of these systems, and briefly describe the two major types of GIS available, especially touching on points that are pertinent to the use of GIS in archaeology.

GIS definitions

A basic understanding of what a geographic information system *is* can be gained by first understanding what it *is not* or rather, *what it is more than*, in terms of other computer based mapping software available. Several researchers have addressed this issue of the definition and distinguishing characteristics of GIS. Cowen (1988) has noted that, 'The basic premise is that a true GIS can be distinguished from other systems through its capacity to conduct spatial searches and overlays that actually generate new information.' By placing the emphasis on the creation of new information, Cowen thus differentiates between GIS and CAD (computer aided design or drafting), and between GIS and DBMS (database management systems). Software systems which automatically draw maps or assign symbols to maps cannot be considered to be true GIS, since they are not creating new information. These systems are essentially only computer driven drafting programs.

Computer mapping programs (CAM) such as GIMMS are also not GIS. The basic difference between CAM and CAD systems is that CAM systems provide a sort of rudimentary linkage between a computer drafting program and a database management system. Cowen (1988) notes, though, that, 'While linking a database to the pictorial representation of geographical entities enables the researcher to address an extensive array of geographical questions, a computer mapping system is still not a GIS'. 'The term [GIS] is restricted to those computer systems which have the capability to interrelate data sets pertaining to different variables and/or to different moments in time. Thus, facilities solely for the manipulation or mapping of individual files are not here considered as geographic information systems' (Rhind, 1981). 'GIS are NOT simple graphics/mapping systems, but are systems that interrelate, manipulate, and analyze a variety of geographically distributed data in addition to mapping' (Kvamme, 1987).

GIS are therefore those which provide for the storage, management, retrieval, display, and of *creation* of geographically referenced data. Crain and MacDonald (1983) have noted that GIS typically evolve from inventory systems to analysis systems to decision support systems. Cowen (1988) emphasizes that, 'a GIS is best defined as a decision support system involving the integration of spatially referenced data in a problem solving environment.'

Data in a GIS are spatially referenced. Twenty-five years ago the geographer Brian Berry (1964) envisioned a geographic matrix containing columns which represent places, and rows which represent attributes or characteristics of those places. By looking at spatially referenced data in this way, we can imagine scanning across a series of locations while looking at the same characteristic, or looking at a number of characteristics applicable to the same place. By adding matrices we can begin to accumulate similar data sets for different times. The information stored in a GIS may therefore be thought of as bits of data related to Spaulding's space, form and time (Spaulding, 1960). Archaeologists have traditionally had difficulty controlling all three of these dimensions at the same time. The use of GIS provides methods for doing so, and, at the same time, storing and manipulating vast amounts of spatially referenced environmental data such as elevation, vegetation, hydrology and land use.

Types of GIS

The two main types of GIS, vector and raster, handle the task of spatial referencing in different ways. Each has advantages and disadvantages for the archaeologist, and will be discussed below.

Vector based systems

Vector based GIS such as ARC/INFO (ESRI, 1986) use a topological structure consisting of points, lines and areas, or polygons, to represent spatial phenomena. We usually perceive the real world as made up of such structures, and at least some of this perception is accurate. For example, geodetic survey stations are point data, roads are lines (this is actually more problematic than it appears, since roads have width as well as length, so at a larger scale they are areas) and regions of homogenous soils are polygons.

Maps have been drawn with vector type data throughout history, and, in fact, have as their basis geometric relationships between points, lines and polygons. The vector approach in GIS therefore has some attributes that make it more satisfying for the display of certain types of features. 'Vectors work well when real world spatial conditions can accurately be defined as lines or edges. Examples might include property lines, the face of a building, or the centre line of a pipeline' (Maffini, 1987). Vector based GIS also tend to produce map output that is more aesthetically pleasing; it looks like the kinds of maps we are used to. As archaeologists we think in terms of irregular boundaries around sites, features, soil types and the like, and lines are well represented in vector based GIS. Drawbacks to the vector approach include slow processing times, difficulties in performing Boolean manipulations between different map layers and, generally, the higher cost of equipment used in these systems (Maffini, 1987; Kvamme, 1988). The process of encoding, or relating points to lines and polygons, and polygons to other polygons and lines is complicated and, depending on how it is done, can produce 'sliver polygons' and other errors in the data structure (Peucker and Chrisman, 1975).

Perhaps for archaeologists, though, the most serious drawback in vector based systems stems from what at first looks like their most attractive feature, their ability to draw accurate lines on computer screens and maps. Maffini explains the problem:

> 'The vector approach has...been used in circumstances for which it is not
> ideally suited. When we look at an image of a region, we see many phenomena
> which have no sharp boundaries. When we impose lines (vectors) on the image
> to bound such phenomena, we introduce a highly precise interpretative element
> into the data which is misleading....Once the line has been drawn it takes on
> a certain immutability.' (Maffini, 1987)

This problem is particularly applicable with archaeological survey work. Although sites may be shown to have boundaries, what do we mean by a site boundary? Is it the place where the surface scatter stops, and if it is, how is that scatter of point data (the artifact positions) recorded as a vector boundary? Usually a ring is drawn on a map, and becomes the site boundary, but it clearly is not drawn by connecting the points of artifact occurrence at the edge of the site. What if we define the site as the limits of human activity associated with a particular locus? Although we take as assumptions that human behaviour is patterned, and that it produces patterned remains in the archaeological record, it is not necessarily so that all kinds of behaviour produce artifactual remains. Under this definition of a site, we might actually never be able to define a boundary. The problem with the vector approach for archaeologists is, then, that it reifies a boundary whose definition is suspect.

Raster based systems

In raster based systems a region is represented by a matrix of grid cells (usually square) forming rows and columns on the X, Y axes, and a numeric Z value that represents some characteristic of the region such as topography, soil type, or slope. Values are assigned to the grid cells in a variety of different methods which are usually under user control. These include a binary switch (presence/absence), extreme value (highest or lowest), average value, predominant value, or centroid of cell (the value at the centre of the grid cell is assigned to the entire cell). Numerous GIS have been developed using raster based data structures, including the Map Analysis Package (Tomlin, 1983), and its derivative, MAPCGI (Cowen and Rasche, 1987).

One advantage of raster based systems is the simple data structure of X and Y locations, with Z values, which make these systems easier to understand and operate. The data structure is easy to manipulate mathematically, making analysis (especially Boolean operations) simple and rapid. For example, two map layers may be combined by simply adding the Z values from each separate layer, on a cell by cell basis, and assigning the sum to a new map layer. The maps may thus be manipulated algebraically. The raster approach is also excellent for handling continuous data, such as elevations, but at the same time can represent such discrete themes as soil types.

If Boolean or algebraic operations are to be performed in vector systems, there is often a hidden conversion of vector to raster data. The operation is performed, and the data are re-converted to vector for display (Maffini, 1987). While there is only one possible result when data are converted from vector to raster, there are many when raster data are converted to vector, so additional problems of interpretation are introduced.

Disadvantages in the raster approach include the large file space required. Typically, a raster system stores arrays of two byte integers (numbers ranging from $-32,767$ to $32,767$ that can be stored in the computer in two character spaces). Even a modest sized map overlay, 100 rows by 100 columns, requires 20,000 bytes of storage, and this is repeated for each map theme in the system. The numeric arrays must be operated upon in computer memory, often two at a time, so hardware memory (and display) limitations do not allow very large matrices. The MAPCGI system, for example, does not allow more than 64,000 grid cells in the database. In operational terms, this means that large areas must be handled at small scales, so detail is lost. As an example, the state of South Carolina is available in the MAPCGI system, but the grid cell size is one kilometer. There is a considerable loss of detail as a result. Increasing the resolution automatically decreases the area that can be examined in a grid cell system, so a compromise must be created between region and scale, which the user of the system is called upon to make, sometimes before enough information is present to make an informed decision. Despite these limitations, Kvamme notes that raster based GIS:

'. . .are particularly well suited for analysis and modelling applications not only because virtually any type of data can be encoded and stored cell-wise, but because these data can be accessed for univariate or multivariate analyses and complex algorithms or decision models can be applied to them. For this reason cell-based GIS have predominately been used by archaeologists for modelling and research purposes.' (Kvamme, 1989)

Archaeological applications of GIS

In this section I examine some of the modelling and research purposes that archaeologists address with GIS methods. As noted above, there are three main areas of research currently conducted via GIS: (1) site location models developed primarily for cultural resource management purposes; (2) GIS procedure related studies; and (3) studies that address larger theoretical concerns related to landscape archaeology through GIS methods.

Site location models and cultural resource management

Kvamme (1989) notes that the use of GIS for 'predictive archaeological location modelling, with its vast data, computational, and cartographic needs, has thus far been the predominant application of GIS in archaeology.' This use of GIS as a cultural resource management tool may be seen to stem directly from Clarke's (1986) emphasis on the development of GIS as a management tool, and Cowen's (1988) definition of GIS as 'a decision support system involving the integration of spatially referenced data in a problem solving environment.' The goal of this approach is to locate areas that are sensitive to the presence of archaeological sites in advance of development, and plan the development phase of a terrain altering project so that it avoids the sensitive archaeological areas. In development projects, 'as in other multipurpose planning, the objective should be to maximize all potential complimentary social benefits at the least social costs' (McHarg, 1971). By presenting archaeological compliance work as a social cost which can be avoided, the actual expenses of archaeological mitigation are reduced, and fewer sites are likely to be destroyed. This method offers results that benefit both archaeologists and developers.

Several studies of this nature have been conducted, some with vector based systems, and others with grid cell or raster systems. The basic approach involves creating a mathematical model and then applying it to the region in question.

A popular methodology by which archaeologists can develop empirical predictions is through quantitative site location studies. To many archaeologists, these site location studies are based upon the assumption that non-cultural aspects (independent variables) of the environment will correlate with and predict site locations (Ebert *et al.*, 1984). Although many archaeologists equate quantitative site location studies with predictive modelling, they will be regarded here as empirical observations which inductively project site location (Ebert and Kohler, 1986). In other words, they are simply correlational models (Marozas and Zack, 1987).

One approach to the creation of site location models involves using logistical regression techniques in a statistical analysis package such as SAS (SAS Institute, 1985). This technique allows a binary, presence/absence indicator of an archaeological site to be used as the dependent variable, and various other environmental factors such as elevation, slope, distance to water and the like, as independent variables (Marozas and Zack, 1987; Warren *et al.*, 1987). It is particularly important to understand the relationship of site locations to these independent variables versus non-site locations. 'The central points [site locations] must exhibit a different set of associations if one is to distinguish between background and potential sites' (Marozas and Zack, 1987). This emphasis on non-site locations should not be confused with non-site archaeology as defined by Thomas (1975), in which the individual artifact is the point of reference, and traditional sites are ignored. Rather, these non-sites are places that are not archaeological sites (that is, where no human cultural remains or activity exists) where environmental data may be collected (Kvamme, 1982).

A problem with this approach is that, in the absence of explicitly collected information about where sites are not, as opposed to where they are, control locations are assumed to be non-sites. A second assumption is that the known site locations are a representative sample of the population:

> '...the location patterns exhibited by the initial site sample used to train the
> pattern classifier (the quantitative model) must be reasonably representative of

the site population under study. The second assumption is that site locations are nonrandomly distributed with respect to the environment or social factors under investigation' (Kvamme, 1986 in Marozas and Zack, 1987).

The basic approach to archaeological pattern recognition requires that the second assumption be made (South, 1977). However, in creating the set of control points against which the site sample is checked, the second assumption is often violated in order to create a set of non-sites: 'While both the SITExxx and the CNTxxx [the control points] coverages were created from the same geographic space, it was assumed that sites occur most infrequently ($p < 0.01$) and with a randomized Poisson distribution' (Marozas and Zack, 1987). Sites cannot be 'nonrandomly distributed with respect to environment or social factors' and at the same time occur 'with a randomized Poisson distribution'.

The contradiction in the operational assumptions which underlie this particular approach to locational modelling render it suspect as far as its utility to predict site or non-site locations is concerned. What is required is firm knowledge that none of the control points are archaeological sites. In cultural resource management reports a more explicit description of areas surveyed, and sites found in them, will help alleviate this shortcoming, but when projects are done in unfamiliar areas, or when explicitly non-site locations are not available, this modelling approach seems risky. This is particularly true since the majority of locational models developed to date appear to use this technique (e.g. Marozas and Zack, 1987; Limp *et al.*, 1987; Warren *et al.*, 1987; Warren, 1988).

An alternative method has been developed by Savage (1989a) that does not require the assumption of non-site locations in the model because the binary logistical regression technique is not used. Instead, site location is used as the dependent variable in a stepwise multiple regression model. The alternative method was developed primarily because information about non-site locations was not available. Under an assumption of a representative site sample, and that sites will occur in areas that are most like those areas in which they have been found (a uniformitarian assumption), this model uses stepwise multiple regression to isolate the various environmental factors which are significant contributors to known site locations. In this case, the site location is created by assigning it a number derived from the cell row and column number where the site is found.

GIS procedure related studies

A few studies have been conducted on the implications of using GIS methods in archaeology, especially with respect to the accuracy of the results obtained. Kvamme's (1988) paper on 'GIS Algorithms and their Effects on Regional Archaeological Spatial Analysis' is an example. Kvamme notes that archaeologists are usually not concerned with the quality of data obtained by computer methods, such as slope, elevation, or aspect. They are often willing to employ such data in GIS models without considering the different outcomes which may result from GIS derived environmental data from different sources. Conclusions reached by these means may be the result of real differences, differences in GIS data sources, or different GIS packages, with different computational algorithms.

Kvamme's study focused on the differences between digital elevation models (DEMs) available to the archaeologist from such sources as the U.S.G.S. and the Defense Mapping Agency. Because the models from these two agencies are available at different scales, and because different smoothing algorithms were used in their creation, different results are likely to be obtained when they are used. In particular, the small scale topographic relief that often seems to influence site location may be lost if the DEM is of low resolution, or if too much smoothing occurred. The results of a locational analysis would therefore be flawed if they are based on such data.

Besides the inconsistencies among different computer derived data sources, problems can exist within a single data source. For example, hydrology data for South Carolina was used by Savage (1989a) to create a projective model of Archaic Period settlement. The data was imported from Digital Line Graph files, which were created by the United States Geological Survey by digitizing their 1:100,000 series map sheets. In this study, where a boundary between two different map sheets was noted, the hydrology data were seen to be inconsistent. Data from the east side of the project study area were much denser in respect to hydrological features than those from the west side. The problem was traced back to the 1:100,000 paper maps, and resulted from different management decisions at the time the maps were drawn. The data inconsistency has since been digitized and incorporated into the Digital Line Graph files. It is unknown at this time just how widespread the problem is, but it impacts major areas in South Carolina, and renders studies based on that data suspect.

Other problems result not from the nature of the data, but from the procedures used to process data within the GIS. In working on the development of demographic models in ARC/INFO, Ezra Zubrow

'...observed while simulating alternative settlement patterns that without changing the parameters differences in resulting migrations would occur. It appeared to be a consequence of the order that one entered the initial centers or population concentrations into the networks of ARC/INFO' (Zubrow, 1988).

The problem occurs when processes which are concurrent in nature must be modelled by computers that operate sequentially. Until some procedure can be developed that will allow concurrent processes to be modelled concurrently, this problem will persist, and modellers that make use of such procedures had best take note of the difficulties. The computer's solution to the problem is not the only one available.

In the absence of truly uniform data quality standards, and in the face of problems related to concurrency, it is necessary for the archaeologist to consider these problems, and to insert prominent caveats in resulting reports (especially CRM reports). Most particularly, the archaeologist should emphasize the danger of reifying the results of locational analyses based on data that may not be accurate enough for the type of predictions created. In essence, the results of our location models represent hypotheses to be tested through archaeological survey, not the end product of a process that creates archaeological 'facts'. Many of the locational analyses using GIS have been undertaken precisely to avoid a large survey, so it appears that the use of GIS in a CRM context may result in more harm than good if compliance is assumed based on the end product of such analyses.

Beyond locational analysis and problems: GIS as a research tool

The emphasis of Clarke (1986) and Cowen (1988) on GIS as management support tools, and the fact that many of the current GIS available have been developed by various government agencies, helps to explain why the majority of initial work with GIS in archaeology has centred on locational analysis and predictive modelling.

There is, however, great potential for using GIS as a research oriented, theory building methodology in landscape archaeology. (I refer to landscape archaeology as the study of spatial relationships among humans and their physical, social and cognitive environments.) Before the development of GIS, many questions related to social organization and spatial clustering or territoriality could only be addressed through such techniques as spatial autocorrelation, cluster analysis, variance to mean ratios and the like. These methods are not only difficult to implement, but even more difficult to interpret. It might well be said that the science of landscape archaeology was at a methodological dead-end (Paynter *et al.*, 1974).

The advent of GIS allows such studies to go forward under a more easily understood and manipulated methodology. Data in a GIS are automatically spatially referenced, and different themes may be explored with reference to other themes through mathematical and Boolean methods. Landscape archaeology and GIS provide a powerful combination of theory and method that promises to advance the study of past social systems in relation to their physical and cultural environments.

For example, Savage (1989b, see Chapter 26) has used GIS to create a social landscape model for the Late Archaic in the Savannah River Valley. Using fifty-one archaeological sites, six minimum band use areas are postulated, based on presumed base camp locations and a distance model created with GIS methods. Population estimates from the six minimum band areas suggest that a single maximum band is represented in the project area.

Adding the third archaeological dimension of time to Berry's (1964) geographic matrix allows us to use GIS to model both diachronic and synchronic social processes. The power of GIS can be harnessed to develop more effective explanations of long-term cultural change. Kvamme (1989) has noted that the time depth of archaeological data is ideal for diachronic simulation studies—each time period may be represented in the GIS as a series of data themes such as archaeological phenomena and environmental information. Spatially referenced data is thereby linked diachronically as well.

The use of GIS as an 'engine' to drive long term processes enabling accurate modelling has been attempted by Allen (1988, see Chapter 25 of this volume). A database was constructed, using the hydrology from New York State, and early historic contact dates. Diachronic aspects of trade patterns were modelled using ARC/INFO. Allen reports that

> 'Alternative trade models are constructed based on formalist and substantive assumptions. These models are combined with the network algorithms of ARC/INFO to predict the distribution of ceramics and other trade goods. The patterns are compared with the archaeological record' (Allen, 1988).

A further application of GIS in long term modelling has been conducted by Zubrow (1988, see Chapter 24). In the application that first drew his attention to the problem of concurrent processes, Zubrow models the spread of colonial population through New York using the various river valleys as migration corridors. A number of different models were constructed, based on different river corridors, and their outputs compared with historical documentation.

The small number of such studies conducted to date, rather than discouraging the researcher, instead point the way toward research potentials in GIS applications in landscape archaeology which are far more exciting than the creation of simple predictive or correlation models. They show that long term temporal, spatial, and cultural processes can be successfully modelled using GIS methodology.

Summary

Although the use of GIS in archaeological research is a new phenomenon, many researchers have begun tapping its enormous potential for storing and manipulating spatially referenced cultural, environmental, and temporal data. The bulk of work currently being done reflects the use of GIS as a management tool, for the prediction of site location in advance of project development. The various problems associated with accepting the results of such modelling episodes as archaeological facts, and the problems related to data variability, can be addressed by treating the generated models as hypotheses. This use promises to allow more effective management of a shrinking cultural resource database if it is coupled with continued archaeological survey and refinement of locational models.

Beyond the modelling, or prediction, of site location lies the potential for using GIS to examine cultural processes synchronically and diachronically, as a methodological tool of landscape archaeology. The methods developed in GIS, when coupled with Geographic Location Theory (Pred, 1967), and the work of current revisionist archaeologists such as Bender (1978), Root (1983), Wobst (1974), Green and Sassaman (1983), and Sassaman (1983) will allow new approaches to the important issues in prehistory.

References

Allen, K. M., 1988, Trade Networks and European Contact: A Case Study Using Geographic Information Systems. Paper presented at the *Third Annual International Conference, Exhibits and Workshops on Geographic Information Systems*, Sydney, Australia.

Bender, B., 1978, Gatherer-Hunter to Farmer: A Social Perspective. *World Archaeology* **10**(2), 204–221

Berry, B. J., 1964, Approaches to Regional Analysis: A Synthesis. *Annals of the Association of American Geographers*, **54**, 2–11

Berry, J. K., 1987, *Introduction to Geographic Information Systems: Management, Mapping, and Analysis for Geographic Information Systems (GIS)*. Special Workshop for South Florida Water Management District

Clarke, K. C., 1986, Advances in Geographic Information Systems. *Computers, Environment, and Urban Systems*, **10**(3–4), 175–184

Cowen, D. J., 1988, GIS vs. CAD vs. DBMS: What are the Differences? *Photogrammetric Engineering and Remote Sensing*, **54**(11), 1551–1556

Cowen, D. J. and Rasche, B., 1987, An Integrated PC Based GIS for Instruction and Research. *Proceedings, Auto Carto VIII*, pp. 411–420

Crain, I. K. and MacDonald, C.L., 1983, From Land Inventory to Land Management: The Evolution of an Operational GIS. *Proceedings, Auto Carto VI*, **1**, 41–50

Ebert, J. I. and Kohler, T.A., 1986, The Theoretical and Methodological Basis of Archaeological Predictive Modelling. Chapter 4 in *BLM Predictive Modelling Draft*, Bureau of Land Management.

Ebert, J. I., Larralde, S. and Wandsnider, L., 1984, Predictive Modelling: Current Abuses of the Archaeological Record and Prospects for Explanation. Paper presented at the *49th Annual Meeting of the Society for American Archaeology*, Portland, Oregon, April, 1984.

ESRI, 1986, *ARC/INFO Users Manual Version 3.2* (Redlands, California: ESRI)

Green, S. and Sassaman, K., 1983, The Political Economy of Resource Management: A General Model and Application to Foraging Societies in the Carolina Piedmont. In *Ecological Models of Economic Prehistory*, edited by G. Bronitsky. (Tempe, Arizona: Arizona State University Anthropological Research Papers, Number 29.), pp. 261–290

Green, S. and Ulrich, T., 1977, The Statistical Analysis of Environmental Variation: A Mapping Approach. Paper presented at the *42nd Annual Meeting of the Society for American Archaeology*, New Orleans, April, 1977

Kvamme, K. L., 1982, Methods for Analyzing and Understanding Hunter-Gatherer Site Location as a Function of Environmental Variation. Paper presented at the *47th Annual Meeting of the Society for American Archaeology*, Minneapolis, Minnesota, April 1982.

Kvamme, K. L., 1986, Developmental Testing of Quantitative Models. Chapter 8 in *BLM Predictive Modelling Draft*, Bureau of Land Management

Kvamme, K. L., 1987, *An Overview of Geographic Information Systems for Archaeological Research and Data Management*. Manuscript on file at the Arizona State Museum, University of Arizona, Tucson.

Kvamme, K. L., 1988, GIS Algorithms and their Effects on Regional Archaeological Spatial Analysis. Paper presented at *Geographic Information Systems Applications in Archaeology*, Society for American Archaeology meeting, Phoenix, April 1988. [See Chapter 10 of this volume]

Kvamme, K. L., 1989, Geographic Information Systems in Regional Archaeological Research and Data Management. In *Method and Theory in Archaeology*, Vol. 1, edited by M.B. Schiffer. (Tucson: University of Arizona Press), pp. 13–203

Limp, W. F., Parker, S., Farley, J.A., Waddell, D.B. and Johnson, I., 1987, *An Automated Data Processing Approach for Natural Resource Management on Military Installations*. Report submitted to U.S. Army Construction Engineering Research Laboratories by the Arkansas Archaeological Survey, Fayetteville.

Maffini, G., 1987, Raster versus Vector Data Encoding and Handling: A Commentary. *Photogrammetric Engineering and Remote Sensing*, **53**, 1397–1398

Marozas, B. A. and Zack, J. A., 1987, *Geographic Information Systems Applications to Archaeological Site Modelling* (Redlands, California: ESRI)

McHarg, I. L., 1971, *Design with Nature*. (New York: Doubleday)

Paynter, R. W., Green S. W., and Wobst, H. M., 1974, Spatial Clustering: Techniques of Discrimination. Paper presented at the *39th Annual Meeting of the Society for American Archaeology*, Washington, D.C., May, 1974.

Peucker, T. K. and Chrisman, N., 1975, Cartographic Data Structures. *The American Cartographer*, **2**(1), 55–69

Pred, A., 1967, Behaviour and Location: Foundations for a Geographic and Dynamic Location Theory. *Lund Studies in Geography, Series B: Human Geography*, Number 27. Lund, Sweden

Rhind, D. W., 1981, Geographical Information Systems in Britain. In *Quantitative Geography: A British View*, edited by N. Wrigley and R. J. Bennett. (London: Routledge and Kegan Paul)

Root, D., 1983, Information Exchange and the Spatial Configurations of Egalitarian Societies. In *Archaeological Hammers and Theories*, edited by J.A. Moore and A.S. Keene. (New York: Academic Press), pp. 193–220

SAS Institute, 1985, *SAS User's Guide*: (Cary, North Carolina: SAS Institute. Inc.)

Sassaman, K. E., 1983, *Middle and Late Archaic Settlement in the South Carolina*

Piedmont. Unpublished M.A. thesis, University of South Carolina, Columbia

Savage, S. H., 1989a, Projecting Settlement Patterns with Geographic Information Systems and the Archaeological Resource Management System. Paper presented at *Interpreting Space: The Use of Geographic Information Systems in Archaeology*, First Annual World Archaeological Congress, Baltimore, January, 1989

Savage, S. H., 1989b, *Late Archaic Landscapes*. Anthropological Studies 8, Occasional papers of the South Carolina, Columbia Institute of Archaeology and Anthropology, University of South Carolina.

South, S., 1977, *Method and Theory in Historical Archaeology*. (New York: Academic Press)

Spaulding, A. C., 1960, The Dimensions of Archaeology. In *Essays in the Science of Culture: in Honor of Leslie A. White*, edited by G. Dole and R. Carneiro. (New York: Thomas Y. Crowell)

Thomas, D. H., 1975, Non-site Sampling in Archaeology: Up the Creek without a Site? In *Sampling in Archaeology*, edited by J.W. Mueller. (Tucson: University of Arizona Press), pp. 61–81

Tomlin, C. D., 1983, *The Map Analysis Package*. Ph.D. Dissertation, School of Forestry and Environmental Studies, Yale University.

Warren, R. E., 1988, Predictive Modelling of Archaeological Site Location: A Case Study in the Midwest. Paper presented at the *53rd Annual Meeting of the Society for American Archaeology*, Phoenix, Arizona, April 1988.

Warren, R. E., Oliver, S. G., Ferguson J. A. and Druhot, R. E., 1987, *Predictive Model of Archaeological Site Location in the Western Shawnee National Forest*. Illinois State Museum Society Quaternary Studies Program Technical Report No. 86. Springfield, Illinois

Wobst, H. M., 1974, Boundary Conditions for Paleolithic Social Systems: A Simulation Approach. *American Antiquity*, **39**(2), 147–178

Zubrow, E., 1988, Concurrence, Simultaneity, and the Buffalo Shuffle: Demographic Modelling with ARC/INFO. Paper presented at the *Third Annual International Conference, Exhibits, and Workshops on Geographic Information Systems*, Sydney, Australia.

4

The diffusion of a new technology: a perspective on the adoption of geographic information systems within UK archaeology

Trevor M. Harris and Gary R. Lock

Introduction

As with the diffusion of any new technology there is a lag-time between the technology becoming available and its uptake by the potential user community. In the case of the application of GIS technology to archaeological data, the UK archaeological community is currently at the very earliest stages of the adoption curve and the lag-time is, as yet, indeterminable. What interests us in this paper is the prospective role which the carriers and adopters of this new technology will have on the rate and pattern of diffusion of GIS through the UK archaeological community. Naturally, the expansion of GIS usage in UK archaeology will be influenced as much by the nature of the archaeological record and the organizational structure of archaeology in this country, as by the many individual decisions concerning whether to adopt or to reject these ideas. In this respect the diffusion of GIS technology may follow a very similar path to that taken during the computerization of the archaeological discipline over the past two decades.

The take-up of GIS technology in the UK at the present time is somewhat limited. We anticipate that within archaeology the nature of the diffusion process itself will be especially important in the way in which the technology is adopted. While the diffusion process will probably contain elements of both contagion and hierarchical expansion, and maybe even relocation expansion from the United States, the ability of important archaeological bodies and other influential institutions in the UK to 'leap frog' the technology through the archaeological structure and enable knowledge to cascade or trickle down to the lower sections of the archaeological hierarchy should not be under-estimated. However, this ability to facilitate the diffusion of GIS is matched by the frictional stress which the same influential bodies can exert on the diffusion process. Clearly, it is possible for the expansion of technological knowledge through a user community to be slowed down, altered course, impeded or otherwise channelled by the attitude of these influential bodies. It is rare for so-called absorbing bodies to be impermeable and completely nullify an innovation pulse for there is an inherent tendency for strong innovation waves to flow around these barriers. Similarly, reflecting barriers, as we indicate later, can actually concentrate and channel the diffusion process in certain areas where more rapid adoption and usage can be envisaged. Nonetheless, in terms of warping the innovation waves in their pure form such hierarchical structures have very important roles to play. The organizational structure of the archaeological community in the UK would suggest that archaeological institutions could be a major force in the adoption process.

In many respects the technology of GIS is now well established (see for example Rhind, 1981; Peuquet and O'Callaghan, 1983; Marble *et al*, 1984; Burrough, 1986). In terms of introducing GIS to the wider archaeological community, however, and demonstrating the application of GIS in archaeological spatial analysis, there is a considerable amount yet to do. We anticipate that GIS will bring demonstrable benefits to the field of archaeology. A regrettable feature of the computerization of the archaeological discipline over the past two decades is that the process has tended to exclude the full locational component from the archaeological record. This is the result of the difficulties encountered in integrating the varying spatial units which define archaeological sites such as point specific data, linear features and polygonal areas, in a computer database. The result has been that archaeological archival systems, as well as archaeological spatial analyses, have been unable to include the total archaeological record within the computing environment. The ability of GIS to integrate disparate spatial and thematic information, both environmental and archaeological, within a computing environment is a powerful tool for archaeologists to explore. The resultant production of composite variables to which extensive spatial and modelling techniques can be applied will open up new vistas in analytical work. We envisage that GIS as a technique will dove-tail within the already existing formal and informal methods of analysis practised by archaeologists. For this reason GIS, once accepted, may find a ready market within the archaeological community. Furthermore, the potential which such systems possess to handle wholesale spatial enquiries of the archaeological record must transform the way in which conventional archaeological database systems have been developed and used. However, as the implementation of Information Technology in general within archaeology demonstrates, the deadening weight of inertia and traditional work practices on the adoption rate of new technology cannot easily be dismissed.

In this paper we do not propose to concentrate on the functional applications of GIS in archaeology *per se* as these are covered elsewhere in this volume. The focus of our interest is to consider the likely areas in which GIS could beneficially be adopted in the UK; to consider the likely rate and extent of its adoption; and to examine those factors likely to constrain or encourage the overall diffusion of this technology within the archaeological community. There are a number of reasons why this emphasis should be so; not the least of which is the current very low adoption and awareness of the technology by archaeologists in this country. A recent government report, (DoE, 1987), commissioned to examine the wider field of geographic information handling in the UK, provides a detailed insight into the current state of GIS in this country and identifies factors thought likely to constrain or facilitate the take-up of the new technology in the near to medium term. In its deliberations the committee drew upon some illuminating comments given in evidence to the enquiry by a North American company long associated with GIS work. We consider that these comments epitomize our concerns in this paper. 'In dealing with a relatively new technology such as GIS' it said

> 'we have found over and over again in North America that the technical problems are minor in comparison with the human ones. The success or failure of a GIS effort has rarely depended on technical factors, and almost always on institutional or managerial ones'. (DoE, 1987, p.154)

Furthermore, it goes on to say

'In short, we believe that the greatest obstacle to greater GIS use will continue to be the human problem of introducing a new technology which requires not only a new way of doing things, but whose main purpose is to permit the agency to do a host of things which it has not done before, and in many cases does not understand.' (DoE, 1987, p.158)

In the context of GIS in the UK at the moment these comments are particularly apposite. Attention has already been drawn to the potential role of archaeological institutions as possible forerunners in the take-up and dissemination of this technology. In reviewing the current low status of GIS in UK archaeology, however, there are a number of areas which are eminently suitable for the rapid and beneficial adoption of GIS. The application of the new technology in these sectors could have a substantial impact, we hesitate to use the word revolution, not only in the ways in which archaeologists handle and use spatial information but in the ways in which archaeological information itself is used within other sectors of society such as planning and cultural resource management.

The development of GIS in UK archaeology

The earliest recognition of the potential which GIS could offer to archaeological analysis was gained at about the same time in both the US and the UK. It would appear though that the initial uptake of the technology has been greater in the former than the latter. The meeting of Commission IV of the UISPP in 1985 as a prelude to the 50th meeting of the Society for American Archaeology, for example, contained just two papers on GIS (Gill and Howes, 1985; Kvamme, 1985a). The same theme was given broader coverage in a symposium entitled 'Computer-based GIS: a tool of the future for solving problems in the past', at SAA50 which covered both methods and principles (Kvamme, 1985b; Ferguson, 1985) and specific regional applications (Bailey *et al.*, 1985; Creamer, 1985). A year later, at the National Workshop on Microcomputers in Archaeology, GIS were again given a high profile with papers on the potential of GIS in archaeology (Kvamme, 1986), the availability of suitable commercial software (Ferguson, 1986; Miller, 1986) and poster displays illustrating GIS applications in North American archaeology at that time. Awareness of GIS and its potential in archaeology has continued to rise in the US (see for example Kellogg, 1987; Kvamme and Kohler, 1988; Kvamme, 1989) even though the number of published applications is still relatively low.

During the same period in the UK similar developments can be identified although they are fewer in number and originate from geographers rather than archaeologists. At the 1986 annual Computer Applications in Archaeology (CAA) conference, Harris (1986) showed the need for computerized spatial data handling procedures at the regional level for archival, educational and research purposes, as well as for aiding the decision-making process within the planning environment. He outlined a GIS application to the archaeology of the Brighton area (see also Harris and Dudley, 1984; Harris, 1985), a theme which was taken up again in a workshop and paper at the next CAA conference (Harris, 1988) where the discussion was broadened to consider GIS facilities for draping archaeological information on to Digital Terrain Models as output from a GIS. Two other geographers have also shown the potential of GIS to archaeology with their work on remote sensing techniques in the Fens of eastern England (Donoghue and Shennan, 1988; Donoghue, 1989). A continental GIS application reported to a UK audience is the work of Wansleeben (1988). Even with the limited power of self-written software Wansleeben was able to apply GIS techniques to six environmental and eight archaeological layers of data representing an area of 4500 sq. km. of the southern Netherlands. Despite the system being restricted to a raster structure and simple two-dimensional graphical output the potential for the integration of environmental and archaeological variables is convincingly demonstrated. With the exception of GIS based intra-site and regional analyses of Danebury Iron Age hillfort by the authors, a recently started survey of wetlands in north-western England, a self-written GIS application in Yorkshire (Powlesland, forthcoming) and some planning-based applications mentioned below, little is currently ongoing in UK archaeology related to GIS, and even less is published. Furthermore, of the many recent UK publications concerned with current issues in archaeological computing, discussion of GIS and spatial data handling techniques is notable by its absence (Martlew, 1984; Cooper and Richards, 1985; Richards and Ryan, 1985; Burrow, 1985; Shennan, 1988; Booth *et al.*, 1989). Encouragingly, some archaeologists closely involved in land-use planning are becoming aware of GIS because of the growing involvement of planning authorities in GIS applications for regional structure planning and development control. These examples, however, remain very much the exception rather than the rule. While applications are progressing in related areas such as archaeological draughting systems for digital plotting (Alvey, 1989, for example), these are undertaken without reference to GIS.

The UK archaeological record

Many of the difficulties associated with the diffusion of GIS within North America over the past two decades will doubtless find their counterparts in the UK. However, some aspects are bound to differ, especially those associated with the nature and extent of the archaeological record itself and the organizational structure of the archaeological community in particular. A number of these factors are specifically highlighted in this paper as being instrumental in the diffusion of GIS. In particular the role of archaeological institutions in the hierarchical diffusion of GIS technology, and of teaching establishments where contagion diffusion could be expected to prevail, are discussed. However, before the discussion focuses upon these factors we should at this point draw attention to an underlying and prevailing reason why we place so great an emphasis upon these institutions as centres of diffusion in UK archaeology.

In many respects our perception of the potential role of GIS in this country goes beyond the specific use of GIS techniques for individual site project work. We anticipate that some archaeologists will, as with the diffusion of computing and quantitative techniques, look to implement GIS in regard to their own specialty areas. The adoption rate will vary in accordance with some of the factors discussed in this paper. Our perspective, however, is to focus not so much upon the adoption of GIS as an additional tool in the archaeologist's analytical armoury for individual project work, important though this is, but on the integration of GIS in the archiving and analysis of the archaeological resource at the regional and national level. This viewpoint is proffered because of the existence in the UK of not only a rich archae-

ological record but one which has been well documented and archived by archaeological bodies for many years. The richness of the archaeological record is such that
any detailed ground survey in the UK inevitably produces the identification of
numerous new sites. This applies also to aerial survey reconnaissance, especially
under favourable ground conditions like the dry summer of 1989 when well over
1,000 previously unknown sites were discovered from the air (Griffith, 1989). The
extent of the record is also evident in the chronological depth of the archaeological
record. Within any geographical area the temporal span is likely to stretch the five
thousand years from the Neolithic period to the medieval period and, indeed, to
more recent remains. What is important in this respect is that the long term recording and inventorying of this heritage by UK archaeologists has resulted in the
development of comprehensive regional and national *computerized* databases of
archaeological sites. The existence of this rich archaeological record and the far-
sighted recording of the sites in regional and national archives suggests that the
advent of GIS in the UK could have an impact at a level greater than that of site
specific applications. The scope exists for GIS technology to be incorporated rapidly
within local, regional and national information storage and retrieval systems, as well
as providing a sophisticated spatial data handling tool for the analysis of archaeological sites at the intra-site and regional levels. This aspect underlies much of the
following discussion concerning the role of UK archaeological bodies who currently
manage and maintain these archive systems.

The UK archaeological organizational structure

It is possible to identify several influential bodies within the organizational structure of UK archaeology which have the potential to act as facilitating gateways, or
obstructing barriers, in the dissemination of GIS to the archaeological community.
The likely response of these bodies to the innovation wave of GIS technology is not
easy to assess at this moment in time, although there are some indications that a
positive response could ensue. The organizations themselves are currently involved
in the initial stages of assessing the technology and formulating an early response
to GIS initiatives such as the 1987 Chorley Report. Clearly there are many factors
involved in the formulation of an institutional response to a new initiative, some of
which will favour early adoption, while others will mitigate against it. Some indication of the possible response of these bodies to the innovation wave of a new
technology can be gained by reference to their past and present performance regarding the adoption of Information Technology in general.

   The county-based Sites and Monuments Records (SMRs) are the most
comprehensive and frequently updated source of archaeological information
available in the UK (Holman, 1985; Chadburn, forthcoming). They provide a unique
potential for regional spatial research in archaeology. Every county in England,
Wales, Northern Ireland and most of Scotland now has SMR cover together with
one or more professional archaeologist(s) specifically dedicated to the management
of the record. All types of information are entered into the record although their
main importance lies in the inventorying of archaeological sites. Despite the potential
of this resource for academic research at all levels, SMRs are virtually dedicated to
cultural resource management activities within the regional planning environment.
This apparent schism between an invaluable archaeological resource and a potential major academic user group is partly due to the unwieldy structure of SMRs and

their failure to fully integrate spatial information within the computerized archaeological database.

The origin of SMRs lies in the strong British tradition of field work and site surveying going back to the beginning of this century (see Burrow, 1985, for a detailed history). The appointment of O.G.S. Crawford as the first Ordnance Survey Archaeology Officer inaugurated a systematic approach to archaeological mapping and recording. In fact, Ordnance Survey (OS) record cards have formed the backbone of most SMRs. The findings of the Walsh Committee (Walsh, 1969), set up to look into the protection of field monuments, recommended that every County Planning Authority should hold a record of all known field monuments and have suitable archaeological expertise on their staff. In view of the threatened, and indeed actual, destruction of the archaeological resource due to rapid urban development, the government of the day reacted favourably to these recommendations. The result is that today nearly the whole of the UK is covered by SMRs with most SMR workers belonging to the Association of County Archaeological Officers (ACAO). Importantly, the emphasis in SMRs is focused on detailed local archaeological information and not just sites of national interest.

The first SMRs of the late 1960s and early 1970s were based on manual card index systems with many soon adopting the Optical Co-Incidence punch card system (Benson, 1974). Although the potential of computer-based SMRs was recognised in the late 1960s it was not until 1974 that the first application emerged (Benson, 1985). Many SMR workers were slow to appreciate the advantages of computerized records. As late as 1978 the ACAO in their *Guide to the Establishment of Sites and Monuments Records* had to incorporate the optical coincidence card system as an option to the mainstream computerization of the Record 'in order to satisfy those members who were perfectly happy to continue with a manual system...and who were unwilling or unable to experiment or invest in the more effective technology then available' (Benson, 1985, p.33). The last decade has seen the slow and somewhat painful adoption of computers by all SMRs with much discussion centred around the standardization of records and terminology (Chadburn, 1988). A recent survey of SMRs (Chadburn, forthcoming) has shown that the 46 in England display a wide range of hardware and software types with 33 being microcomputer based while the rest employ either minicomputers or mainframe systems. It is perhaps surprising that in the late 1980s 'a few SMRs still rely heavily on Optical Co-Incidence Cards for retrieval, and they can only undertake limited searches of their records'. *(ibid)*

Although the present situation is the result of *ad hoc* development with very little standardization in hardware, software or content, one theme common to all SMRs has been their inability to integrate satisfactorily the full locational component of an archaeological record within the computerized archaeological database. SMR database structure very closely mirrors the earlier card index structure. Within the computerized database the main spatial unit is the parish and usually an OS National Grid reference is given. Operating parallel to the computerized database, the SMRs have a series of 1:10,560 scale (or metric equivalent) OS maps, sometimes with various overlays, old maps of interest or coverage at different scales for sensitive areas, onto which each database entry is plotted by hand and referenced by a unique SMR number. Some SMRs also include a collection of aerial photographs which may or may not be referenced to the database and/or the maps. The spatial information associated with an archaeological record is therefore split between three different storage media and, not uncommonly, different physical locations within the same building. This rather inflexible structure greatly restricts the range of questions that can be asked of the data and the quality and nature of the response from an SMR. Archaeological enquiries with an emphasis upon location must be phrased in terms of either the fields within the database record structure (usually related to the parish) for computer-based output, or map sheet number for manual cartographic output. A request to a computerized SMR, for example, for information concerning all prehistoric sites in an area of five by ten kilometres will typically result in a five centimetre thick printout and two large photocopies of maps (each covering a five kilometre square) covered in hand-drawn black markings and corresponding SMR reference numbers.

This aspect of SMRs highlights a number of general problems relating to the handling of spatial information in the archaeological record. The single grid reference and corresponding symbol on the map are applied to a site regardless of the actual size of either. A small barrow, for instance, or even a single artifact is represented by a symbol which is probably hundreds of metres in size if taken to scale. It is the same in reverse for large sites; even if they are shown to size on the map, in the database they are recorded by a single point grid reference which is usually an estimation of the site centroid. A well preserved Late Prehistoric field system, for example, which may be many hectares in extent, is carefully drawn onto the OS base maps. In the database, however, it is represented by a single grid coordinate with no indication of size or shape. The spatial component of an archaeological record is therefore effectively divorced from the computerized database and stored independently on hand-drawn maps. There is an obvious need to be able to record digitally the extent, nature and shape of each site. A basic premise in archaeology is that human activity involves the ordering and use of space and that such activities are likely to be represented by patterning within the archaeological record. As such, archaeological phenomena are underpinned by their unique position in space and time and by the latent relationships existing between them. The spatial component in archaeological analysis is thus particularly important and yet has been excluded from the computing environment because of the difficulties previously associated with integrating archaeological information which differ in their basic spatial unit. For these reasons the potential which SMRs have for archaeological analysis remains virtually untapped by archaeologists. Recent advances in computer hardware and in software in the form of GIS, now make it possible to record and integrate archaeological information according to the basic spatial primitives of point, line, zone and pixel and thereby overcome these fundamental problems.

Besides the problems associated with storing archaeological information outside the computerized database on hand-drawn maps, there is a further problem in that maps are notoriously difficult to update and tend to fossilize information. While this does not necessarily apply to the archaeological information on the SMR maps, which can be added or altered manually, it does apply to the underlying OS topographic base maps which are often years out of date; a last major revision in the 1950s is not unusual. When a new version of a map is released by the OS the whole SMR would have to be copied by hand from the old to the new maps if the latest topographic information were to be utilized. Further, the digital encoding of these maps is presently being undertaken by the OS to provide national topographic coverage. This digital resource will form a major topographic database for the UK which may be linked with other environmental databases. Without a full locational reference, however, the SMRs will only be able to tap a small part of the potential offered by the availability of this digital map information.

Not only are the computerized databases unable to provide archaeological information in a form which satisfies many archaeological enquiries, but in their present computerized form they also fall short of satisfying the requirements of the overseeing body to which they are currently dedicated. Because of the ever present pressure on the historic environment in the UK from urban and rural development, the main application of SMRs has been geared firmly toward servicing planners and the extensive planning system in the UK. Given the origins of the SMRs and the fact that most registers are actually located in County Council offices this is perhaps not surprising. Most enquiries originating from planning authorities are concerned less with the archaeological content of the record *per se* and more with assessing the significance and possible impact on the cultural resource arising from the granting of planning permission to a developer to build at a given location. Many planning enquiries therefore are spatially oriented and geared toward knowing about what exists, or often what might exist, at a given location or the area immediately adjacent to it. In this case the computerized system is abandoned at the first hurdle to play a secondary role to that of the primary visual inspection of the hand-drawn maps. Similarly any possibility of modelling the computerized database or establishing a predictive capability based upon multivariate relationships for example, is severely curtailed because of the lack of a sufficiently accurate locational component.

Integration with other environmental data is also severely limited. Archaeology has been painfully slow to integrate with other areas of landscape conservation despite the fact that the archaeological record is a finite and rapidly disappearing resource. The greening of British politics over the last few years has included only a minimal archaeological input. A trend toward a coordinated and integrated approach to the many different strands of landscape conservation could be a major motivating factor for the adoption of GIS in archaeology on a large scale. Much of this integration would undoubtedly occur at the county level and the SMRs could be central to this process. It may be that many SMRs are introduced to GIS technology on the coat-tails of county planning departments. GIS certainly offers the opportunity for archaeology to raise its public profile and integrate with not only other environmental databases but with conservation and resource management concerns as well. This potential has not gone unnoticed by some archaeologists working within planning departments such as in Hampshire and Kent. In the former the current intention is to establish a GIS based 'Environmental Record' which will include the county SMR together with other environmental aspects of the landscape including woodlands, historic buildings as well as officially designated areas such as Sites of Special Scientific Interest and Environmentally Sensitive Areas. Kent has the most recently established SMR in England and this is currently being added to a county database using GIS software. Other county councils are also considering feasibility studies for establishing regional databases covering their statutory areas of responsibility, including that of the cultural landscape. Certainly, GIS based SMRs enabling the integration of the spatial and thematic components of the archaeological record as well as other environmental data offer an attractive view of the future in which improved and more diverse uses of these archaeological databases are thereby encouraged.

After the loose confederation of SMRs, the second major organization in UK archaeology has just recently taken a faltering but significant step toward such a coordinated approach to the computerization of the archaeological record. The

Royal Commission on the Historical Monuments of England (RCHME) was established by Royal Warrant in 1908 and has a variety of duties. Its brief is national in scope (there are equivalent bodies in Wales and Scotland) and includes the creation and curation of a national archive of archaeological monuments and historic buildings. This corpus of information is housed in the National Monuments Record (NMR), the archaeological component being the National Archaeological Record (NAR) and in a library of aerial photographs. The NAR is the now defunct archaeology section of the OS, together with its collection of record cards, which was transferred in its entirety to the Southampton office of the RCHME a few years ago, showing a certain continuity in the recording of the heritage at the national level. It is, in fact, from these records that the new Kent SMR was compiled and supplied as an ORACLE database. Other duties of the RCHME include large-scale fieldwork in the form of land and aerial surveying to update the NMR and its more recently acquired role of providing archaeological information for publication on OS maps. To a certain extent the RCHME appears to have a more enlightened attitude towards computing than is the case for many SMRs. This is significant because a recent policy review has recommended that the RCHME should adopt a leading role in the coordination and development of SMRs. The implications of such a policy are important for the possibility now exists that the RCHME could develop a coordinated national approach to the computerized handling of archaeological records. In a coordinated reorganization under the aegis of the RCHME GIS could be expected to play a central role.

The computerization of the NAR has been progressing for several years (Leech, 1986; Hart and Leech, forthcoming) and the awareness of the potential offered by a fully integrated GIS system within the RCHME has been recognised (Grant, pers. comm.) although specific GIS applications have yet to be established. The RCHME has acknowledged the fundamental importance of applying computer technology to many of its statutory functions and to this end has recently created a Computer Services Department to which it has made a commitment of continued expansion (RCHME, 1988). The NAR currently operates within an ORACLE relational database environment linked to GIMMS digital mapping software using OS 1:625,000 series digital maps. Areas of aerial photograph coverage can be overlain onto a digital map background using GIMMS/Photonet to provide a map-based interactive search facility. Other areas of RCHME computer usage include CAD production of drawings from electronically captured survey data and the possibility of the NAR being available for on-line interrogation (Hart and Leech, forthcoming). It would seem that the RCHME has recognised the importance of the spatial element within the data it manages, and while its current computer systems are not ideal as far as analysis of this component is concerned, awareness within the organization is sufficiently developed already to consider adopting GIS technology.

The third pertinent organization within UK archaeology is the official office of state archaeology. In England this is the Historic Buildings and Monuments Commission (HBMC), usually known under its popular name of English Heritage, which has statutory responsibilities under the Ancient Monuments and Archaeological Areas Act 1979 and the National Heritage Act 1983. Again, there are equivalents in Wales and Scotland operating from the Welsh and Scottish Offices and within the Department of the Environment for Northern Ireland. As with the RCHME, English Heritage has a range of responsibilities which are national in character and mainly concerned with the preservation of buildings and monuments of national importance and with their promotion to the public. An important task of English Heritage, especially in the eyes of the archaeological profession, is the administering of central government's annual financial grant to archaeology. The system is one whereby individual archaeological bodies can submit bids for project funding. In the year 1987/88, for example, 561 applications were made, of which 277 were successful in getting a share of the £3.5 million ($5.6 million) distributed through English Heritage (English Heritage, 1987). This process is important within the present context because these grants are the means by which money has been specifically allocated for the computerization of the SMRs. The distribution of grants in this way has enabled English Heritage to exert a not inconsiderable influence over the development of the SMRs. This is reflected in the number of SMRs that use SAMSON which is software written by English Heritage based on the package Superfile. Out of the 33 microcomputer-based SMRs, 25 use SAMSON (Chadburn, forthcoming) despite it being a flat file database lacking much of the sophistication of modern database packages. In 1987/88 nearly £350,000 ($560,000) went to SMRs, while in the year 1988/89 a further £240,000 ($384,000) was given to 25 SMRs with a further commitment to another two years funding. In return English Heritage has been able to use the computerized SMR databases in the performance of its statutory duties. One such obligation is to keep a schedule of monuments of national importance which are protected under the 1979 Act. This schedule currently stands at about 12,800 monuments in England (approximately 2 per cent of the 635,000 known) but under the recently initiated Monuments Protection Programme (MPP) it is due to increase to about 12 per cent of the known total over the next few years. The identification of new sites to be added to the schedule is to be based on interrogation of SMRs.

While English Heritage has supplied considerable aid and advice to SMRs concerning computerization of the register, this has been aimed almost exclusively at establishing standard database applications with no attempt to integrate the spatial component. English Heritage is aware of digital spatial data systems because its own computer-based Schedule of Ancient Monuments (SAM) has recently been re-plotted onto a series of 6000 raster scanned OS maps at 1:10,000 scale (Clubb, 1988). In its dual role as provider and advisor to SMRs English Heritage could have played a key role in raising the awareness of the archaeological community to the new technology and speeding the diffusion of GIS through an important user community. It now seems, however, that the future lead role is to be with the RCHME and whether this is in the form of advice and/or financial help remains to be seen. A review of SMRs has already started, involving the RCHME and the ACAO, in which possible replacements for SAMSON are to be discussed. Despite this English Heritage remains an influential body within English archaeology and it is of interest to note its awareness of GIS, despite having to justify and demonstrate to central government the need for GIS to perform tasks which present systems cannot do (English Heritage, pers. comm.). One immediate application could be the ongoing assessment by English Heritage of the impact on archaeological sites of the massive programme of road building carried out by the Department of Transport during the late 1980s. This assessment is currently being done manually using the map sheets of SMRs. Such an exercise using GIS-based SMRs and buffering the appropriate corridors is one application area where the benefits of introducing GIS technology to the SMRs is clearly demonstrated. English Heritage is also currently supporting some GIS work in the form of financial support for the North West Wetlands Project, a large-scale survey and assessment of the wetlands of north-western England. The estimated cost of this project is £1.4 million ($2.24 million) over a period of 8 years of which between £5000 ($8000) and £10,000 ($16,000) per annum is designated for computing facilities including GIS.

Diffusing the technology: increasing archaeologists' awareness of GIS

Such a limited uptake of GIS by UK archaeologists, in spite of the obvious advantages of such techniques, may initially seem disappointing. As we suggest, however, it actually belies a number of potential application areas where the technology could make a substantial impact. Much of this response can be directly attributed to the widespread lack of awareness of GIS capabilities and of potential application areas in archaeology. Indeed, this lack of current awareness of GIS amongst archaeologists closely mirrors that of many other discipline areas in the UK. In raising this awareness level the recent publication of the government report into *Handling Geographic Information* (DoE, 1987) has important implications for the archaeological community, as well as the wider GIS user community (for a discussion of this report see Masser, 1988; Chorley, 1988; Rhind and Mounsey, 1989). The Chorley Report, named after the chairman of the enquiry, was commissioned by the government in 1985 on the recommendation of a House of Lords Select Committee enquiring into *Remote Sensing and Digital Mapping* (HMSO, 1983). The report is important for acknowledging the rapid technical developments currently taking place in the handling of geographic information (for the government's responses see DTI, 1984; DoE, 1988). It also acknowledged the considerable commercial importance now attached to the field; a factor which contributed in no small measure to the setting up of the enquiry. The report is further important for focusing attention on those factors likely to influence the diffusion and take-up of GIS in the UK.

The Commission of Enquiry certainly had no doubt about the importance of GIS and its anticipated impact on society. The development of GIS, they claimed, was as significant to spatial analysis as

> 'the invention of the microscope and the telescope were to science, the computer to economics, and the printing press to information dissemination. It is the biggest step forward in the handling of geographic information since the invention of the map.' (DoE, 1987, p8)

The report highlighted what the committee felt were the principal barriers to obtaining the full benefits from this new technology. These barriers were related to problems arising from the great diversity of users and uses of GIS, and the failure by many, through lack of awareness of the potential benefits, to recognise the central importance of these systems (DoE, 1987, p1). The report recognised the continuing fall in the relative cost of computer processing power and memory, as well as that of GIS software which the committee anticipated would contribute considerably to the adoption rate of GIS. Furthermore, recommendations in the report to lessen existing restrictions placed on access to government data, as well as a more aggressive approach to marketing geographical data in the hands of government departments were further proposals designed to encourage greater use of geographical information. Likewise, improved data documentation and exchange standards for geographical information, the adoption of a basic spatial unit where possible, and the more rapid transfer of topographic information into digital form by the OS as mentioned earlier, were all considered important in the general goal of increasing the take-up of GIS. Most of these problem areas have their counterparts in archaeology though they have been little documented over the past few years.

Two aspects, highlighted by the report, are singled out here as being central to the adoption of GIS by the archaeological community in the UK. The first of these, touched upon earlier, concerns the current lack of awareness amongst archaeologists of the benefits to be gained from GIS technology. The commission considered this lack of awareness to be one of the most important reasons for the low take-up of GIS in the UK to date (DoE, 1987, p96). Certainly within the wider community the UK lags behind developments in North America with respect to the level of GIS awareness and, as a result, the range of application areas involving GIS. This lack of user awareness occurs at several levels and can range from ignorance of the techniques themselves and the benefits to be derived from using them, to how best to actually use them (DoE, 1987, p96).

This lack of cognizance is important not just in terms of the immediate take-up of the technology but also in the longer term because of the financial investments currently being made by archaeological organizations. Investment decisions concerning the computerization of the archaeological record which are made independent of GIS technology could delay the widespread diffusion of GIS in archaeology for many years. Certainly the actions of organizations, such as English Heritage and the RCHME, have important implications for the diffusion of GIS because these organizations directly influence the direction of much archaeological work including the purchase of hardware and software. As the report points out:

'The full benefits of sharing geographic information and Geographic Information Systems cannot be realised unless all the potential sharers are aware of them.'

Thus while some sections of the archaeological community may adopt GIS relatively quickly, the main benefits of adopting such technology could remain limited because of the lack of awareness or the rejection of GIS by others. For example, rejection or prolonged resistance to GIS by the SMRs would certainly affect the ability of other archaeologists, GIS users or not, to tap the full potential of these regional databases, undertake full spatial analysis, or exchange archaeological information.

The underlying need is clearly to increase the awareness of archaeologists as a whole to the new technology. The Commission of Enquiry and the publication of the Chorley Report have begun this process by bringing GIS to the attention of government agencies, including those with statutory responsibility for the historic and archaeological heritage. Indeed, the anticipated hierarchical diffusion of GIS through influential organizations is actually acknowledged by the committee in that they proposed targeting the heads of organizations at the chief executive and senior management levels as the focus for promotional efforts. It is envisaged that these will form 'seeding points' which will stimulate the uptake of GIS in the UK (DoE, 1987, p97).

While both the RCHME and English Heritage have considered the Chorley Report and its recommendations in the light of their current working practices, neither organization is formulating an official response for external publication. Within the RCHME, GIS is a 'live issue forming a part of the formulation of a new information strategy document' and is likely to comprise 'the core part of future retrieval systems' (Grant, pers. comm.). One major problem that the RCHME foresees, and a potentially serious constraining factor on its uptake of GIS, concerns the availability and cost of OS topographic digital map data for the country as a whole. The OS has released its 1:625,000 digital map series which, although adequate down to a scale of around 1:100,000, is not sufficient for most archaeological work. The OS is also releasing the 1:250,000 series of maps in digital form which could provide a suitable working scale for archaeological work. Due to a policy of self-financing imposed by government upon the OS, however, the cost to organizations of leasing the raw topographic digital data is likely to be in the region of £60,000 ($96,000) per annum. The initial response of many archaeological organizations, such as the RCHME, is that such a charge would be a prohibitive expense, although they are still to formulate a formal response to this proposed level of charging. Solutions could include forming an interest group to lobby for a change in OS costing policy, or alternatively, to produce independent digital topographic datasets. Because of the RCHME's recently acquired responsibilities for the SMRs, the latter option could be undertaken on a regional cost sharing basis in cooperation with local bodies such as County Councils.

The role of educators and training establishments in the diffusion of GIS

A vital factor determining the take-up of GIS in the UK concerns the necessary training of personnel in GIS technology. The ability of the archaeological community to implement GIS technology on even a modest scale, even if a strong desire to do so existed, would be seriously hampered by a shortage of people skilled in GIS. Rapid adoption may therefore founder on the rocks of impracticability because the necessary skilled personnel to establish and maintain these systems were not forthcoming. At best this could result in the adoption process being a lengthy affair; at worst a skills shortage could form an almost impermeable barrier from which only minor seepage of the technology occurred. It is likely that until this shortfall is met, the diffusion of GIS technology will be heavily constrained. The focus here is clearly upon educational and research institutions to provide a lead. It might be expected that these institutions would provide the early innovators, carriers and adopters of this technology, as well as undertake the necessary basic research and development facilitating GIS implementation within archaeology. In their role as educators it might also be expected that they would undertake to train future archaeologists in the use of GIS and provide the core of technicians to operate the systems. The development of demonstration projects to illustrate the benefits of GIS applications in archaeology would similarly increase the general awareness of archaeologists to GIS.

Assessing how educational institutions and archaeology departments in UK higher education might respond to the innovation of GIS is best done by examining their responses to the impact of earlier similar technological innovations. It is of more than passing interest to examine how successful universities have been in encouraging archaeology students in the use of computers in general, even though such adoption has been undertaken from an initial very low awareness base. The teaching of archaeology in British universities is very largely non-vocational in the sense that the practical requirements of being a professional archaeologist have a minor influence on course content. Because of the traditional autonomy of universities the structure and content of archaeology courses are very diverse (Austin, 1987) and computing fits into these more often as a result of the research interests of particular members of staff than from any overall teaching strategy (Greene, 1988, for example). There is also a distinct lack of communication between the major archaeological employers within the UK and university departments, which serves only to emphasize that a qualification in archaeology is a general measure of ability rather than an indicator of specific skills. The Institute of Field Archaeologists (IFA), through its Training Working Party, has been investigating the whole area of training within the profession and, although it has yet to prove itself in the eyes of many UK archaeologists, the IFA is rapidly becoming one of the few bodies capable of bringing together trainers and employers in curriculum formation.

A recent survey of the teaching of archaeological computing in British universities (Martlew, 1988) has shown that within 18 undergraduate courses, an introduction to computing can vary from a mention within a single lecture to a compulsory one-year course of several hours per week. Many students embarking upon a degree course in archaeology have no computing skills at all and are taught basic skills as a corollary to undertaking an archaeological project. This applies equally to graduate students who are invariably dependent upon their own interests and those of their supervisors or other members of staff within their department. Only two British universities have lectureships specifically in computer applications and quantitative methods in archaeology, and only two offer post-graduate courses in this area. Because of a lack of resources and expertise some departments rely on their archaeology students attending general computing courses or being taught by staff from computing departments. This has generally met with disapproval from the students who feel their endeavours to be too far removed from the archaeological source of their interest. In general though this situation could easily prove analogous to a future situation where archaeology students are taught GIS applications by geographers. Indeed, some archaeology departments already have close links with geography departments and the emulation of geographical techniques is certainly not new to archaeology (see for example Hodder, 1977; Bintliff, 1986; Grant, 1986). The establishment of a national curriculum in GIS, as currently undergoing beta-testing in the United States, may provide a basis for such inter-disciplinary teaching to proceed. It is beyond doubt, however, that the most successful archaeological computing is taught by archaeologists using relevant archaeological data and, if possible, involving current on-going research which fixes the emphasis firmly on the archaeological content. The rather confused and uncoordinated situation as outlined here is not an ideal foundation on which to contemplate teaching a new technology such as GIS to archaeologists. While increasing financial demands for the required hardware and software would present a major problem in the present climate of financial retrenchment, what is perhaps more worrying is the perceived lack of intent by faculty. It is noticeable that in many departments the teaching of archaeological computing falls to the younger members of staff. The present response of more senior and influential academics to general computing skills in archaeology is problematic and not promising for inaugurating the teaching of specialisms such as GIS. The realistic conclusion at this stage must be that the early carriers and adopters of GIS, and the provision of trained technicians, are unlikely to be produced by university archaeology departments in the short, or even medium, term.

GIS and archaeological spatial analysis

In addition to the twin issues of GIS awareness and training, a further important factor influencing the adoption of GIS technology by archaeologists is the anticipated position of GIS within the existing approaches to spatial analysis practised by archaeologists. Since early this century archaeologists have used a variety of techniques to analyze and interpret spatial patterning in the archaeological record. In essence these techniques can be generalised as being map-based or statistically oriented approaches. We contend that GIS provides a further phase in the evolution of archaeological spatial analysis by permitting much greater flexibility to be exercised in structuring the raw data and enabling both map-based and quantitative approaches more closely to complement each other.

Map-based approaches are inherently descriptive in nature and provide a basis for interpretation through informal and intuitive methods of analysis. This usually involves comparing and commenting on the distributions of archaeological phenomena plotted on maps, often in conjunction with background contextual information. This approach has evolved in recognition that at both the intra-site and regional levels such graphical representation is a simple and rapid method of displaying and summarizing large quantities of complex spatial and thematic data. Since the early 1970s quantitative spatial analysis has followed a very different line of enquiry fuelled partially from a concern about the perceived subjectivity of the map-based approach. Based on techniques mainly borrowed from disciplines such as geography and plant ecology (Hodder and Orton, 1976) archaeological spatial quantitative methods are concerned with quantifying the relationships between points and producing some form of summary statistic with an associated confidence limit. Quantitative spatial analysis has developed rapidly over the last few years with many new techniques being introduced by archaeologists for specifically archaeological problems (Hietala, 1984) and even resulting in a dedicated software package (Blankholm, 1989). Discontent with these approaches, however, has centred on their reliance on point data or cell counts and their inability to incorporate essential background information in the analysis thereby forcing a severe reductionist approach to the data. In 1982 Kintigh and Ammerman argued for 'heuristic spatial analysis combining the intellectual sophistication of intuitive approaches with the information processing capacity and systematic benefits of quantitative treatments' (Kintigh and Ammerman, 1982, p33). Further criticism of standard spatial quantitative methods has been listed by Whallon (1984) who sees the main problems as centring on basic incongruencies between the statistical methods and the nature of archaeological data, as well as the type of questions asked of the data. This has resulted in 'a large measure of dissatisfaction with available approaches and their results' (Whallon, 1984, p242).

There has always been a perceived lack of empathy between practitioners of the two respective approaches. We suggest that integration of the two approaches using GIS provides is a valid way forward. The strengths of such an integrated approach were recognised many years ago by Clarke in his model of the Iron Age settlement at Glastonbury (Clarke, 1972). This work remains a landmark in spatial analysis in archaeology because it embodies the major strengths of map-based methods yet reflects Clarke's quantitative approach to analysis. This approach permitted the combination of a variety of data types such as artifacts, structures, features and environmental variables, so as to build up as complete an archaeological picture as possible. The data were then subjected to a developing lattice of analyses combining vertical spatial relationships, horizontal spatial relationships, structural relationships and artifact relationships. As Clarke states

'the spatial relationships between the artifacts, other artifacts, site features, other sites, landscape elements and environmental aspects present a formidable matrix of alternative individual categorisations and cross-combinations to be searched for information'. (Clarke, 1972, p363)

The more relevant information that can be integrated within an analysis will increase the range of questions that can be asked and improve the resulting interpretations. It is this flexibility and power for manipulating, restructuring and analyzing the raw archaeological data and favouring the integration, rather than the separation, of map-based and quantitative approaches to spatial analysis, that are being offered by GIS.

General summary and conclusions

The main theme of this paper has been to isolate the likely rate and pattern of diffusion of GIS within the UK archaeological community. It remains to reiterate and emphasize some of the more important elements of the preceding discussion. We see two major factors which are likely to influence the adoption of GIS in the UK; the nature of the archaeological record and the organizational structure of archaeology. The record is not only extremely rich and varied throughout the UK but is also well documented through the system of county-based SMRs, either in a computerized form or as maps. Although the organizational structure of archaeology is very fragmentary across the UK, certain influential bodies can be identified and the RCHME in particular should be mentioned in its new role involving the coordination and development of SMRs. These two factors combine together to suggest that the SMRs are likely to be the springboard for the *widespread* adoption of GIS in UK archaeology.

Although it is very early days yet to comment on this new role of the RCHME, its success or failure in encouraging the adoption of GIS could depend on the relationship established with the ACAO and whether or not financial backing is made available to replace or enhance the funding traditionally obtained from English Heritage. If the RCHME does advocate a more strident push towards GIS and away from the standard database solution of English Heritage, potential strains within the existing funding structure can be visualized. Many SMRs cannot rely on GIS facilities being supplied by their planning department and it must be remembered, in any case, that the take-up of GIS has been limited in all areas in the UK and not just in archaeology. The latent potential exists for good GIS applications and the need for GIS in UK archaeology is easily demonstrated;

- The spatial component is vital within archaeological data and yet it is currently divorced from computer-based archival systems such as the SMRs;
- A major drive towards GIS is beginning within planning departments and it is essential that SMRs are incorporated within it;
- The two approaches to spatial analysis in archaeology, map-based and quantitative, could move some way towards unification with GIS;
- Regional and national archives would become more accessible to regular analytical and research work rather than being dominated by planning applications and the occasional *ad hoc* research project;
- OS topographic digital databases are becoming available at scales that are useful to archaeology.

Within these influencing factors two major drawbacks to the adoption of GIS need to be emphasized. The first is the raising of awareness of the potential of GIS and, again, this applies to all application areas in the UK and not just archaeology. The quotes from the Chorley Report cited above show that in the North American experience it is human and not technological problems which form the greatest barriers to GIS adoption. The potential of sharing archaeological and geographical data will be severely limited if all of the potential sharers are not aware of the benefits. This returns us to the importance of the SMRs in this context for if they are not included in GIS applications in UK archaeology they will leave a large hole to fill. This lack of awareness has probably been the main reason for the slow adoption of GIS in the UK although since the publication of the Chorley Report, which is important as the official 'Government seal of approval' amongst other things, the situation is beginning to change.

The second major drawback is in the training of technicians and users of GIS, again generally but especially in archaeology. It is difficult to avoid an element of pessimism judging by the precedent of the adoption of IT within university archaeology courses. A large proportion of qualified archaeologists in the UK have never used a database package in anger which is not encouraging when the complexity of GIS software is considered together with the underlying cartographic, statistical and spatial concepts. It is unlikely that universities will play a major role in the *general* training of GIS archaeological users in the short and medium term although they are likely to produce isolated exemplar research projects.

Despite these problems we do see universities providing the early adopters and disciples of GIS in UK archaeology whose projects will play an important part in raising the general awareness. This combined with hierarchical diffusion, potentially from the RCHME through the SMRs, suggests that while GIS applications in UK archaeology are not about to soar to the heights of those in North America, at least they have taken off.

References

Alvey, B., 1989, Hindsight. *Archaeological Computing Newsletter*, **19**, 4–5

Association of County Archaeological Officers, 1978, *A guide to the establishment of sites and monuments records*. (Association of County Archaeological Officers)

Austin, D., 1987, Archaeology in British Universities. *Antiquity*, **61**, 227–238

Bailey, R., Howes, D., Hackenberger, S. and Wherry, D., 1985, Geographic Information Processing in land use modeling and testing in the Columbia River basin. Paper presented at the 50th Annual Meeting of the Society for American Archaeology, Denver, USA

Benson, D., 1974, A Sites and Monuments Record for the Oxford region. *Oxoniensia*, **37**, 226–237

Benson, D., 1985, Problems of data entry and retrieval. In *County Archaeological Records: progress and potential*, edited by I. Burrow, (Somerset: Association of County Archaeological Officers) pp 27–34

Bintliff, J.L., 1986, Archaeology at the interface: an historical perspective. In *Archaeology at the Interface: studies in archaeology's relationships with history, geography, biology and physical science*, edited by J.L. Bintliff and C.F. Gaffney (Oxford: British Archaeological Reports), International Series 300, pp.4–31

Blankholm, H.P., 1989, ARCOSPACE. A package for spatial analysis of archaeological data. *Archaeological Computing Newsletter*, **19**, 3

Booth, B.K.W., Grant, S.A.V. and Richards, J.D., (eds), 1989, *Computer usage in British Archaeology*, 2nd Edition (Birmingham: The Institute of Field Archaeologists)

Burrough, P.A., 1986, *Principles of Geographical Information Systems for land resources assessment*. (Oxford: Clarendon Press)

Burrow, I., (ed), 1985, *County archaeological records: progress and potential*. (Somerset: Association of County Archaeological Officers)

Chadburn, A., 1988, Approaches to controlling archaeological vocabulary for data retrieval. In *Computer and Quantitative methods in archaeology 1988*, edited by S.P.Q. Rahtz, (Oxford: British Archaeological Reports) International Series 446, 2 Volumes, pp.389–398

Chadburn, A., forthcoming, Computerized county Sites and Monuments Records in England: an overview of their structure, development and progress. In *Quantitative methods in archaeology 1989*, (Oxford: British Archaeological Reports) International Series

Chorley, R., 1988, Some reflections on the handling of geographical information. *International Journal of Geographical Information Systems*, **2**(1), 3–9

Clarke, D.L., 1972, A provisional model of an Iron Age society and its settlement system. In *Models in archaeology*, edited by D.L. Clarke, (London: Methuen), pp. 801–869

Clubb, N., 1988, Computer mapping and the Scheduled Ancient Monument record. In *Computer and Quantitative methods in archaeology 1988*, edited by S.P.Q. Rahtz, (Oxford: British Archaeological Reports) International Series 446, 2 Volumes, pp. 399–408

Cooper, M. and Richards, J.D., (eds), 1985, *Current issues in archaeological computing*. (Oxford: British Archaeological Reports), International Series 271

Creamer, W. 1985, The upper Klethia Valley: computer generated maps of site location. Paper presented at the 50th Annual Meeting of the Society for American Archaeology, Denver, USA.

DoE, 1987, *Handling Geographic Information*. Report to the Secretary of State for the Environment of the Committee of Enquiry into the Handling of Geographic Information, chaired by Lord Chorley, (London: H.M.S.O.)

DoE, 1988, *Handling Geographic Information: the Government's response to the Report of the Committee of Enquiry Chaired by Lord Chorley, Department of the Environment*. (London: H.M.S.O.)

Donoghue, D.N.M. and Shennan, I., 1988, The application of multispectral remote sensing techniques to wetland archaeology. In *The exploitation of wetlands*, edited by P. Murphy and C. French, (Oxford: British Archaeological Reports), British Series 186, pp. 47–59

Donoghue, D.N.M., 1989, Remote sensing and wetland archaeology. In *The archaeology of rural wetlands in England*, edited by J.M. Coles and B.J. Coles, (Exeter: University of Exeter, Department of History and Archaeology) WARP Occasional Paper 2, pp.42–45

DTI, 1984, *Remote Sensing and Digital Mapping: The Government's reply to the First Report from the House of Lords Select Committee on Science and Technology*. (London: H.M.S.O.), Cmd 9320, Department of Trade and Industry

English Heritage, 1987, *Rescue archaeology funding in 1987–88*. (London: English Heritage)

Ferguson, T.A., 1985, Use of Geographic Information Systems to recognize patterns of prehistoric cultural adaptation. Paper presented at the 50th Annual Meeting of the Society for American Archaeology, Denver, USA.

Ferguson, T.A., 1986, The Professional Map Analysis Package (pMAP) and its archaeological application. In *National Workshop on Microcomputers in Archaeology*, (New Orleans: Society for American Archaeology)

Gill, S.J. and Howes, D., 1985, A Geographical Information System approach to the use of surface samples in intra-site distributional analysis. Paper presented at UISPP Commission IV Symposium on Data Management and Mathematical Methods in archaeology, Denver, USA.

Grant, E., (ed), 1986, *Central places, archaeology and history*. (Sheffield: University of Sheffield, Department of Prehistory and Archaeology)

Greene, K., 1988, Teaching computing to archaeology students. *Archaeological Computing Newsletter*, **14**, 21–24

Griffith, F., 1989, 1989 bumper year for AP. *British Archaeological News*, **4**(5), (London: Council for British Archaeology) pp. 65–66

Harris, T.M., 1985, GIS design for archaeological site information retrieval and predictive modelling. In *Professional archaeology in Sussex: the next five years*. Proceedings of a one day conference by Association of Field Archaeologists, Seaford, Sussex

Harris, T.M. and Dudley, C., 1984, A computer-based archaeological site information retrieval system for the Brighton area: a preliminary report. *Sussex Archaeological Society Newsletter*, **42**, 366

Harris, T.M., 1986, Geographic Information System design for archaeological site information retrieval. In *Computer Applications in Archaeology 1986*, edited by S. Laflin, (Birmingham: University of Birmingham), pp.148–161

Harris, T.M., 1988, Digital Terrain Modelling and three-dimensional surface graphics for landscape and site analysis in archaeology and regional planning. In *Computer and quantitative methods in archaeology 1987*, edited by C.L.N. Ruggles and S.P.Q. Rahtz, (Oxford: British Archaeological Reports), International Series 393, pp.161–172

Hart, J.S. and Leech, R.H., forthcoming, The computerization of the National Archaeological Record: progress and the future for online access. In *Computer and quantitative methods in archaeology 1989*, (Oxford: British Archaeological Reports), International Series

Hietala, H., (ed), 1984, *Intrasite spatial analysis in archaeology*. (Cambridge: Cambridge University Press), New Directions in Archaeology Series

H.M.S.O., 1983, *Remote Sensing and Digital Mapping*. Report of the House of Lords Select Committee on Science and Technology, chaired by Lord Shackleton, 2 volumes, (London: H.M.S.O.)

Hodder, I., 1977, Spatial studies in archaeology. *Progress in Human Geography*, **1** (1), 33–64

Hodder, I.R. and Orton, C.R., 1976, *Spatial Analysis in Archaeology*. (Cambridge: Cambridge University Press), New Studies in Archaeology Series

Holman, N., 1985, Evaluating the contents of Sites and Monuments Records: an alternative approach. *Archaeological Review from Cambridge*, **4**(1), 65–79

Kellogg, D.C., 1987, Statistical relevance and site locational data. *American Antiquity*, **52**(1), 143–150

Kintigh, K.W. and Ammerman, A.J., 1982, Heuristic approaches to spatial analysis in archaeology. *American Antiquity*, **47**, 31–63

Kvamme, K.L., 1985a, Geographic Information Systems techniques for regional archaeological research. Paper presented at UISPP Commission IV symposium on data management and mathematical methods in archaeology, Denver, USA

Kvamme, K.L., 1985b, Fundamentals and potential of Geographic Information System Techniques for archaeological spatial research. Paper presented at the 50th Annual Meeting of the Society for American Archaeology, Denver, USA

Kvamme, K.L., 1986, An overview of Geographic Information Systems for archaeological research and data management. In *National Workshop on Microcomputers in Archaeology*, (New Orleans: Society for American Archaeology)

Kvamme, K.L., 1989, Geographic Information Systems in regional research and data management. In *Archaeological Method and Theory*, Volume 1, edited by M.B. Schiffer, (Tucson: University of Arizona Press), pp. 139–203

Kvamme, K.L. and Kohler, T.A., 1988, Geographic Information Systems: technical aids for data collection, analysis, and displays. In *Quantifying the present and predicting the past: theory, method and application of archaeological predictive modeling*, edited by J.W. Judge and L. Sebastian, (Washington D.C.: U.S. Government Printing Office), pp. 493–547

Leech, R.H., 1986, Computerization of the National Archaeological Record. In *Computer Applications in Archaeology 1986*, edited by S. Laflin, (Birmingham: University of Birmingham), pp 29–37

Marble, D.F., Calkins, H. and Peuquet, D., (eds), 1984, *Basic readings in geographic information systems*. (New York: SPAD Systems)

Martlew, R., 1984, *Information systems in archaeology*. (Gloucester: Allan Sutton Publishing)

Martlew, R., 1988, New technology in archaeological education and training. In *Computer and Quantitative methods in archaeology 1988*, edited by S.P.Q. Rahtz, (Oxford: British Archaeological Reports) International Series 446, 2 Volumes pp 499–504

Masser, I., 1988, The development of geographic information systems in Britain: the Chorley Report in perspective. *Environment and Planning B: Planning and Design*, **15**, 489–494

Miller, L., 1986, MIPS: A Microcomputer Map and Image Processing System developed for easy use by individual natural scientists. In *National Workshop on Microcomputers in Archaeology*, (New Orleans: Society for American Archaeology)

Peuquet, D. and O'Callaghan, J., (eds), 1983, *Proceedings of a United States/ Australia workshop on the design and implementation of computer-based geographic information systems*. (Amherst, New York: International Geographical Union Commission on Geographical Data Sensing and Processing)

Powlesland, D., forthcoming, From the trench to the bookshelf: computer use at the Heslerton Parish Project. In *Computing for archaeologists*, edited by S. Ross, J.C. Moffett and J. Henderson, (Oxford: Oxford University Committee for Archaeology) Monograph No. 18

Rhind, D., 1981, Geographical Information Systems in Britain. In *Quantitative Geography*, edited by N. Wrigley and R.J. Bennett, (London: Routledge)

Rhind, D.W. and Mounsey, H.M., 1989, Research policy and review 29: the Chorley Committee and 'Handling Geographic Information'. *Environment and Planning A*, **21**, 571–585

Richards, J.D. and Ryan, N., 1985, *Data processing in archaeology*. (Cambridge: Cambridge University Press), Cambridge manuals in archaeology

Royal Commission on the Historical Monuments of England, 1988, *Annual Review 1987–1988*. (London: RCHME)

Shennan, S.J., 1988, *Quantifying archaeology*. (Edinburgh: Edinburgh University Press)

Walsh, D., 1969, *Report of the Committee of Enquiry into the arrangements for the protection of field monuments 1966–8*. Reprinted 1972, Command 3904 (London: HMSO)

Wansleeben, M., 1988, Geographic Information Systems in archaeological research. In *Computer and Quantitative methods in archaeology 1988*, edited by S.P.Q. Rahtz, (Oxford: British Archaeological Reports) International Series 446, 2 Volumes, pp.435–451

Whallon, R., 1984, Unconstrained clustering for the analysis of spatial distributions in archaeology. In *Intrasite spatial analysis in archaeology* edited by H. Hietala, (Cambridge: New Directions in Archaeology Series, Cambridge University Press), pp.242–277.

5

GIS, archaeology and freedom of information

Linda F. Stine and Roy S. Stine

'Free people are, of necessity, informed; uninformed people can never be free', (Senator Long, in O'Brien, 1981, p. 28)

In 1966 President Johnson signed the Long-Dirksen bill (Public law number 89–554). It has officially been in effect since 1967 (Houdek, 1981; Sherick, 1978). This law, more commonly known as the Freedom of Information Act (FOIA, as amended 1975, 5. U.S.C. 552) is based upon the premise that information generated by federal agencies belongs to the people of the United States. Prior to the passage of the FOIA the burden of proof was upon the citizen to provide need or cause to examine government documents. Now the process has shifted from a 'need to know' to a 'right to know' basis (O'Brien, 1981; United States Congress, House of Representatives serial no. 13174–2).

The FOIA specifically pertains to documents held by the administrative agencies and the executive branch of the Federal government. For example, certain federal agency management studies would be included under this act. It would also cover specific types of digital information, such as geographic information system (GIS) files. The law does not apply to the material maintained by the legislative and judicial branches (O'Brien, 1981, pp. 1–28; United States Congress, House of Representatives serial no. 13174–2). FOIA does contain nine major exceptions (see Sherick, 1978, pp. 27–37). Some of these are national security or foreign policy items covered by an executive order and items which are specifically exempted from disclosure by a statute. As discussed in O'Brien (1981) these exemptions also include 'Trade secrets', personnel, medical or other records whose disclosure would 'constitute a clearly unwarranted invasion of personal privacy' (O'Brien, 1981; see also Sherick, 1978, pp. 28–32). Additional exempted data include current law enforcement investigation records, some materials concerning financial institutions, and certain geological and geophysical information.

Archaeological site locations are usually protected by specific statutes, under FOIA heading #3, 'Matters that are specifically exempted from disclosure by statute' (Sherick, 1978, pp. 29; see also O'Brien, 1981). This has been interpreted rather broadly. As a result, over one hundred statutes are used by agencies to defer the FOIA (Sherick, 1978). Archaeological materials have apparently been protected under the general preservation statutes, such as Public Law 93–291, the Archaeological and Historic Preservation Act of 1974 (e.g. Keel 1988; United States Congress, Interagency Archaeological Services, 1979). In particular, under the National Historic Preservation Act of 1966, title one, section 101, subsection (a): the Secretary of the Interior is authorized

> 'to withhold from disclosure to the public information relating to the location of sites or objects listed on the National Register whenever he determines that the disclosure of specific information would create a risk of destruction or harm to such sites or objects'
> (King *et al.*, 1977, pp. 205–206).

Under the provisions of this act, the location of National Register sites, regardless of state, are currently protected. Passage of the Archaeological Resources Protection Act of 1979 (Public Law 96–95) has ensured protection of federal site locations, under section 9(a), 'Confidentiality'. This states that

> 'Information concerning the nature and location of any archaeological resource for which the excavation or removal requires a permit or other permission under this Act or under any other provision of Federal Law may not be made available to the public under subchapter II of chapter 5 of Title 5 of the United States Code...,'
> (Public Law 96–95—October 31, 1979, 96th Congress).

State Freedom of Information laws tend to conform to the Federal FOIA (United States Government, Subcommittee on Administrative Practice and Procedure, 1978). As a result, individual states have specific formal exemptions for archaeological data, usually under a state archaeological protection act. For example, Massachusetts and Virginia have laws and regulations that protect archaeological sites. In addition, they both have exemptions in their respective FOIAs that protect their archaeological site files from open, unsupervised access (Massachusetts Historical Commission, personal communication 1989; Virginia Department of Historic Preservation, personal communication 1989). State and territorial FOIAs do not always protect archaeological site files. Guam's and South Carolina's FOIAs, for example, do not exclude archaeological data. Guam has a 'Sunshine' Act, Public Law 19–5, that includes archaeological records with all other public documents, requiring open access (Guam Historic Preservation Office, personal communication 1989). Last year, the Governor's Office vetoed a bill that would have excluded the archaeological site files from South Carolina's FOIA. To the surprise of some archaeologists, the strongest lobby against passage of the bill came from the South Carolina Press Association. This association is against any exemptions from the state FOIA, regardless of individual merit. Instead of formal state legislation, the association prefers agencies to use the existing system of regulations (Steven Smith, South Carolina Deputy State Archaeologist, personal communication 1989).

Those states that have no specific legal stand on access to archaeological site files have left interpretation of the FOIA to the agency in charge. When surveyed this year, the majority of State Historic Preservation Offices (SHPO) indicated that they have developed specific, if informal, policies limiting access to archaeological site files.

Archaeologists can use the federal FOIA exemptions to help protect federal archaeological sites. In most cases, they can also use state regulations to help protect state sites. However, certain federal agencies have learned to manipulate the FOIA to deny scholars access to certain planning decisions. Although this has been mitigated somewhat by the 1975 amendments to the FOIA, agencies still have some leeway in their interpretation of 'right to know'. Scholars can use the FOIA to gather data concerning the existence, condition, and protection of archaeological sites under federal jurisdiction.

There are three major reasons for archaeologists to be aware of the FOIA. These are as follows: (1) concerns of archaeological site protection; (2) issues relating to the access of agency documentation concerning sites; and (3) consideration of the effect of the digital revolution on current policies and interpretations of the FOIA. Currently, general interpretation of the FOIA exemptions allows for the protection of archaeological sites. This is managed through the exclusion of certain types of archaeological data from public access. The effects of GIS development and implementation have sparked debate about continuation of this policy. In addition, protection of site location information through relating existing preservation laws to FOIA exemption 3 is occasionally challenged (Patricia Cridlebaugh and Nancy Brock, South Carolina State Historic Preservation Office [SCSHPO], personal communication 1989).

Archaeologists can obtain management information from recalcitrant federal agencies under the FOIA (see King *et al.*, 1977, pp. 179, 317–318). Archaeologists can determine if any previous archaeological work had been undertaken in any federally mandated area. Documentation demonstrating the agency's compliance with cultural resource management laws, and use or misuse of data for planning purposes may be requested (King *et al.*, 1977). Of course, precise, detailed wording of any request under the FOIA is needed to ensure speedy and complete agency compliance (King *et al.*, 1977). For example, one archaeologist requested information from a Tennessee agency. After long delays, she was allowed access to certain files. However, she was not informed that the files had already been purged. If not for a few ethical workers, she would never have known that the files had been changed (Patricia Cridlebaugh, SCSHPO, personal communication 1989).

This cautionary tale has serious implications. Archaeologists need to have some assurance that the data they receive are complete and accurate. This will be just as true for digital information as for printed matter. Scholars requesting information under the FOIA should take the time to peruse existing guides to using the act (e.g. Sherick, 1978). This saves unnecessary delays, complications, and unknowing loss of information.

The increasing use of digitized data by government agencies, as well as by archaeologists, has led to debate concerning access to digital information. The use of digital systems has resulted in the ability to transfer information inexpensively and simply. Scholars and legislators are currently debating how these electronic databases relate to the FOIA. Archaeologists need to have a voice in this debate, as it concerns balancing the need for site protection with the desire for access to electronic information for planning and research purposes.

Actually, site protection may be enhanced by allowing planners and engineers access to site locational information. Once in a management database, such information would allow planners to 'avoid' adverse effects to archaeological sites. In cases where a site would be destroyed or damaged, planners would have more time to contact the correct preservation agency to work out a plan of action. Ignorance of site locations would no longer be an issue in any planning debates.

Federal law, replicated by many state laws, prohibits access to data such as site location without a proven need to know. As a result, professional pot hunters can be denied access to information concerning the location of specific types of sites. Imagine such a person gaining access to a GIS, showing both real and potential site locations, ranked by density and intactness! However, this does open up a Pandora's box of questions concerning how to decide 'need to know'. Generally, scholars with a proven research interest and those involved in a cultural resource management project are allowed supervised access to site file data.

In a recent survey of fifty-one United States Territorial and State Historic Preservation Offices, 96 per cent ($n = 49$) stated that they did have a set policy concerning access to archaeological information (1989 survey, see Appendix A). The majority indicated that they followed an in-house, informal policy. In all, 98 per cent ($n = 50$) indicated they allow professional archaeologists and their students access, and 96 per cent ($n = 49$) allow cultural resource management archaeologists admittance. Other scholars are given access to site file information at 92 per cent of the offices surveyed ($n = 47$).

Only 27 per cent, or 14 of those surveyed, indicated that they allow the public supervised access to site information. For example, in South Carolina, members of the general public have to write a brief research proposal discussing why they need access to archaeological site files. In a recent interview, the Deputy State Archaeologist stated that informal, written guidelines concerning site file usage were followed. Office personnel can describe research proposal guidelines to interested people by referring to an 'in-house guide'. These written guidelines are not available to the public as yet, nor have they been formally presented as set policy (Steven Smith, personal communication 1989).

The United States Congress Federal Interagency Coordinating Committee on Digital Cartography (FICCDC) has published the results of a survey of federal agencies and their views on GIS (1988). They discovered that although 84 per cent ($n = 37$) of the agencies surveyed were using or planning to use a GIS, only 27 per cent ($n = 10$) have any formal written policy concerning access and use (United States Congress, FICCDC, 1988). These include those agencies most likely to deal with archaeological materials, such as the Forest Service, Bureau of Land Management, and the National Park Service. The Army Corps of Engineers, as of 1988, was still investigating how best to write their policy (United States Congress, FICCDC, 1988).

Information management requires that access policy be set. Agency directors need to identify information management activities, and to plan for acquiring, controlling, pricing, budgeting, and evaluating information. Each agency must determine their data needs, who has jurisdiction over the data, and who sets policy about the data. Information relationships between agencies need to be known, and standard guidelines should be established (Kettinger *et al.*, 1988). This is especially true for those developing and using a GIS.

The final consideration of the FOIA and archaeology pertains to digital data systems. Archaeologists are increasingly using computers to manage their site files. In a recent survey (summer 1989) of SHPOs and related governing agencies, respondents were asked if they had computerized their site files. From a total of 51 respondents, 84 per cent ($n = 43$) stated that they were already in the process of doing so (see example of survey form in Appendix A). An additional 16 per cent ($n = 8$) responded that they were planning on computerizing their files in the future. As one can see, every responding agency ($n = 51$) stated that they planned to eventually have all their files in digital format.

A total of 86 per cent ($n = 44$) of those responding to the survey, stated that they were not planning, through choice and/or funding constraints, on developing a GIS.

Only 14 per cent ($n = 7$) of the agencies managing archaeological data were already employing a GIS. This has important implications for archaeologists and planning agencies. Future planners should have some access to archaeological site information. Site locations and other attributes need to be digitized into shared state and perhaps federal level GIS. Information on destroyed sites' locations can be shared without adverse effect. Existing sites could be protected and their locations incorporated for planning purposes. GIS operators could also be used automatically to exclude archaeological site locations, if warranted.

Guidelines for interagency access to archaeological information need to be clearly stated. Such cooperation raises many issues concerning archaeological data. How much information should preservation agencies impart about specific site locations to other planning agencies? Would other agencies see the need to protect the location of archaeological sites? Would site protection laws pertain to other agencies? Would the digital data, once included in another agency's database, be subject to full disclosure under the FOIA? One envisions certain interagency personnel, seeing nothing wrong with collecting artifacts, using data to pot hunt sites. Government agencies differ in their views of archaeology and the goal of archaeological site protection. As a result, decisions concerning access to automated archaeological data will have to be jointly made at all levels of government.

In a 1988 survey of federal agencies, it was found that 32 per cent ($n = 12$) are already using a 'broad scoped' GIS system. In all, 45 per cent ($n = 20$) of those responding stated that they were planning on implementing a full GIS by 1990 (United States Congress, FICCDC, 1988). The Bureau of Land Management is already using such as system, as is the National Park Service. The Forest Service and the Army Corps of Engineers are using a limited GIS at present, but are planning on expanding these systems as soon as possible (United States Congress, FICCDC, 1988).

These agencies are already sharing layers with other departments. For example, the Forest Service shares data with international, federal, state, and local governments. The National Park Service trades data with both federal and state agencies. The Bureau of Land Management and the Army Corps of Engineers cooperates with other federal agencies (United States Congress, FICCDC, 1988). This raises the issue of who determines what GIS or database information can be shared, and whether access should be limited to other government entities.

When GIS data are shared, a hierarchy of information levels must be discussed. For example, agreement upon data interface, availability, and security must be reached (Harper and DeCario, 1988; see also Bell, 1988; Marshall, 1988). Data interface considerations include the ability of other agencies to alter, knowingly or unknowingly, an agency's data. A system could be constructed that gives free access to all. However, some are calling for interacting systems that only allow an agency to '...view, plot or report on...' another department's data (Harper and DeCario, 1988).

The availability of data within a department is a matter of priority. System technicians need to have access to each individual data layer. Section managers need to have ready access to maps that reflect policy decisions.

The availability and security of such information in regards to other departments, researchers, and the public must be determined (Harper and DeCario, 1988; Bell, 1988; Marshall, 1988; Chrisman, 1987). Should researchers, including archaeologists, be allowed to purchase specific layers? Should there be differential access for university researchers as opposed to private consultants? Managers must also decide if the general public should be allowed the same access. Agency personnel must determine if the general public requires access to enough digital information to actually manipulate data.

A related topic of interest is the cost of copying such information. As described elsewhere (Scholtz and Chenhall, 1976), the cost of initial digitization of archaeological data is extremely high. However, once the database has been developed, corrected, and tested, the costs of upkeep remain relatively low. Those obtaining copies of such information in a government database should pay a nominal fee for replication, but should not have to pay recovery costs for the creation of the database.

In 1975 amendments to FOIA (Public law number 93–502) were passed over President Ford's veto. These amendments require that the government agencies affected by FOIA publish comprehensive indexes. Agency fees for locating and copying the records must be uniform and moderate as well. Importantly, the amendments prevent agencies from withholding an entire document if only part of the information is exempt, a procedure often evoked in the past. Instead, the document must be edited and released. The amendments also state that the agencies do not have to search materials for an individual, analyze documents, and/or collect information the agency does not have (United States Congress, House of Representatives serial no. 13174–2).

Certain state archaeological agencies have had a policy requiring that contractors pay for site file information, while other researchers pay no user's fee. Perhaps this was somehow justified in the past, but the amendments to the FOIA especially caution against such differential treatment of users. The FOIA does affect all agencies connected to archaeology. For example, all printed and mailed documents generated by the South Carolina SHPO are available to the public under the state FOIA (Patricia Cridlebaugh, personal communication 1989).

Two of the most important aspects of the FOIA amendments pertain to fees and how long it takes an agency to respond to a request. The amendments specifically state that 'fees should not be used for the purpose of discouraging requests for information or as obstacles to disclosure of requested information' (United States Congress, House of Representatives serial no. 13174–2, p. 8). Furthermore, fees may not exceed the actual cost of searching for and copying the requested documents. Search fees are apportioned at approximately five dollars per hour and copy fees at approximately ten cents per copy. (Note these are 1974 cost estimates.) Agencies are not allowed to charge for review of the information. If they deem a portion of the data to be exempt from FOIA they may not charge the individual with the time it takes to make this decision. Secondly, the agency must respond to a request for information within ten working days. If the agency cannot respond to a request they have the same time period to let the individual know this fact (O'Brien, 1981; United States Congress, House of Representatives serial no. 13174–2).

Many government agencies are beginning to require both hard and digital copies of cultural resource compliance reports (e.g. Army Corps of Engineers). The development of large information databases and GIS by government agencies has resulted in an increasingly large body of digitized archaeological data. Although articles on the creation of databases and certain computer methods can be found in the archaeological literature (e.g. Dibble and McPherron, 1988; Hobler, 1982; LeBlanc, 1976; Scholtz and Chenhall, 1976), little discussion has been generated concerning the effect of digitization on the dissemination of archaeological information (Cummings, 1981).

Archaeologists have been debating about the 'grey literature' in our profession, with primary emphasis on the lack of dissemination of cultural resource management reports (Brose, 1985; Roberts, 1989). In 1979, Somers stated that archaeologists had a responsibility to disseminate their compliance reports. He advocated use of the National Technical Information Service's (NTIS) system to give the profession greater access to such studies (Somers, 1979).

Anyone who has paid the high cost of NTIS xeroxed reports understands the limitations of using this system. Many find it difficult to justify purchase of a microfiche reader. Regardless, a researcher can only afford so many studies during any particular year. Furthermore, the copies rarely have an enticing appearance, and may be difficult to read. With the digital revolution, scholars could send an inexpensive disk to a particular agency and request a copy of the compliance report. Maps and other illustrations could be faxed or, if on a GIS, simply copied and sent in disk format. The agencies in question should be allowed to charge duplication costs. With full copyright warnings, this system should allow for wider distribution of reports. Archaeologists could greatly increase their reference libraries, at a fraction of today's cost.

One of the great powers of a GIS is its ability to generate new information (Cowen, 1987). Agencies using a GIS have been creating additional types of aggregate data layers. These myriad layers are being created by archaeologists and other scientists. This information, concerning attributes of the natural and cultural landscape, has been painstakingly collected and entered into the database at a high cost. Should scholars be allowed to purchase digital copies of these data, at cost? This raises additional legal and ethical questions. For example, one wonders how the information should be referenced and updated. Researchers at all levels need to keep careful track of each layer's lineage, often a difficult task with existing systems.

Who is the ultimate author of digital, aggregated information? Some type of use guidelines will have to be promulgated for digital archaeological data. These guidelines can be produced solely by members of government agencies, or jointly constructed by diverse members of the profession. The information revolution has reached archaeology, and researchers need to become involved in these types of management decisions.

Do archaeologists 'need to know' information such as how many Archaic sites are found on elevations ranging from 200–400 ft in a particular region? Should individual researchers have to duplicate the tremendous data entry activities required to develop and apply a GIS? At all levels, the United States government is turning to automation. Archaeologists, as citizens, pay taxes to help support digitization of various databases. Access to aggregate data should be allowed, and at cost. As a result, many more researchers could use a GIS to explore various questions in archaeological method and theory.

Conclusion

The courts have been investigating liability issues between the manufacturers and users of GIS data (Epstein and Roitman, 1987). As of mid-December, 1989, no such cases were found that pertain particularly to archaeology and GIS. In a search of the LEXIS legal database, no state or federal case was found under the following parameters: a search was made, within fifty words on either side of the terms 'FOIA' or 'site file access', 'archaeology', 'archeology', and 'GIS'. No such cases were found (Ezra Zubrow, personal communication 1989). However, one suspects that such cases will be forthcoming with increasing use of GIS in archaeology.

The 95th through 98th congress has passed over 200 laws that relate to the information field. Yet in spite of this legislation there has been no central theme, no unifying goal (Day, 1986, pp. 215–219). As burgeoning experts in information systems in general, and geographic information systems in particular, archaeologists need to become the appraisers of relevant information.

It is our responsibility, both as individuals and as members of professional organizations, to let the government know what data archaeologists need and how much the information is worth. Geographic information systems allow for collection, integration, and creation of rich archaeological data sets. Researchers need to help decide who should be the caretakers of this information, as well as who should enjoy information access (Godspiel, 1986, pp. 264–266). In this time of budget crunches and increasing reluctance of the executive branch to enforce the FOIA, archaeologists need to make their concerns known.

References

Bell, K. M., 1988, Land information integration: the Hub System in Greenland. *URISA*, **3**, 165–180.

Brose, D. S., 1985, Good enough for government work? A study in 'grey archaeology'. *American Anthropologist*, **87**, 370–377

Chrisman, N. R., 1987, Design of geographic information systems based on social and cultural goals. *Photogrammetric Engineering and Remote Sensing*, **51**(10), 1367–1370

Cowen, D., 1987, GIS vs. CAD vs. DBMS: What are the Differences? *GIS/LIS Proceedings*, Vol. 1, San Francisco

Cummings, C. R., 1981, Cultural resources management: a statement of concerns from a conservation archaeology perspective. *Journal of Field Archaeology*, **8**, 95–98

Day, M. S., 1986, Factors related to a national information policy. In *Government Information: An Endangered Resource of the Electronic Age* (Washington, D.C.: Special Libraries Association) pp. 211–228

Dibble, H. and McPherron, S., 1988, On the computerization of archaeological projects. *Journal of Field Archaeology*, **15**(4), 431–440

Epstein, E. F. and Roitman, H., 1987, Liability for information. *URISA*, **4**, 115–125

Godspiel, S., 1986, Conclusion. In *Government Information: An Endangered Resource of the Electronic Age*, (Washington, D.C.: Special Libraries Association) pp. 264–266

Harper, G. W. and DeCario, V.N., 1988, Monroe County Florida: a joint agency GIS. *URISA*, **4**, 94–102

Hobler, P., 1982, Catalogs of artifacts and the computer, report on REPORT. *Journal of Field Archaeology*, **9**, 536–538

Houdek, F. G., 1981, *The Freedom of Information Act: Comprehensive Bibliography of Law Related Materials*, (Austin, TX: Tarlton Law Library, University of Texas)

Keel, B. C., 1988, Preface. In *Advances in Southeastern Archaeology 1966–1986 Contributions of the Federal Archaeology Program*, edited by Bennie C. Keel,

Southeastern Archaeological Conference Special Publication No. 6

Kettinger, W. J., Jefferson, R.D. and Marchand, D.A., 1988, Information resource management: a policy guide for local government. In *Information Management Series*, (Columbia, SC: Bureau of Government Research and Service, University of South Carolina)

King, T. F., Hickman, P.P. and Berg, G., 1977, *Anthropology in Historic Preservation, Caring for Culture's Clutter*, (New York, NY: Academic Press)

LeBlanc, S. A., 1976, Archaeological recording systems. *Journal of Field Archaeology*, **3**, 159–168

Marshall, M. W., 1988, A local government's approach to developing an integrated land information system, Prince William County's experience. *URISA*, **4**, 195–206

O'Brien, D. M., 1981, *The Public's Right to Know: The Supreme Court and the First Amendment*, (New York, NY: Prager)

Roberts, D. G., 1989, The obligation to document: a consultant's view of archaeological reporting and publishing. *The Society for Historical Archaeology Newsletter*, **22**(3), 9–10, edited by N. F. Barka (Williamsburg, VA: College of William and Mary)

Scholtz, S. and Chenhall, R.G., 1976, Archaeological data banks in theory and practice. *American Antiquity*, **41**(1), 89–96

Sherick, L. G., 1978, *How to use the Freedom of Information Act (FOIA)*, (New York, NY: Arco)

Somers, G. F., 1979, Using NTIS, or how to disseminate archaeological reports. *American Antiquity*, **44**(2), 30–32

United States Congress, Federal Interagency Coordinating Committee on Digital Cartography, Office of Management and Budget, 1988, *A Summary of GIS Activity in the Federal Government*, (Washington, D.C.: A Reports Working Group)

United States Congress, House of Representative Reports, volume 3-2, 1978, Report of Committee on Government Operations **1–16**, 106–107 with Exceptions, 95th Congress 1st Session, Serial no. 13174-2. (Washington, D.C.: U.S. Government Printing Office)

United States Congress, Interagency Archaeological Services, 1979, *Archaeological and Historical Data Recovery Program 1979*, (US Department of the Interior, National Park Service, Washington, D.C: U.S. Government Printing Office)

United States Congress, Subcommittee on Administrative Practice and Procedure (compiler), of the Committee on the Judiciary of the United States Senate, 1978, *Freedom of Information: A Compilation of State Laws*, (Washington, D.C.: U.S. Government Printing Office)

Appendix A

GIS survey form
June 12, 1989

Institution _____

_____

_____

Please check the appropriate space: YES   NO

1. Are you automating your site files?   ___   ___

2. Are you considering automation of your files?   ___   ___

3. Are you automating your site files using a Geographic Information System (GIS)?   ___   ___

4. If not, are you considering using a GIS to automate site maps?   ___   ___

5. Do you have a policy set concerning access to your site files?   ___   ___

6. Does this policy allow access to:

   a. Archaeology professors/students?   ___   ___

   b. Contract archaeologists?   ___   ___

   c. Other scholars?   ___   ___

   d. General public?   ___   ___

7. Additional comments:

_____

_____

_____

PART II

Theory and methodology

6

Contemplating space: a commentary on theory

Ezra B. W. Zubrow

The research of any field is bounded by the types and quality of its theory and methodology. When one begins to use a new analytical tool such as geographic information systems (GIS), one rapidly discovers that it is not equivalent to a mechanic changing wrenches. The changes are more profound. New analytical tools such as GIS are not just better tools. The methodology evolves to take advantage of the new tool. New problems develop because new solutions are possible (Marble, Chapter 2). Even the theory changes. Archaeology is replete with such innovations. Radiocarbon dating is an example. The result was not only substantive changes in better dated sites. There were also methodological changes in the way one excavated sites or examined soils and artifacts. Changing tools had a ripple effect and soon the nature of the questions being asked was altered. Questions of artifactual function and site formation became as important as stylistic or cultural identification. We expect similar consequences from the introduction of geographic information systems to archaeology.

Ideally, the road will be smooth and archaeologists will be able to use GIS easily and effectively. Stine and Lanter's chapter in this volume provides a template for a smooth transition. However, there are analytical and procedural implications which archaeologists should consider. The road may have intellectual potholes for the unwary.

The redefinition of theory

Archaeology is a discipline complete in itself. On the other hand, geographic information systems is a technique and a tool from another discipline. There are always problems when one takes a tool from one discipline and applies it to the problems and data of another. Often these are problems of semantics. Terms in one discipline, such as 'coverage' or 'strata' (meaning area sampled and type of soil level in archaeology) may have quite different meanings in another (meaning thematic mapping layers in geography).

At the central core of the archaeological endeavour is time. Archaeology's theory, methodology and tools are devised to show how human behaviour changes over time. Geography's theory is often synchronic and is clearly not geared toward diachronic and historical studies which cover exceptionally long time spans. What is true of geography is even more true of GIS. In the case of geographic information systems, there is a distinct lack of analytical tools which are designed for the complexity of temporal data. Indeed, there is a continuing debate on how to construct temporal data structures (Marble, Chapter 2). The result is that even non-spatial databases have problems with temporal data.

Yet archaeologists and geographers do have common goals and common concepts—for example the need to understand the landscape. Past landscapes influence present landscapes and the landscapes of the prehistoric past will impact the landscapes of the future. Thus, there is an increasing tendency for archaeologists and geographers to cooperate and develop interpretations of prehistoric physical and social landscapes. Crumley and Marquardt (Chapter 7) suggest that landscapes can be used as a unifying concept and are a better alternative than traditional regional analysis in archaeology. They claim the landscape orientation allows one to answer new questions made operationally possible through GIS.

It is a tool redefining theory

The tool can redefine theory in opposite ways. Crumley and Marquardt (Chapter 7) and Green (Chapter 27) use the landscape approach to reduce the diversity of categories in archaeology. In a very real sense, Stine and Lanter (Chapter 8) suggest a 'deconstructing' of the typical archaeological region and archaeological site. They suggest that one of the advantages of GIS is the ability to examine all of the data in point form. One can 'deconstruct' the traditional categories such as site, region and culture area. Then, one regroups or reconstructs the data into new categories or the traditional categories. More categories are possible. Viewing a complex landscape as an integrated set of individual points, lines and areas is a type of 'pointillism'. GIS and its 'pointillism' may help change the theory of archaeology as Seurat and his 'pointillism' helped change the theory of art from representational forms to expressionist.

Many readers will have considerable experience with computers and with a variety of programs. Computer 'jocks' are a notoriously anti-theoretical group. Thus, when one is faced with a chapter on the theory of such computer-based systems as geographic information systems there is a strong tendency to quip 'Theory? Well young man, theoretically it was supposed to work ...', or that 'the theoretically better upgrade which Ezra Fuzzwhistle put on the system crashed my GIS losing 16 months of archaeological data'.

Yet even the most anti-theoretical archaeologist has been faced with the problem of accurately estimating the number and location of archaeological sites in areas which have not been surveyed. Warren (Chapter 9) makes a clear statement about how to develop a predictive model for finding archaeological sites and how to test it. He integrates GIS with logistic regression analysis. The GIS is part of the verification tool. Zubrow (Chapter 24) points to a wider use of the GIS in science. He provides an example of how the processes which are being studied are themselves modelled by the GIS. Thus, the theoretical characterization of migration is actually built into the GIS and is reflected in the data structure.

The two perspectives

The reader should understand that this book is written from an archaeological perspective. Most of the contributors are archaeologists with the exception of D. Marble, T.D. Decker, D. Lantner and R. Stine. It is appropriate, therefore, to say a few words about archaeological theory and compare it to the use of theory in geographic information systems. Theory in archaeology has the characteristics of generality and abstraction and is closely associated with various well-defined and often debated schools of thought. Thus, archaeological theory is related to substantive concepts such as culture history, social ecology, reconstruction, evolutionary process and post-processual interpretation. Significant time and considerable professional literature has been devoted to developing theory and attempting to test it in the archaeological record. Whatever theoretical 'flavour' an archaeologist is today, there is no question that theory plays a central role in the types of questions one undertakes as well as the types of answers one accepts.

In GIS the concept of theory is less mature. The very existence of theory is in question. The concept of theory in GIS has been a matter of considerable debate (Cowen, 1988). From this second perspective, there are substantial numbers of practitioners and writers who believe GIS is a technique and thus has no theory. Others believe that the theoretical aspects of GIS are clear but take differing positions on a range of issues. Some believe the theory of GIS is closely related to concepts of the relationships of space to mapping. Others think it is related to the theory of algorithm development such as the nature of self correction or floating stacks. The reader might consider the following exposition. Underlying GIS are principles of computer science. Hardware, software, and algorithms are based upon digital concepts and thus there are significant biases toward Boolean logic and standard set theory. Hence GIS are unable to handle such concepts as 'maybe', 'fuzzy sets' or 'phenomena which are sometimes located in a place and sometimes are not.' All of these 'fuzzy phenomena' are well known to any archaeologist who looks at a fractured rock and tries to decide whether or not it was used as a tool. Similarly, the actual definition of location or space in GIS is not a fixed reality. As Kvamme's article (Chapter 10) points out, different GIS packages use different algorithms for interpolating spatial data. The location of phenomena such as artifacts and sites and their contextual relationships may be partially a result of the algorithm chosen as much as the cultural or even analytical reality.

The theoretical implications of raster, vector and object systems

GIS are divided generally into raster, vector and object oriented systems. Each of these puts limitations on the way you define your units, their spatial relationships, and ultimately the type of meaning which can be ascribed to the GIS representation of a phenomenon. It ultimately limits the type of analysis which is possible and constrains research.

Raster systems (such as PC-MAP, MAP, GRASS, ERDAS, OSU Map-for-the-PC) divide space into grids. There are numerous advantages to these systems such as compact data structures, smaller memory requirements, and being relatively simple and easily accessible, which make modification possible. All of these systems set the level of resolution at grid cell size.

Ultimately, raster systems 'live and die' by the grid system. The characteristics of grid systems not only provide the strength of the system, but they also provide the weakness. If the object is larger or smaller than the grid cell, arbitrary decisions must be made to represent the object with one unit more or one unit less. This is the graphic equivalent of the rounding problem in statistics. In fact, this could be called the graphic rounding problem. Although one may adjust the grid cell size and even use hierarchical grid units in various nesting designs, the problem is that lines are inherently one-dimensional and continuous while the grid cells are inherently two-dimensional and discontinuous. In some sense the limitation is such that 'never the twain shall meet'.

Furthermore, the definition of the grid is dependent upon the number or value given to it. This means the definition is dependent upon the nature of the ascription process. All attribute characterization of the database takes place outside the raster part of the data structure. This has some inherent weaknesses which will become more obvious when contrasted with vector GIS. *The critical concept for theory is that for raster systems, meaning is independent of boundaries.*

Vector systems are a second type of GIS. These systems use points, nodes and lines to represent spatial phenomena. Such systems as ARC/INFO have a variety of theoretical advantages. Since they are not grid-based, the incomparability problems of one-dimensional line and the two-dimensional grid are solved. Furthermore, since points, lines and polygons are the basic three ingredients of the analytical package, vector GIS more closely match cartographic methods. Although attribute analysis must take place outside the vector structure, vector systems generally have a much broader set of hooks which allows one to use other databases. Finally, the rounding problem is solved. One does not need to worry that the phenomenon being represented does not correspond exactly to the grid units.

The disadvantages of vector systems are also quite numerous. They foster false and phoney precision. They tend to be highly inefficient in terms of memory usage. The code is very difficult to find and even more difficult to get into. Often this is because they are proprietary and, even more importantly, because as the systems get larger, management réquirements limit the amount and type of access of even the owners of the system.

Generally, meaning is limited to nodes and lines in vector systems. Attributes are related to the nodes and lines and then polygons are constructed from these without any necessary meaning. One might look at this from another perspective and observe that one breaks spatial phenomena into nodes and lines and then attaches meaning to them. For example, consider a lake with an island. The meaning is given to the lake boundary and to the island boundary. If one removes or loses the island boundary, the island takes on the characteristics and the attributes of the lake since it does not exist independent of its own boundary line. In a raster system, the boundaries are made up of a line of grids. The interior areas of the island still retain their identity even if the boundary is broken. This is the case even if the researcher who uses the raster system does not ascribe a separate number to the island boundary.

Object oriented systems, such as System 9, are based upon object oriented programming languages. The fundamental advantage is that one defines objects such as roads, houses, or artifacts in the GIS data structure. These objects can act as primitives to be combined into other objects, i.e. a region can be made up of roads, lakes, houses, etc. These objects are replicable and can be placed in any location. These systems tend to be window oriented, icon driven and easy to program for applications. Often these data structures are hierarchical and unidimensional. One disadvantage seems to be that once the object hierarchy is developed, it is usually difficult if not impossible to add new objects into the structure at the primitive level. On many systems (but not all) the program code is difficult to reach. On the other hand, many systems have the advantage of code referencing so that from any place in any application one can reuse a piece of code without replication or restructuring.

For archaeologists, one problem with object oriented systems is just the fact that objects must be defined. This adds increased complexity because often we do not know the spatial definition of the objects we are studying. The attributes may vary as one moves through space, time or even the study. Critical concepts such as site and activity area are arbitrarily defined.

These definitional problems have major implications for archaeological theory. Consider, for example, a ridge with archaeological objects strewn across it. There are real problems in defining the ridge and the distribution of archeological objects, as well as with defining the site. Traditionally, the archaeologist has been forced to make arbitrary decisions on these topics simply because he or she had no way to incorporate, access or use the data without these constructions. GIS allows access to large amounts of data without these constructs while also making possible a regional siteless archaeology in which the artifact is the unit of analysis, or, in the GIS sense, the 'object.'

We can use the 'site' concept to point out some generalities about the GIS process. Since the site is such a fuzzy concept which varies in definition from archaeologist to archaeologist, and indeed by the same archaeologist in different contexts in the same study, it is difficult to define in a non-arbitrary manner. On the other hand, artifacts and features are more concrete. In general, the less concrete and more arbitrary the entity, the more difficult it is to use and develop an object oriented system.

What are the implications of these three systems for studying the prehistoric or historic cultural landscape? Raster-based systems force one to examine the densities and distribution of objects in space. For example, if there are 15 points in a raster grid unit, the most feasible way to represent this phenomenon is to give the grid unit a value. The vector system's boundary and point structure corresponds to the combination of line and polygon to form objects. This allows both a site and siteless (object and pointillistic) approach to archaeology. Object language systems accomodate all approaches (point, polygon and object) at the expense of flexibility.

To query or not to query: Standard Query Language

GIS has taken the concepts of verification and Standard Query Language (SQL) from computer science. There is a movement in database creation and GIS to use a standard language for questioning the database. One advantage is that it makes cross-database questioning easy. However, one limitation is that all questions must fit the particular format of the Standard Query Language. For example a typical archaeological use of SQL would be: 'find Mesolithic Bann flakes and Neolithic hollowscrapers where the location is greater than north 520 east 320 and strata equals four in ascending order'. These languages are standard for INFO, DBASE, RBASE, INTERGRAPH, System 9, etc.

Conclusion

There clearly is a relationship between the theory of databases, the theory of GIS and the theory of archaeology. In some sense, archaeology is the quintessential candidate for the use of GIS. All of our data are spatially referenced, often in three dimensions. We deal with data at many levels of spatial resolution from the artifact to the culture area. Context, the spatial relationships of artifact to artifact, is the bread and butter of our knowledge gaining enterprise. Yet, the authors believe that the theoretical implications that GIS has for archaeology are just beginning to be realized. GIS will change the way archaeology is done and, as the work changes to make use of these systems, it will both expand and limit the types of archaeological research which are possible. The direction archaeology has taken in the last twenty years has been towards more theoretically oriented research. This has tended to mean less data oriented archaeology. The utilization of GIS will mean that theoretically sophisticated, complex data oriented work will not only be possible but the norm for the next decade and perhaps into the next millenium.

Reference

Cowen, D.J., 1988, GIS vs. CAD vs. DBMS: what are the differences? *Photogrammetric Engineering and Remote Sensing*, **54**, 1551–1558

7

Landscape: a unifying concept in regional analysis

Carole L. Crumley and William H. Marquardt

Introductory concepts

LANDSCAPE is the spatial manifestation of the relations between humans and their environment. Included in the study of landscapes are population agglomerations of all sizes, from isolated farmsteads to metropolises, as well as the roads that link them. Also included are unoccupied or infrequently occupied places, such as religious shrines, resource extraction sites, river fords, passes through mountains and other topographical features that societies use and imbue with meaning.

Landscapes are real-world phenomena. In interacting with their physical environment, people project culture onto nature. That is, they make decisions and expend energy according to their own mental models of how the world operates. To the extent that such models are always partial, unintended consequences may result from human action. For example, we consciously extract coal to provide energy for domestic consumption and industrial production; in the process, we unintentionally produce acid rain, which may have a long-term deleterious effect on agriculture and, eventually, on the health of our own society.

Contradictions inevitably arise as humans interact with their cognized environments. *Within* human groups contradictions result from the differential participation of people in the development of models of reality. *Between* human groups contradictions emerge because people occupying particular localities develop models of their environments based on their specific needs and experiences; these models may be at variance with those of other groups, leading to competition over scarce resources, religious conflicts, and the like. Contradictions constitute the raw material of change, which occurs in the resolution of conflicts and tensions between and among human groups, and between humans and their physical environment.

Even though people may share a large number of conventional understandings about the real world, they participate differentially in the production and application of models of reality. A dynamic tension between the infrastructure (the realm of material production and social relations) and the superstructure (the realm of ideas) characterizes human life. Landscapes are manifestations of that totality; hence they are appropriate objects of anthropological investigation.

As abstractions whose components vary between individuals and among groups, landscapes cannot be studied directly. The concept of *SCALE* activates both human-environmental relations and our study of those relations. Just as specific models of reality are conceived, negotiated among human groups and applied at specific scales, so are our investigations of landscapes undertaken—and the results of our studies applied—at specific spatial and temporal scales. When we choose a particular scale during one moment of our analysis, we do so because at that effective scale we can comprehend patterns: functional centres and the connections between them. A *REGION* is a unit that we recognize at a certain scale in its distinctiveness from and interrelations with other such units, both spatially and temporally. Our recognition of pattern and distinctiveness at one scale simply begins the analysis, uncovering contradictions whose resolution leads us to other scales, elements and structures.

Two types of structures determine landscape. *SOCIOHISTORICAL STRUC-TURES* include class, inheritance, descent, political liaisons and interest groups, defence, trade and laws, along with the administrative units through which people draft and enforce them—in short, sociohistorical structures are political, legal and economic. *PHYSICAL STRUCTURES*, although not without societal meaning, are those that nonetheless are relatively independent of human control, such as climate, topography and geology. In our view, the sociohistorical and physical structures *and their interpretations* (aesthetic, symbolic, religious, ideological) are determinative and mutually definitive of landscape. Change in the order of importance of certain structures (and ultimately their modification or replacement) is not only a function of adaptation to physical environmental challenges, but also is a function of the resolution of conflicting and contradictory interpretations of the meaning of sociohistorical structures. For example, when considering questions of defence in the history of our research area in Burgundy, relief (elevation) can be seen to have been more important during the Iron Age, when the inhabitants relied on hillforts, than during the Pax Romana, when hillforts were abandoned for romanized cities and settlements at lower elevations, military threat shifted to a supraregional boundary and an improved military and administrative road network facilitated import and export trade. Priorities were re-ranked and the landscape reflected those decisions.

BOUNDARIES are worthy of study because they often serve simultaneously as both *edges* and *centres* within the landscape under investigation. For example, the quantity of information and/or goods moving *along* a boundary may often be significantly greater than the quantity moving *across* that boundary. From the standpoint of the groups divided by the boundary, that boundary is an edge, a periphery. From the point of view of participants in commerce and communication, the boundary is in fact an important kind of functional centre. An area peripheral to two (perhaps antagonistic) groups is also a neutral territory, a no-man's land that has the potential to serve as a meeting ground for commercial and social activities between the groups divided by it, and for others passing through.

We consider boundaries to be of substantive interest because of their inherent duality, from the point of view both of the researcher and of the societies under study. In general we eschew drawing arbitrary boundaries around an area of interest that are intended to last for the duration of the study. Instead we have approached a research area at a number of different scales, observing a multiplicity of not necessarily coincident boundaries and the fertile ground for contradictions those divisions generate.

Finally, and of central importance to researchers interested in applying GIS technology, is the concept of *HETERARCHY* and its distinction from the more familiar structural concept of hierarchy. Heterarchy denotes a structural condition in which elements have 'the potential of being unranked (relative to other elements) or ranked in a number of ways, depending on systemic requirements' (Crumley, 1979; Marquardt and Crumley, 1987). In contrast, hierarchical structure is one in which some elements, on the basis of certain factors, are in the condition of being

ranked subordinate to others. We contend that a common error, not just in settlement archaeology but in ecology, biology and elsewhere, is that researchers
uncritically 'nest' levels of analysis. While hierarchical structures are of considerable
utility in the organization of data at a variety of scales, the researcher risks forget-
ting that hierarchy is an analytical strategy imposed on the heterarchical face of
nature. These hierarchical links, whether programmed or conceptual or both, bias
analysis and render inoperable a dynamic concept of region in which it is quite
possible (perhaps even probable) that events which occur at 'subordinate' levels have
major systemic effects; they change parameters (boundaries) or 'control-levels'
(centre-periphery shifts, scale changes) or the ranking of various elements. For
example, defence becomes more important than commerce which causes elevation
to be a more significant factor in explaining changes in settlement patterns than clay
deposits. This will require more sophisticated analytical techniques than are currently
used in most GIS analysis.

We argue that administrative hierarchies, while eminently worthy of study, are
only one version of the way elements might be ranked. If we are to study change
in human concepts of structure over time, across space and in the human mind, we
must include other, more flexible concepts of organizational structure. The concept
of heterarchy reminds us of a natural, multidimensional fluctuation in the importance of elements.

Discussion

Any area of the earth may be termed a 'region' for purposes of the study of human-
environmental relations, so long as one can recognize within it demonstrable
homogeneity. But such a region is inadequately conceptualized in the sense that both
its temporal relations (connections with the past and the future) and spatial relations
(connections with other areas at the same scale and at larger and smaller scales) are
unspecified. This deficiency, prevalent in spatial approaches ranging from traditional
culture area studies to modern regional marketing analyses, can be rectified, we
believe, by utilizing a multiscalar (hence, multiregional) and multitemporal analytical
strategy.

We concentrated our initial research efforts toward learning how administrative
boundaries change. This was appropriate for two reasons. First, the boundaries
themselves are often recorded, even though records of the decision-making process
are generally lost. Second, shifts in boundaries signal shifts in the order of priorities
and thus reveal changing conceptions of environment. Thus, by studying the relations that influence structural change in one continuously observable but fluctuating
parameter (administrative boundaries), one can begin to study the dynamics of a
region. A recognizable region emerges when there is consensus both about what
characteristics are important and about their concomitant spatial representations.
When we define a region, we do so because we can comprehend, identify and select
it as a unit in its relationships with other units; thus, the use of the term region is
always with respect to a certain perceptual size. It is defined at a scale at which the
researchers believe they can distinguish pattern. To find an appropriate scale of
analysis one must search for (1) a measure of the connectivity (at different scales)
of the area under consideration with contiguous areas and (2) areas that seem to
exhibit a high degree of overlap of a variety of boundaries. Viewed in this way, a
region never has the same meaning, nor does it occupy the same boundaries,

throughout its history. Region, defined spatially and temporally, is a system, but the system is not confined to the limits of an administrative or political region. By *region* we mean a spatial configuration at a scale at which certain phenomena exhibit recognizable areal distribution. In France, regions have historically been seats of power and influence, representing impressive consensus but frequently failing to expand that consensus to a broader scale. This is an important and recurrent theme in the history of France, and a cornerstone to the understanding of the emergence of the French state.

Understanding the relationship between centre and periphery is equally essential. What constitutes boundary and centre administratively? A boundary that is a river not only divides two territories and serves as a limit to them both, but centralizes interaction between them and in turn links both territories to areas up- and downstream. In similar fashion, cities may aggregate, integrate and mediate varieties of custom and opinion, serving a function also served by some boundary areas. Some centres and some boundaries are sparsely populated, yet charged with meaning, e.g., 'no-man's land', 'ceremonial centre'; some teem with human mental and physical activity, e.g., 'gateway cities', 'markets'; all gather their importance because of the spatially localized coincidence of various scales of activity and thought on the face of the earth.

As with regional analysis generally, an extremely important principle in the study of boundaries and centres is that of scale. What may constitute a centre at one scale is a boundary at another. For example, Vichy, a spa and symbol of provincial elegance in east-central France since ancient times, was established as a Roman centre of administrative control and also as a centre of healing and revelation; but it was also used as a wedge between the two great, rival Celtic powers, the Arverni and the Aedui, who had in preconquest times enjoyed the greatest power in the east of Gaul and whose territories lay in two major mountainous areas, the Massif Central and the Morvan. Reached easily from the east, Vichy served a similar purpose in World War II as the seat of government for German-occupied France. The Massif Central and the Morvan, where the Arverni and the Aedui had once held sway, were now strongholds of the Resistance; the Germans wisely knew the risks of being trapped in Paris if this mountainous region, which also has contained since pre-Roman times some of the most important commercial activities and routes in western Europe, were to be regained by the French. Similarly Dijon, the administrative and cultural centre of modern-day Burgundy, is situated at the juncture of the southeast-facing slopes of the Saone river valley and the rolling uplands of Champagne, and at the break in terrain between the low country west of the Belfort Gap and the Paris Basin. Dijon's strategic importance as a break-in-bulk point, or gateway city, in both north-south and east-west commerce, and its central location relative to a wide variety of resources, has assured its continuing importance since the time of the Dukes of Burgundy in the twelfth century. The city of Autun is similarly situated at a more local scale. It would seem, then, that both centres and edges may be functional centres—that is, any place that serves a function or functions not equally available elsewhere, be it border crossing or marketplace. The essential difference between concepts of boundary and centre would then turn on questions of scale, context, and perception: a heterarchy of values.

Some places are more consistently chosen as centres—for whatever reason—than others; some characteristic of a place causes many to recognize, out of an infinite geographic array, the possible advantages of that place over others. Examples are the place on a steep path where most climbers fall short of breath, or the place on a hillside that catches the breeze on a still, hot summer day but is protected from winter winds. Local directions have long been given by reference to such homely places; the farmer's directions, seemingly vague to the traveller, are so only because the traveller is bent on thinking at the broader scale of his journey. He is not accustomed to seeing the familiar landmarks at a more local scale. The traveller considers himself nowhere, between two somewheres; the farmer is in the middle of his somewhere, and his mental map of home territory is dotted with places of note, obvious, so far as he is concerned, to all.

Many such places are given meaning as a result of their daily function in society at some scale or scales. A particularly large tree may shelter from the sun both the farmer on errands between house and field and the traveller between towns. However, such practical concerns may reveal only one of the tree's numerous functions at a variety of scales. If the tree is an oak and the farmer Celtic, the tree is given a certain mystical significance as well. If, additionally, it has grown tall and beautiful because its roots benefit from a spring at its base, the farmer and his neighbours are likely to consider the tree and the spring sacred, and activities around the tree may be significantly more spiritual than practical. Suppose that local legend has it that cures for barrenness are effected there, and women come from some distance away, to visit the spot and to dip their breasts in the spring's waters. The power of the place may become more widely known, and travellers may begin to stop there for spiritual reasons, beyond the practical considerations of shade and refreshment. Then suppose that at a later time, others with different political and religious ideas (which are in turn related to differing vested interests and perceptions of the environment) find the story of the power of the tree offensive and the customs paganistic, and seek to Christianize the spot by erecting a cross. Some such sequence of events forms the history of many places in Burgundy and elsewhere in the world. To assume that there is not a continuity of sacred meaning despite the ostensible claim to the spot by those of one religion or another is to deny the essential centrality and importance of places deemed sacred partly because of their physical characteristics and partly because of the meaning attributed to them by the people for whom they are a portion of the daily, visible landscape. One such spring, at Certenue, in Burgundy, retained this continuity of sacred meaning as late as the end of the nineteenth century.

Similarly, prominent places such as Mont Dardon have for centuries been held in reverent esteem. Assuredly, Mont Dardon has long served more practical defensive functions: it was first fortified more than two thousand years ago, in the late Iron Age, and has been regularly reused since, most recently in World War II by the Germans as well as by the Resistance. But its voluptuous contours, the magnificent view it affords, and its high visibility from every direction must have provided local inhabitants a point of inspiration in the distant as well as the more recent past. From this perspective, it is not at all difficult to understand why Mont Dardon has had a succession of temples, chapels and churches, as well as fortifications, on its summit, and why the people from the three contemporary communes that touch near its highest point still attend rural festivals there, where they erected three monumental concrete crosses. Similar wooden crosses marked the summit in earlier centuries.

Administratively speaking, for the people of the communes of Uxeau, Ste.-Radegonde and Issy-l'Eveque, Mont Dardon marks a boundary, an edge. It is also—and has long been—a centre as well, for defence, inspiration, and social gatherings. During peaceful times its sacred functions would have been more obvious; during times when local defence was critical, its name would have been a rallying cry for both attacker and attacked.

The intentional destruction of at least one of the architectural manifestations of its sacredness, a tenth-century church that seems not to have stood more than perhaps one hundred years, illustrates the lightning-rod effect of a holy place and the essential relationship of politics and religion in geographical space. The incorporation of one group's sacred space into that of another's by the political redefinition of boundaries is devastating, but commonplace. The West Bank of the River Jordan affords a striking contemporary example.

GIS applications

Since 1983, an important aspect of our project has been the construction and analysis of a GIS database (see Madry and Crumley, Chapter 28). We consider the dynamic study of region, utilizing the concepts of landscape, scale, sociohistorical and physical structures, and boundaries to have been made practical with the advent of GIS. The concept of heterarchy is a powerful research tool in that it necessitates critical review of underlying analytical strategies and forces the researcher to separate structures characteristic of the group being studied (e.g., sociohistorical, administrative) which may or may not be hierarchical—or hierarchical at one level and heterarchical at another—from both physical structures (e.g., regional geological facies) and from research analytical categories. The data entry of landscape elements (e.g., sociohistorical structures: maps showing administrative boundaries or resource extraction sites and physical structures, soils and geology maps) at various scales (pedestrian survey, aerial photography, satellite-derived remotely sensed imagery) enables us to analyze temporal and spatial connections with far greater rapidity and accuracy. Changes over time in boundaries and site locations, and shifts in centre/periphery relations, which may be documented and analyzed readily with GIS, have forced us to rethink previous assumptions and enabled us to pose new research questions which direct subsequent investigation.

We have been able to document changes in settlement and land use in the Arroux valley for the past 2500 years, from Celtic times, through the Roman and medieval periods to the present. When the archaeological data are juxtaposed with the geology and soils, topography (especially elevation and slope) and meteorology, it becomes clear that the warming climate of the Roman Optimum (approximately 250 B.C. to A.D. 250) drove Celtic settlement to higher elevations—where moist conditions continued to prevail—and opened lower elevations to Mediterranean vegetation and concomitant Roman agricultural practices. In the aggregate, we have been able to document the effect on human settlement and land use of the northward movement of the ecotone between temperate and Mediterranean ecological zones. We intend to use these and other historical analogies to model appropriate contemporary responses in Europe to global warming.

Summary and conclusions

In conclusion, we argue that ecological, economic, social and historical factors must be integrated if the dynamism of culture is to be understood. Through the analysis of space, especially settlement and land use, archaeologists have long attempted to document both material and social change. We argue that to increase our ability to model successfully, cognitive and historical features must be added to the more familiar environmental analyses. Furthermore, theoretical and practical approaches must be thoroughly integrated. Landscape is the spatial manifestation of the dynamic relations between humans and their environments. Landscape as a concept can combine theory and practice and facilitate interdisciplinary comparison and synthesis which in turn can lead to broader and more useful understandings of both regional development and cultural change. GIS is the tool which allows its practical application.

Acknowledgements

Portions of this chapter are reprinted, with the permission of Academic Press, Inc., San Diego CA, from Crumley, C. L., and Marquardt, W. H., (eds), 1987, *Regional Dynamics: Burgundian Landscapes in Historical Perspective.*

References

Crumley, C. L., 1979, Three locational models: an epistemological assessment for anthropology and archaeology. In *Advances in archaeological method and Theory,* edited by M. B. Schiffer, Volume **2**, (Orlando: Academic Press, Inc.), pp. 141–173.

Crumley, C. L., and Marquardt, W. H., (eds), 1987, *Regional Dynamics: Burgundian Landscapes in Historical Perspective*, (San Diego: Academic Press, Inc.)

Marquardt, W. H. and Crumley, C. L., 1987, Theoretical Issues in the Analysis of Spatial Patterning. *In Regional Dynamics: Burgundian Landscapes in Historical Perspective*, edited by C. L. Crumley and W. H. Marquardt, (San Diego: Academic Press, Inc.), pp. 1–18

8

Considerations for archaeology database design

Roy S. Stine and David P. Lanter

Introduction

A geographic information system (GIS) 'is a decision support system involving the integration of spatially referenced data in a problem solving environment' (Cowen, 1988). Its software is designed to collect, store, retrieve, manipulate and display objects defined as points, lines, or areas (Clarke, 1986; Dueker, 1987). What makes a GIS unique is its power to synthesize. GIS manipulation creates new spatial entities from existing spatial data. Deriving new geographic features from old ones is a function of the software. The software transforms implicit spatial relations in existing geographic databases into explicit relations. This is analogous to the functioning of statistical packages that extract measures of central tendency, strength of relationship, and significance from statistical samples and populations.

Geographic entities (of which archaeological entities are subsets) are uniquely and individually represented within databases by the spatial data structures. By querying the data structures it is possible to determine the location and thematic value of the various features represented in the database. However, queries concerning spatial relations such as distance, proximity, connectivity, juxtaposition, and contiguity are not possible if they are not explicitly coded within the data structure. GIS functions are designed to deduce spatial relationships of the various geographic entities in the database and explicitly represent and store them (the relationships) as new geographic entities.

The ability to analyze existing spatial data and synthesize new information has profound implications for the archaeologist (Peregrine, 1988). These capabilities can speed descriptive archaeological analysis by detecting and mapping distributions of artifacts, features, and sites. It can aid sophisticated predictive archaeological modelling. GIS operators can be applied to a spatial database to determine settlement patterns within an area of interest. More specifically, a GIS can be used to generate potential feature locations within a site and site locations within a region.

The ability to compare the contents of squares, sites, or regions depends on the design of the database. This paper discusses spatio-temporal GIS database design considerations. It then discusses the impact of this design on descriptive and predictive archaeological modelling.

The geographic matrix

GIS organize data in three ways: (1) least common geographical unit (LCGU), (2) objects, and (3) layers. Least Common Geographical Units, also known as Integrated Terrain Units (ITUs) integrate all pertinent spatial data records into a single set of all classes (Chrisman, 1975; Dangermond, 1988). Wild's System-9 is an example of a modern LCGU-based GIS. Object-oriented approaches integrate individual geographical entities and attributes into semantic objects with inheritable properties (Kjerne and Dueker, 1986; Charlwood *et al.*, 1987; Herring, 1987; Gahegan and Roberts, 1988; Egenhofer and Frank, 1989). Intergraph's TIGRIS is one example of a commercially available object-oriented GIS.

While the LCGU and object-oriented approaches are finding favour among some users, the layer-based geographic information system remains the most widely used. This is because theme layers allow users to visualize a cartographic database as a set of registered map separations. Spatial relationships between relevant archaeological and geographic entities are analyzed, in such systems, with the help of neighbourhood and overlay functions (Chrisman and Niemann, 1985; Kjerne and Dueker, 1986; Aronson, 1987; Bracken and Webster, 1989). Examples of layer-based GIS include Environmental Science Research Institute's vector-based ARC/INFO and the raster-based Map Analysis Package (Tomlin, 1980).

Regardless of organizational approach, geographic phenomena have three basic characteristics: (1) the spatial location of the phenomena, (2) descriptive information or attributes about the phenomena, and (3) the time and duration of the phenomena (Berry, 1964). Berry's spatial, aspatial, and temporal dimensions serve as a framework for organizing archaeological information. Once organized in this fashion the archaeological data are more readily analyzed (Figure 8.1).

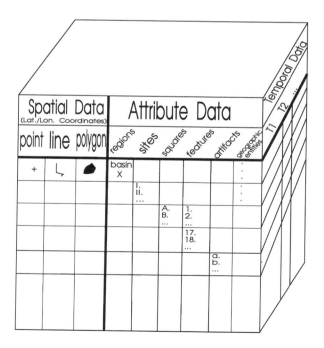

Figure 8.1. Three characteristics of archaeological information (after Dangermond, 1983).

The spatial dimension

Vector and tessellation models provide two successful approaches for storing and processing spatial data in GIS (Peucker and Chrisman, 1975; Peuquet, 1984; Maffini, 1987). Vector approaches capture graphical representations of archaeological and geographic entities with strings of coordinate X, Y pairs organized as points, lines and areas. The evolution of vector data models has witnessed the use of alternative techniques for storing points, lines, and areas. Examples include spaghetti, topologic, GBF/DIME, POLYVRT, and chaincodes.

In contrast to vector data models, tessellation models organize point, line, and polygonal entities within a mosaic of geometric tiles. Each tile represents a homogeneous graphical region characterized by the presence of some archaeological or geographic entity. The geometry of tiles may be regular squares, triangles, or hexagons which may be nested in hierarchies subdividing large non-homogeneous areas. Of the three, square (raster) tessellations are the most popular spatial data organization in non-vector GIS.

Vector and raster models have their own strengths and weaknesses. Peuquet (1984) suggests that vector data models

'are direct digital translations of the lines on a paper map. This means that the algorithms also tend to be direct translations of traditional manual methods. The repertoire of vector-mode algorithms is thus both well-developed and familiar. The primary drawback of vector data models is that spatial relationships must be either explicitly recorded or computed. Since there is an infinite number of potential spatial relationships, this means that the essential relationships needed for a particular application or range of applications must be anticipated.'

Spatial relationships, according to Peuquet,

'are 'built-in' for regular, tessellation type models. Grid and raster data models are also compatible with modern high-speed graphic input and output devices. The primary drawback is that they tend to be not very compact. Regular tessellations tend to force the storage of redundant data values.... Another drawback is that the algorithm repertoire is less fully developed than raster data models.'

Peuquet concludes

'neither type of data model is intrinsically a better representation of space. The representational and algorithmic advantages of each are data and application dependent, even though both theoretically have the capability to accommodate any type of data or procedure.'

Whether handled within the context of vector or raster based GIS, the two fundamental archaeological entities are *artifacts* and *features*. In a GIS context an artifact is defined as any discrete, transportable material that shows evidence of past human activity. A feature is viewed as any spatially fixed remnant of past cultural activities. In order to relate artifacts and features to any of the relationships described below, all aspects need to be placed within a spatial context.

The spatial dimensions of archaeological information revolve around the relationships between (1) artifacts to features, (2) artifacts and features to squares, (3) squares to sites, and (4) sites to regions. If one prefers the approach of Dunnell and

Dancey (1983), one may view the relationship as solely between artifacts and regions. In utilizing a siteless or non-site approach (Dunnell and Dancey, 1983) each artifact and surface feature would need to be spatially referenced. In this way the artifact densities are located within geographic and environmental zones. Then the relationships between different artifact densities and between the artifact clusters and natural and cultural landscape are analyzed. These relationships are then viewed within the matrix of the surrounding region. The *square* provides the arbitrary, spatial context for excavating artifacts and features. The coordinate system of the square is typically cartesian. The resolution of the archaeological database is therefore determined by the smallest unit of measure needed to identify either discrete artifacts or the individual squares. These resolutions allow one to differentiate spatially between individual artifacts in the three dimensions of the unit, or the three dimensions of each square. As each artifact or feature is found, it needs to be placed within its vertical and horizontal space, i.e., the X and Y coordinates within the square, and a subsurface Z. In turn, each unit is located within the coordinate system of the site.

The *site*, or artifact cluster, is an area that holds evidence of artifacts or features. As stated by Dunnell and Dancey (1983, pp. 267-287), 'Anyone who has done much fieldwork is aware that distinguishing a site and setting its boundaries is an archaeological decision, not an observation'. While the exact location of a site border may be difficult to determine, arbitrary borders do provide a starting point for determining the cartesian grid of the site. The origin is used to delineate the X, Y starting point, generally arbitrary numbers like 0, 10 or 100. The Z is generally a value below surface level or an arbitrary site elevation. The site's coordinate system permits subsurface configurations, mapped within squares, to be tied to the site as a whole. In this manner, the spatial relations between artifacts and features found within squares can be analyzed and mapped. Once mapped, sites may be studied and compared to determine how they relate to each other and how they fit within regions.

The *region* is an area that provides a context for intersite comparisons. The '...concept of region requires that recognition of boundaries, as well as of the functions of centers must remain flexible over both space and time if the investigator is to grasp the manner in which the region functions and changes.' (Crumley *et al.*, 1987). Regions may be delineated by distinctive geographic characteristics on the surface of the earth. The size of a region may vary as a function of the area at which intersite comparisons are to be made. A researcher may choose to compare sites within a drainage basin, between several drainage basins, or across continents. Such site comparisons may serve as the basis for determining past regional settlement patterns.

When detailing information about archaeological sites, it is not enough to build a database from artifacts and features. Each entity must be placed within a regional and geographic context. That is, the site must relate to an environment of geographic entities such as lakes, streams, roads, hills, valleys, and population centres. The interrelationship of artifacts, features, squares, sites, geographic entities and regions requires the use of a common world grid coordinate system such as latitude and longitude or universal transverse mercator. An additional requirement is a mapping scale that is large enough to capture the smallest artifact excavated, or the individual squares.

The ramifications of this approach for archaeology, in terms of GIS, is that the data could rapidly be 'clumped' or aggregated into quantitative entities. For instance, artifactual data, within a site containing certain types of pottery sherds, may be generated into a density map from the data structure. Feature data, such as a house, could have the refuse disposal patterns at twentieth century farmsteads (South, 1979) rapidly generated. Using numbers from Moir and Jurney (1987), one would generate a 'buffer' of 30 to 50 feet to create the adjacent refuse pattern and a buffer of 30 to 70 feet for the peripheral refuse pattern.

After researching spatial structure within a site one may turn to the analysis of structure between sites. The primary decision to make at this juncture is, at what scale should the sites be overlaid? Many GIS software packages have scale changing potential. This means that sites digitized into a database at different scales can automatically be regenerated into a common scale. Once a common scale is decided upon the sites may be overlaid. The research potential of rapidly overlaying similar sites from differing areas is vast. One could analyze commonality between site structure, project average boundaries among sites of a similar temporal period, and undertake various other innovative procedures.

In summary, a world coordinate system and adequate spatial resolution are the spatial requirements for an archaeological database capable of supporting: (1) spatial queries that interrelate archaeological features and geographic entities, and (2) the use of GIS functions to derive spatial relations implicit in a database of artifacts, squares, sites, features, geographic entities and regions.

Aspatial dimension

Spatial data are linked with their aspatial attributes to describe important real-world characteristics of archaeological or geographic phenomena. Separating aspatial attributes from the graphic representation permits retrieval and display of geographic and archaeological entities meeting a combination of aspatial and spatial query criteria. Aspatial characteristics such as colour, type, material, texture, weight, etc. can be stored in either the hierarchical, network, or relational data models (Atre, 1980).

The hierarchical model organizes data in a tree structure. It is made up of nodes and branches. Each node represents an entity and is made up of a collection of descriptive data attributes. The oldest database systems are based on the hierarchical data model; examples include IBM's Information Management System (IMS), Intel's System 2000, and Informatic's Mark IV. Hierarchical systems are based on rigid tree structures that are not easily reorganized and promote redundant storage and attributes describing entities stored in different branches of the tree. They do provide for rapid partitioning of data sets and retrieval of stored data.

The network model organizes nodes of data with an interconnection network of directed labelled links. Examples of commercially available database systems based on the network data model include Cullinane's IDMS, Cincom's TOTAL, Honeywell's IDS/II, UNIVAC's DMS 1100, and Digital Equipment Corporation's DBMS-10 and DBMS-20. Network data systems are more powerful than hierarchical systems in modelling real work data relationships without the necessity of data redundancy. This power is balanced by the inherent complexity in navigating and updating tangled webs of pointers connecting related entities and their attributes.

The relational data model uses tables to organize attributes of entities and relationships between entities. Organizing attributes in tables makes the relational model easy to understand focusing attention on the information content of the database. This stands in sharp contrast with the hierarchical and network data models that force users to be conscious of the physical storage of data in order to access the information content. Another strength is that the relational model supports the normalization or removal of redundant information from the database and is based upon a solid mathematical theory of relations not existing for the other data models.

The power that a simple conceptual model provides users is traded off with a high cost in computational overhead that threatens to bog down the speed of processing in large databases. However, most commercially based GIS use the database management systems based on the relational model to provide flexible organizational and querying capabilities. Examples of relational database systems used with GIS are Relational Technologies' INGRES, Henco's INFO, and Oracle Corporation's ORACLE.

Regardless of which database model is used, archaeological attributes describe entities encountered in the field. Much of field archaeology consists of excavation and filling out forms. Indeed, there are survey forms, level forms, feature forms, unit forms, site forms, state forms, company forms, school forms, and forms about the forms. In each case a decision about artifact taxonomy must be made before the project truly begins.

Archaeologists interested in the historic period would have to decide if they are going to group artifacts according to a taxonomy such as the South Functional Classification Scheme (South, 1977), or the Sprague Functional Classification Scheme (Sprague, 1981). There are also descriptive classification taxonomies to choose from. These consist of placing objects into groups by attributes, such as material of manufacture. Archaeologists may also create their own typologies. Classification is an extremely important consideration if a GIS is utilized. For the artifacts to be compared to one another, they must be entered into the database under the same classification system. Artifact comparisons between two sites are, therefore, limited to sites that have utilized the same system.

When grouping artifacts the accustomed thought process is to aggregate discrete items (artifacts) into arbitrary units (squares). It is possible to pull artifact types into their smallest component parts. Abler (1987) has stated that we can now accomplish analysis on a pointillism basis. This means that researchers can analyze and formulate questions about sites in their disaggregated form. Sites may now be compared to other sites as a discrete series of artifacts and attributes. The proper methods and statistics for fully utilizing this pointillism approach are only now beginning to be researched.

With a pointillism approach the artifacts and features may be analyzed by attribute. One could, for instance, research a single artifact type within a site. One could also analyze an artifact group between sites. In addition, one could study a single artifact type within a region. For pointillism to be possible, however, the database must be built from points, or artifacts, located with explicit X, Y, Z coordinates. This type of approach also has vast implications for intricate spatio-temporal modelling.

Temporal dimension

Generally, most vendors of a GIS work with time in its proper sequence; in other words, Friday the 6th would be followed by Saturday the 7th. Archaeologists view time differently. They observe recent litter on the surface of the ground overlaying various strata, usually dating to earlier periods. For example, after initial excavation one may find historic period artifacts. After further excavation, following the law of superposition, one may find prehistoric artifacts. In other words, the archaeological process of data recovery moves backwards rather than forwards, chronologically.

Langran (1988) suggests that a goal of contemporary GIS design efforts should be to enable the analysis of changes in spatial information through time. Such a temporal GIS would store information about the changing nature of geographical features as they occur outside the database in the real world. This is important to producers of map products that have strict requirements for currency and historical accountability (Hunter, 1988; Vrana, 1989). Archaeological research would greatly benefit from these developments.

Langran and Chrisman (1988) indicate 'just as the topological data structure would provide a means of navigating from an object to its neighbor in space, the corresponding temporal data structure would provide a means of navigating from a state or a version to its neighbor in time.' This suggests that the archaeologist could study the transformation of a site or a state, as it interrelates to its neighbours and environment over time.

Armstrong (1988) provides a conceptual model for creating a time topology by tagging each geographic feature in a GIS with temporal information. Tagging each geographic feature with a time stamp attribute facilitates the tracking of how that feature changes across many dates (Figure 8.2). With a time attribute it is possible to retrieve temporal information about a given object, or retrieve features associated with a particular time. In Armstrong's conceptual model a duration of a particular feature's valid spatial representation can be explicitly stated or calculated from the equation:

$$\text{Duration} = \text{Time}(n) - \text{Time}(n-1)$$

Basoglu and Morrison (1977) illustrate how a system of feature-based time stamping is implemented; '...the most logical approach was to start with up-to-date and precise data for modern county boundaries (1970) and then work back to 1790, deleting and adding and changing lines.' This implies that various time slices can be selected to view the temporally changing configuration of historical county boundaries.

Langran, Chrisman, Basoglu and Morrison have paved the way for a GIS database designed around spatio-temporal references. Contemporary GIS databases, however, cannot be optimized for space, time, and spatio-temporal retrieval. Instead, they must be ranked by importance to choose which dimension will dominate (Langran, 1989).

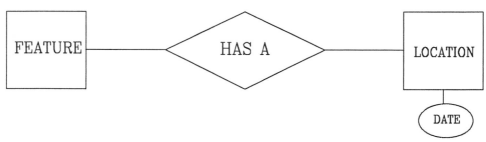

Figure 8.2. Temporal attribute of a feature's location (after Armstrong, 1988).

Spatio-temporal queries

As with attribute queries, temporal queries cut across the different levels of spatial analysis. The researcher may divide a site into its temporal components by searching the database for all attributes between time A and time B. From that information one could generate a map of just those attributes. Single period attribute maps from a variety of sites could then be overlaid and analyzed. One could also generate a chronological site map of a region. By selecting a specific time period, a map of all the sites in the area that fit the time criteria would be generated. A GIS designed to analyze changes in historical geographic states will provide functions for '...detecting trends, cycles or other patterns in spatio-temporal data that might explain underlying processes or be used in forecasting future states'. (Langran, 1988). Archaeological queries based upon such functions would focus on inventory and analysis. Inventory queries would involve accessing stored descriptions of the area of interest. This would necessitate analyzing the history of archaeological sites with one or more geographic features. Examination of individual time slices could then provide a view indicating the state of the study area at various times.

To accomplish temporal queries, archaeological entities need to be endowed with several time attributes. First they would need the dates they were discovered during excavation, then they would need the dates of cultural affiliation. These time stamps could then be used to view the chronological history of a particular artifact or feature. In this manner the temporal aspects for areas of interest could be traced from prehistoric to historic to the date of excavation. Analysis of spatio-temporal archaeological data would involve navigating across time and space to 'explain, exploit, or forecast a regions's components and processes' (Langran, 1988).

Conclusion

An integrated spatio-temporal archaeological database can be analyzed with GIS functions. Time stamps, a single classification system and a common coordinate system for squares, sites, and regions will serve as a method for integrating archaeological and geographic databases. This would allow for scientific inquiry into underlying archaeological spatial structures. These structures link archaeological entities such as artifacts and features to sites, and they place sites within regions.

References

Abler, R. F., 1987, The National Science Foundation National Center for Geographic Information and Analysis, *International Journal of Geographic Information Systems*, **1**(4), 303–326

Armstrong, M.P.,1988, 'Temporality in Spatial Databases', *Proceedings of GIS/LIS '88*, Vol. 2, pp. 880–889

Aronson, P., 1987, Attribute Handling for Geographic Information Systems. *Eighth International Symposium on Computer-Assisted Cartography*, pp. 346–355

Atre, S., 1980, Data Base. In *Structured Techniques for Design, Performance, and Management*, (New York: John Wiley and Sons) pp. 83–123

Basoglu, U. and Morrison, J. L., 1977, The Efficient Hierarchical Data Structure for the U.S. Historical County Boundary File. *Papers of an Advanced Study Symposium on Topological Data Structures for Geographic Information Systems*, Laboratory for Computer Graphics and Spatial Analysis, Harvard University, Cambridge, MA

Berry, B., 1964, Approaches to Regional Analysis: A Synthesis. *Annals of the Association of American Geographers*, **54**, 2–11

Bracken, I. and Webster, C., 1989, Toward a Typology of Geographical Information Systems. *International Journal of Geographic Information Systems*, **3**(2), 137–152

Charlwood, G.I. *et al.*, 1987, Developing a DBMS for Geographic Information: A Review. *Eighth International Symposium on Computer-Assisted Cartography*, pp. 302–315

Chrisman, N.R., 1975, Topological Information Systems for Geographic Representations. *Proceedings of the International Symposium on Computer-Assisted Cartography*, pp. 346–351

Chrisman, N.R. and Niemann, B., 1985, Alternative Routes to a Multipurpose Cadastre: Merging Institutional and Technical Reasoning. *Proceedings of the Seventh International Symposium on Automated Cartography*, pp. 84–93

Clarke, K. C., 1986, Advances in Geographic Information Systems, *Computer Environment Urban Systems*, **10**(3/4), 175–184

Cowen, D. J., 1988, GIS vs. CAD vs. DBMS: What are the Differences? *Photogrammetric Engineering and Remote Sensing*, **54**(11), 1551–1555

Crumley, C. L. and Marquardt, W. H. (eds.), 1987, Theoretical Issues in the Analysis of Spatial Patterning, *Regional Dynamics Burgundian Landscapes in Historical Perspective*, (San Diego, CA: Academic Press Inc.)

Dangermond, J., 1983, A Classification of Software Components Commonly Used in Geographic Information Systems. In *Design and Implementation of Computer-Based Geographic Information Systems*, edited by D. Peuquet and S. O'Callaghan, IGU Commission on Geographical Data Sensing and Processing, Amherst, N.Y.

Dangermond, J., 1988, A Review of Digital Data Commonly Available and Some of the Practical Problems of Entering Them into a GIS. *Proceedings of the 1988 Annual Conference of the American Congress on Surveying and Mapping*, Vol. 1

Dueker, K. J., 1987, Geographic Information Systems and Computer-Aided, *APA Journal*, Summer.

Dunnell, R. C. and Dancey, W. S., 1983, The Siteless Survey: A Regional Scale Data Collection Strategy. In *Advances in Archaeological Method and Theory*, edited by M. Schiffer, Vol. 6, (San Diego,CA: Academic Press Inc.) pp 267–287

Egenhofer, M.J. and Frank, A.U., 1989, Object-Oriented Modeling in GIS: Inheritance and Propogation. *Proceedings of the Ninth International Symposium on Computer-Assisted Cartography*, pp. 588–598

Gahegan, M.N. and Roberts, S.A., 1988, An Intelligent Object-Oriented Geographical Information System. *International Journal of Geographic Information Systems*, **2**(2), 101–110

Herring, J.R., 1987, TIGRIS: Topologically Integrated Geographic Information System. *Eighth International Symposium on Computer-Assisted Cartography*, pp. 282–291

Hunter, G. J., 1988, Non-current Data and Geographical Information Systems: A Case for Data Retention, *International Journal of Geographical Information Systems*, **2**(3), 281–286

Kjerne, B., and Dueker, K.J., 1986, Modelling Cadastral Spatial Relationships using an Object-Oriented Language. *Proceedings of the Second International Symposium on Spatial Data Handling*, International Geographical Union Commission on Geographical Data Sensing and Processing, Williamsville, NY, pp. 142–157

Langran, G. 1988, Temporal GIS Design Tradeoffs, *Proceedings of GIS/LIS 1988*, Vol. 2, pp 890–899

Langran, G., 1989, Accessing Spatiotemporal Data in a Temporal GIS. *Proceedings of the Ninth International Symposium on Computer-Assisted Cartography*, pp 191–198

Langran, G. and Chrisman, N. R., 1988, A Framework for Technical Geographic Information, *Cartographica*, **25**(3), 1–13

Maffini, G., 1987, Raster versus Vector Data Encoding and Handling: A Commentary. *Photogrammetric Engineering and Remote Sensing*, **53**(10), 1397–1398

Moir, R. W. and Jurney, D. H., (eds.), 1987, *Richland Creek Technical Series*, Vols. II–IV, Archaeology Research Program, Institute for the Study of Earth and Man, Southern Methodist University, Dallas, Texas

Peregrine, P., 1988, Geographic Information Systems in Arhcaeological Research: Prospects and Problems. *Proceedings of GIS/LIS '88* pp. 873–879

Peucker, T.K. and Chrisman, N.R., 1975, Cartographic Data Structures. *The American Cartographer*, **2**(1), 55-69

Peuquet, D.J., 1984, Conceptual Framework and Comparison of Spatial Data Models. *Cartographica*, **21**(4), 66-113

South, S., 1977, *Method and Theory in Historical Archaeology*, (New York: Academic Press)

South, S., 1979, Historic Site Content, Structure, and Function, *American Antiquity*, **44**(2), 213-236

Sprague, R., 1981, A Functional Classification for Artifacts From 19th and 20th Century Historical Sites. *North American Archaeologist*, **2**(3), 251-259

Tomlin, C.D., 1980, The Map Analysis Package (MAP). *Papers in Spatial Information Systems*, Yale University School of Forestry and Environmental Studies.

Vrana, R., 1989, Historical Data as an Explicit Component of Land Information Systems, *International Journal of Geographical Information Systems*, **3**(1), 33-49

9

Predictive modelling in archaeology: a primer

Robert E. Warren

Introduction

Prediction plays an integral role in the scientific method. When a scientist discovers patterning in a set of observations and develops an hypothesis to explain the observed patterns, the hypothesis, if useful, has predictive implications for future observations. The implications can then be tested with new or independent data. If the new data conform to the predictions, the test provides support for the validity of the hypothesis. In the scientific method, then, prediction is a mechanism for testing explanations.

Prediction is also a key component of many spatial models that have been developed by archaeologists in recent years, although it generally plays a somewhat different role in these models. Predictive models are deductive or inductive formulations that predict unknown trends or events. Deductive models are deduced from theory, and are analogous to the kind of prediction that we usually associate with the scientific method. However, most predictive modelling in archaeology is inductive. Inductive predictive models are simply composite patterns or uniformities that are detected in empirical observations. The patterns are described in such a way that they can provide expectations concerning the archaeological characteristics of unknown areas. Such models have no necessary connection with the testing of hypotheses, although hypotheses may be advanced to explain the patterns upon which they are based. It follows that predictive models in archaeology may be either explanatory or descriptive. There is nothing inherently unscientific about either approach, although the most useful models will be those that lead to a scientific understanding of the uniformities they portray.

This chapter presents an introduction to the topic of predictive modelling in archaeology. The focus is on spatial models of archaeological site location created with the aid of geographic information systems (GIS). The first section expands on some of the ideas presented above and provides a conceptual background for predictive models. The second is an overview of the approaches to predictive modelling that have been popular among archaeologists during the past few years. Finally, the third section describes a particular approach that integrates GIS with a multivariate statistical procedure called logistic regression analysis. This approach, developed by Kvamme (1983a, 1983b, 1989), has been used successfully at the Illinois State Museum and a number of other institutions. It forms the basis for several of the models described below in the applications section of this volume, including a model of site location in the Shawnee National Forest of southern Illinois (Warren *et al.*, 1987; Warren, this volume, Chapter 18).

Conceptual framework

Predictive models are devices that make use of existing knowledge to forecast trends or events. Models of this genre have a wide variety of topical applications, ranging from projections of future stock prices to predictions of where hidden natural resources may be found. Models may also differ greatly from one another in terms of their form and function. However, all predictive models have three basic elements in common: information, method and outcome. In essence, a predictive model uses method to transform information into a predicted outcome.

The first element is the available knowledge or body of information from which a model is derived. Two basic kinds of information can be used to develop predictive models. These are (1) theories that explain the processual effects of variables on the trends or events of interest in a cause-effect relationship, and (2) empirical observations, which ordinarily consist of (a) the observed interactions among independent and dependent variables at earlier times, in previously studied areas, or in sampled parts of an area of interest, and (b) information on the variables and conditions that may influence a predicted outcome, either in the future or in a sampled area of interest.

In concept, it is possible to develop a predictive model either by pure deduction from theory or by pure induction from empirical observations. In practice, however, most models make use of both theory and observation. For example, models of archaeological site location could be developed using a deductive approach that stresses the theorized cultural and biological needs of a society. Such needs would be used to guide the selection of independent variables. Models like this can be developed in the absence of information on archaeological site location, but they cannot be implemented or tested without such observations. Similarly, one could develop a purely inductive model of site location, but in the absence of theory the process of variable selection would be inefficient and the resulting model would run the risk of being weak and uninterpretable.

One of the most powerful approaches to empirical prediction is a family of procedures called probability models (see below). A crucial information requirement of these models is data on positive and negative responses to stimuli. Hence, the dependent variable in a probability model consists of two or more mutually exclusive and exhaustive groups, which may be coded as success vs. failure, response vs. no response, presence vs. absence, and so on, with respect to one or more independent variables. This procedure is analogous to the 'null model' approach to data analysis, which allows an observer to track the response of a dependent variable to different stimuli and thereby measure the differential effects of multiple independent variables (Conner and Simberloff, 1986). Spatial applications of these procedures, such as models designed to predict the locations of archaeological sites, require that independent variables are measured not just for places that are known to contain sites, but also for places where sites are known to be absent (i.e., nonsites).

The second element of a predictive model consists of the method or body of methods used to transform information into a prediction. In the development of a deductive model, one could use theory to build a hierarchy of conditional arguments. The hierarchy would be designed to translate the readily observable properties of some phenomenon into predictions as to whether the phenomenon contains unobservable properties that are of interest or value. For example, mineral prospectors in eastern Washington developed a hierarchical inference network that can be applied to maps of geological, geophysical and geochemical variables to predict the locations of hidden mineral deposits (Campbell *et al.*, 1982).

Methods analogous to those used by mineral prospectors have been developed for predicting archaeological site location. In a pioneering application, Limp used theory and observation to select a set of 13 environmental variables that may have been important to prehistoric peoples in southern Arkansas when they chose places to conduct activities that left traces in the archaeological record (see Limp and Carr, 1985). Each of the environmental variables was then dichotomized as a yes-no choice (favourable vs. unfavourable for settlement) and the dichotomies were assembled and reassembled like building blocks to create a vast series of hierarchical decision trees. Each tree differed from all others in terms of the number or position of its constituent blocks. Next, a sample of the many possible decision-tree structures was applied to environmental data in the region to identify 'viable locales' for settlement. These locales were compared with the locations of known archaeological sites to evaluate the potential predictive power of each tree. One advantage of this approach is that its structure may mimic the decision-making processes of prehistoric peoples. However, Limp and Carr found that the method is cumbersome to implement and it is difficult to test the results. It is limited further by the fact that when continuous variables are dichotomized at branchings, the number of possible trees becomes infinite.

Empirical predictive models usually rely on statistical methods. There are a wide variety of appropriate techniques, but it is probably fair to say that most of the procedures in current use are variants or analogues of regression analysis. At the elementary end of the regression family is bivariate linear regression (Blalock, 1979). This procedure can be used to define the relationship between two variables for a sample of cases. One can then predict the score of a new case on one measurement (dependent variable) when its score on another measurement (independent variable) is known. Multiple regression is an extension of bivariate regression for problems that have one continuous dependent variable and two or more independent variables (Lewis-Beck, 1980).

A popular analogue of multiple regression is discriminant function analysis. In its simplest form, this is a multivariate procedure that uses observations made on members of two distinct groups to create an axis of discrimination that statistically separates the two groups in such a way as to minimize the number of misclassifications of group membership (Anderberg, 1973). Once defined, a discriminant function can be used to predict the group membership of a new case (see Klecka, 1980).

Related to both multiple regression and discriminant function analysis is a family of procedures called probability models (Aldrich and Nelson, 1984). The most popular of these is the logit or logistic regression model (Stopher and Meyburg, 1979). Logistic regression is a robust and flexible procedure. It can accommodate one or more independent variables, and it operates on either tabulated data, where there is more than one subject per case, or on data for which each case is an individual subject (Engelman, 1985).

The key element of the logistic regression procedure is the fact that the dependent variable is dichotomous or polytomous, which means that it is comprised of two or more mutually exclusive and exhaustive groups. Instead of predicting the scores of cases on a continuous dependent variable as in multiple regression, logistic regression predicts the probability that each case is a member of each group. The family of probability models is well suited for developing predictive models of archaeological site location, as the appropriate dependent variable in such models is a natural dichotomy: the presence versus the absence of an archaeological site.

Finally, the third element of a predictive model is the predicted outcome. There are four basic kinds of predictions, each of which corresponds to one of the four levels or scales of measurement: nominal, ordinal, interval, and ratio scales.

Nominal predictions are categorical; they assign cases to groups or classes that have no assumed relationships to one another. For instance, an archaeologist may wish to predict the raw material type of a lithic artifact based on the distances between its find spot and the source areas of different kinds of raw material in a region. Depending on the method used to make the prediction, the outcome would be a determination of the most probable category of raw material.

Ordinal predictions are at a higher level than nominal ones, as ordinal categories can be ordered or ranked along a common scale. However, ordinal scales are limited by the fact that the relationships among categories are monotonic. On an ordinal scale we can say that one category is greater than or less than another category in some respect, but we cannot measure the magnitude of the difference between them. For instance, an archaeologist may want to rank the environmental settings of an area in terms of their potential for containing sites. Lacking evidence from the area in question, the analyst could examine data from a well-known area to assign the array of settings to three groups. The predicted sequence would provide no indication of the exact magnitude or abundance of sites, but the outcome could be useful as a ranking of high, medium and low potential. Note that it is difficult to imagine a deductive predictive model of archaeological site location achieving anything greater than ordinal-scale predictions. This is the scale used by Limp in the decision-tree model described earlier (see Limp and Carr, 1985).

Interval-scale predictions have the capacity not only to order cases or categories relative to one another, but also to determine the exact distance between them. However, interval scales also lack absolute zero points, so it would not be possible using such a scale to say that one event is twice or three times as probable as another. For example, the mineral prospectors mentioned earlier developed an interval scale of 'favourability' for predicting where porphyry deposits occur (Campbell *et al.*, 1982). Their scale ranges from -5 to $+5$, with higher values representing a measurably greater favourability than lower ones. The scale served its purpose quite well, although its users could not interpret a score of $+4$ as being twice as favourable as a score of $+2$. Note that the family of ordinary least-squares regression procedures, including multiple regression, generates interval-scale outcomes that can range from negative to positive infinity. Scores may be useful on scales such as this, but they cannot be manipulated as ratios or interpreted as probabilities.

Ratio-scale predictions have all the qualities of interval predictions, plus the added benefit of an absolute or nonarbitrary zero point. The outcomes of discriminant function analysis and probability models are ratio-scale measures that are constrained not only by basement values of zero, but also by ceiling values of one. The results of these procedures are not like the predictions of size or weight that would be obtained from ordinary regression, which, as I have just noted, are interval-scale measures. Instead, the outcomes are predictions of the probability of group membership.

As noted earlier, outcomes of this genre are ideal for archaeologists seeking to develop predictive models of site location. The two groups in this case are locations that do and do not contain sites (i.e., sites vs. nonsites). Each outcome is a prediction of the probability that a site occurs at a given location. Probability-based predictions have several advantages over predictions on other scales. For instance, they are readily interpretable (values range between 0 and 1), they can be treated as ratios (a probability of 0.6 is twice as high as a probability of 0.3), and their accuracy can be tested with sample data (see below).

Archaeological approaches

Prediction is an important focal point of archaeological research, as it is in all sciences. Most archaeologists interested in the contexts of sites have probably developed intuitive models of where sites may be expected to occur in the places they have studied. However, only in recent years has predictive modelling emerged as a formal component of site location research in archaeology. In the past, the focus of most locational studies was patterns of settlement and subsistence. These studies examined the spatial distributions of sites to gain insight into the cultural and environmental adaptations of societies. The immediate goal was interpretation rather than prediction. Today, both approaches thrive and there is a basic dichotomy among the proximate objectives of locational research. Superficially, the two types of models are very similar. The results of prediction studies are certainly of interest to settlement pattern studies, and vice versa. However, in most instances it would be difficult to transform a traditional settlement pattern model into a useful predictive model without a significant restructuring of method.

One of the most important of the many recent developments in archaeological predictive modelling is a more widespread recognition of the methodological implications of these dichotomous goals. Developers of predictive site location models are abandoning many of the familiar methods of settlement pattern research in favour of methods that are more appropriate for making predictions. This does not mean that settlement pattern studies are necessarily flawed or are declining in importance. Nor does it mean that predictive models lack interpretive potential. Rather, archaeologists are becoming aware of the appropriate conceptual framework of predictive modelling, and the methods being used in predictive research are improving.

In two recent review articles, Carr (1985) and Kohler and Parker (1986) present critical appraisals of archaeological predictive models. Both articles highlight key elements of the transition away from the traditional methods of settlement pattern research.

Perhaps the most important change is the abandonment of the 'site' and the adoption of the 'land parcel' as the basic unit of analysis. As noted earlier, this approach stems from the fact that probability models require information on both positive and negative responses to stimuli. In the terminology used here, a positive response is a land parcel that contains a site and a negative response is a parcel that lacks a site (nonsite). The land parcel approach also is important because it is the only way to measure the relative effects that different independent variables have on site location (Kvamme, 1983a, 1985). In essence, land parcels provide background or control data on environmental distributions in an area, against which the distributions of site locations can be compared (see Kellogg, 1987).

Another important change is the move toward appropriate statistical procedures that are capable of converting information on site and nonsite locations into formal predictions of site-presence or site-absence probability. Two of the procedures mentioned above—discriminant function analysis and logistic regression—produce formulae that yield probability estimates. Comparisons of these procedures usually show that logistic regression outperforms discriminant function, particularly when independent variables do not have normal distributions. Logistic regression also has several practical advantages that make it the procedure of choice for many archaeological applications (see below).

As noted by Carr (1985), a third component of the shift away from the methods of settlement pattern research is an expansion of the types of independent variables used in locational analysis. In most predictive models these variables are measures of the natural environment. They differ from the variables emphasized in settlement pattern studies in that they are not limited to factors related to subsistence. For example, important determinants of site location may include such variables as soil moisture, surface slope, and access to crucial technological resources.

In his critical commentary, Carr (1985:117–120) describes five often-used empirical approaches to predictive modelling and outlines their limitations.

(1) *Density transfer*. This approach consists of simply transferring or projecting the density of sites in a surveyed area into an environmentally similar area that has not been surveyed. Limitations of the approach are: (i) it is restricted to the use of categorical independent variables, (ii) it cannot discriminate the effects of different independent variables on site location, (iii) it requires the use of large land parcels, so it cannot provide specific indications of site location, and (iv) tests of predictions would require the field survey of extensive tracts of land.

(2) *Density regression*. This approach uses the results of a multiple regression of site density vs. environmental variables in a surveyed area to predict site densities from environmental variables in an unsurveyed area. Limitations (iii) and (iv) of the density-transfer approach also apply here.

(3) *Significance regression*. The best-known example of this approach used multiple regression to predict the culture-resource significance of small land parcels in Colorado (James *et al.*, 1983). The dependent variable is a complex interval-scale measure of site significance, and the independent variables are environmental measurements. The primary methodological limitation of this approach is that the dependent variable is a composite index of unrelated factors, which Carr (1985, p.119) feels is 'an inappropriate application of multiple regression procedures.'

(4) *Discriminant function analysis*. This approach uses a discriminant function analysis of site locations, nonsite locations, and environmental variables from sampled areas to predict site-presence probabilities in unsampled areas. The main limitations of this approach involve the statistical procedure itself: (i) it assumes multivariate normality of sample data, which is unlikely in locational problems, (ii) it assumes equal covariance matrices, and (iii) it does not readily accept mixtures of categorical and interval-scale independent variables (Parker, 1985).

(5) *Logistic regression analysis*. This approach is similar to discriminant function analysis in its use of site locations, nonsite locations, and environmental variables from sampled areas to predict site-presence probabilities in unsampled areas. It differs, however, in that the logistic regression procedure is less constrained by statistical assumptions. Also, most tests indicate it provides more powerful and consistent predictions, particularly in situations where the assumptions of discriminant function analysis are violated (Kvamme, 1983b; Press and Wilson, 1978). Further, it readily accepts mixtures of nominal, ordinal, interval, and ratio-scale

independent variables, and it operates on either tabulated data or on data
with one subject per case. Its output is a formula that provides readily
interpretable predictions of site-presence or site-absence probability.

Considering the various limitations and benefits of these approaches, logistic
regression analysis is clearly the method of choice for empirical predictive modelling
of archaeological site location. Several predictive models have been developed in the
United States using this procedure. Kenneth L. Kvamme—who is credited with
developing this approach for archaeology—has created several models in western and
southeastern Colorado (Kvamme, 1983a, 1983b, 1984, 1985). Kvamme's models are
almost unparalleled in terms of their innovative definition of variables, their persistent methodological rigour, and their exposition of the vast research potential of
predictive modelling. Similar in design to the Kvamme models is one developed in
south-central Arkansas by Sandra C. Parker (Scholtz, 1981; Parker, 1985).

An integrated approach

The approach to predictive modelling described in the remainder of this chapter
integrates the use of GIS (for creating data and displaying results) with logistic
regression analysis (for generating predictive models).
    Sophisticated methods for predicting archaeological site location place heavy
demands on the processes of data acquisition and analysis. The basic unit of observation in the approach outlined here is the land parcel. Land parcels can be of any
dimension, but studies have shown that as the size of parcels decreases, there is a
corresponding increase in the accuracy and resolution of the information portrayed
by a group of parcels on a map (see Kohler and Parker, 1986). Small parcels are
thus more desirable than large ones, as they yield high-resolution data which can be
transformed into high-resolution predictive models. However, small parcels also have
an important negative benefit; every time we divide the linear dimensions of a parcel
in half, the total number of parcels quadruples. The trade-off for high resolution
is potentially vast numbers of data points. For instance, land parcels in the Shawnee
National Forest study area are square grid cells measuring 25×25 metres (0.06 ha),
and there is a total of 145,162 cells in the study area (see Warren, Chapter 18).
    To develop a true probability model, it is necessary to identify all the cells that
have been surveyed in a study area and mark each cell as either a site or a nonsite.
There are 1238 site cells and 17,945 nonsite cells in surveyed portions of the Shawnee
study area. It is also necessary to assign values to cells that designate their
environmental characteristics. Each cell is assigned a separate value for each of the
environmental variables used in the analysis. Twenty-six environmental variables
were defined for the Shawnee study area, which yields a total of 3,774,212 environmental values. To these are added sequential cell labels, column and row coordinates, and a utility variable. Altogether, then, the Shawnee data set consists of
more than 4.5 million values.
    Obviously, it is not feasible to accumulate such a vast amount of information
by hand, and it would be very difficult to analyze the data by hand. Computer-
assisted approaches are crucial to the implementation of the logistic regression
procedure, both for the collection and the analysis of data.

Data collection

Recent advances in computer technology, combined with the development of sophisticated computer software, have made it possible for archaeologists to develop complex predictive models. Computers can be used not only to analyze large sets of data, but also to help capture, interrelate and manipulate data for analysis and display. Systems of integrated computer hardware, software and peripheral equipment, which together are capable of gathering and processing spatial data, are referred to as geographic information systems (GIS).

The Illinois State Museum uses a GIS maintained by the Illinois Natural History Survey. As is discussed elsewhere in this volume (Warren, Chapter 18), this system was used to generate the data from which the Shawnee predictive model was derived. A generalized flow chart of the steps required to create the data set is shown in Figure 9.1(a). The first step in the process is to transfer information from original maps to computer storage using a coordinate digitizer. The data enter the computer as digitized lines (e.g., elevation contours) and polygons (e.g., soil series boundaries), where each constituent datum is identified by column and row coordinates. Next, the images are edited to make corrections. Then values are assigned to the polygons (e.g., soil series code), and a regular grid of cells is imposed over the data. Finally, an interpolation program is applied to the elevation coverage to assign values to cells that fall between the digitized contours.

The processed images are referred to as primary coverages. Each coverage is a grid of cells with basic environmental and archaeological values assigned to each cell. The next step is to manipulate and interrelate the data through further processing to create a series of secondary coverages that are potentially useful predictors of site location. For example, the primary elevation coverage is transformed into a variety of new variables, including surface slope and several measures of local topographic relief. A geometric formula is used to measure the distance from each cell to the nearest stream. Also, soil series that have shared characteristics with respect to certain variables are recoded in various ways, so that distinct coverages of such variables as surface runoff and biome of soil formation can be created.

Data analysis

Once the full suite of secondary coverages has been created, it is possible to proceed with the predictive modelling analysis, Figure 9.1(b). Although the central element of this process is the logistic regression statistic, which is described below, other steps in the procedure are just as important to the ultimate success of the model.

Before the model can be developed it is necessary to segregate the surveyed cells, which contain sites and nonsites, from the unsurveyed cells, whose archaeological statuses are unknown. Data for the surveyed cells are then assembled onto a new tape file for analysis.

In the initial stage of analysis, summary statistics are calculated for comparing the sample area with the remainder of the study area. This is done to determine whether the survey sample is representative of environmental variation in the study area as a whole. In a similar vein, grouped summary statistics and univariate statistical tests are run on site versus nonsite locations to determine whether there are significant environmental differences between the two groups.

The second stage of analysis consists of running the logistic regression program on a random subsample of the sites and nonsites. The program I use is a stepwise procedure that measures the predictive power of each independent variable and

(a)   Creation of GIS Data Sets

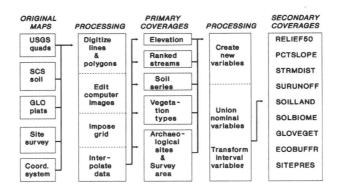

(b)   Predictive Modeling

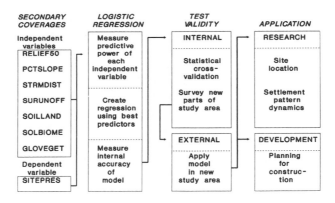

Figure 9.1. Generalized flow charts of the procedures used to (a) create computer files of geographical data using a geographic information system (GIS), and (b) create, test, and apply probability-based predictive models of archaeological site location from GIS data sets.

enters only an optimum subset of these as actual predictor variables in the model. It then produces a formula that can be applied to the predictor variables to predict the probability of site location in known or unknown areas. It also measures the internal accuracy of the model using procedures described below.

Once the model is created, it is necessary to test its validity with independent data. This can be accomplished with a cross-validation procedure, which consists of applying the model to the subsample of sites and nonsites that were withheld at random from the process of model development. The results are checked to find the proportions of correct and incorrect predictions.

In the last stage, we evaluate the differential effects of predictor variables on site probability to determine which variables have the strongest impacts on the results. We also apply the model to cells throughout the entire study area to produce shaded contour maps of predicted site-presence probability.

After a predictive model has been developed, it should be tested further with new fieldwork aimed at previously unsurveyed areas. It is desirable to direct these surveys toward environmental settings that may be underrepresented in the original survey area, as the results of new surveys can also be fed back into the model to improve its accuracy and reliability. Models can also be tested in new study areas to evaluate the geographic and environmental limits of their predictive power.

Finally, if tests of a model indicate that its performance meets or surpasses desired thresholds of validity, it may be appropriate to use the model for management, planning and research purposes.

Logistic regression analysis

Logistic regression or logit analysis belongs to a family of statistical procedures called probability models. These models were developed for special regression problems in which the dependent variable is a categorical measure rather than an interval or ratio-scale measure. In most applications of logistic regression, the dependent variable is a dichotomy that is comprised of two classes or groups. The groups may be distinguished from one another by the presence or absence of a certain characteristic, by their success or failure with respect to some criterion, or by their response or lack of response to a stimulus. The independent variables consist of one or more predictors, which can be measured on nominal, ordinal, interval or ratio scales. The regression function is a nonlinear relationship between the independent and dependent variables that predicts the probability that any given case is a member of either group. When plotted, the function defines a symmetrical, S-shaped ogive curve with asymptotes at probability values of 0 and 1.

The logistic regression model was originally developed by Berkson (1944). Interest in the model has proliferated in recent years with the advent of high-speed computers and the development of appropriate software. It is becoming an important procedure in economics (Johnston, 1984), transportation planning (Stopher and Meyburg, 1979), remote sensing (Maynard and Strahler, 1981), medicine (Gong, 1986; Prentice, 1976), and the social sciences (Aldrich and Nelson, 1984). The model is compatible with certain microeconomic assumptions of behaviour, where it is assumed that one can maximize some outcome by choosing from among a set of possible alternatives or courses of action.

In this section I present an informal description of the logistic regression model with reference to a simple hypothetical example. The intent is to provide a heuristic guide to the procedure that will aid in the development of models and the interpretation of results. Formal descriptions of the procedure can be found in Aldrich and Nelson (1984), Cox (1970), Neter *et al.* (1983), and Stopher and Meyburg (1979).

To illustrate the operation of the logistic regression model, let us consider a hypothetical example. Suppose that we have compiled information on 2000 locations in a study area. We know that half of the locations contain archaeological sites and half do not, and we have measured the distance from each location to a stream that runs through the area. Our problem is to fit a logistic regression function to the data that will allow us to predict the probability that a given location contains a site. The independent variable X is the ratio-scale measure of distance to stream. The dependent variable Y is binary and consists of two dichotomous groups: locations where sites are absent (Group A), which are coded as 0, and locations where sites are present (Group B), which are coded as 1. To examine the data we segregate the cases by group and then tabulate the scores at regular intervals along the independent

Table 9.1 Univariate logistic regression of hypothetical tabular data

Variable (X_i)	Frequency Group A A_i	Group B B_i	Proportion Group B p_i	Logistic transformation p'_i	Least square regression p'	Cutpoint probability Group B p(B)	Group A p(A)
0	0	0	—	—	−17.08	0.0000	1.0000
10	0	0	—	—	−15.37	0.0000	1.0000
20	0	0	—	—	−13.66	0.0000	1.0000
30	1	0	0.000	—	−11.96	0.0000	1.0000
40	10	0	0.000	—	−10.25	0.0000	1.0000
50	44	0	0.000	—	−8.54	0.0002	0.9998
60	117	0	0.000	—	−6.83	0.0011	0.9989
70	205	1	0.005	−5.32	−5.12	0.0059	0.9941
80	246	10	0.039	−3.20	−3.42	0.0318	0.968
90	205	44	0.177	−1.54	−1.71	0.1534	0.8466
100	117	117	0.500	0.00	0.00	0.5000	0.5000
110	44	205	0.820	1.5	1.71	0.8466	0.1534
120	10	246	0.960	3.2	3.42	0.9682	0.0318
130	1	205	0.995	5.32	5.12	0.9941	0.0059
140	0	117	1.000	—	6.83	0.9989	0.0011
150	0	44	1.000	—	8.54	0.9998	0.0002
160	0	10	1.000	—	10.25	1.0000	0.0000
170	0	1	1.000	—	11.96	1.0000	0.0000
180	0	0	—	—	13.66	1.0000	0.0000
190	0	0	—	—	15.37	1.0000	0.0000
200	0	0	—	—	17.08	1.0000	0.0000

Variable (X_i)	Correct predictions Group A	Group B	Total	Percent correct Group A	Group B	Total
0	—	—	—	—	—	—
10	—	—	—	—	—	—
20	—	—	—	—	—	—
30	0	1000	1000	0.00	100.00	50.00
40	1	1000	1001	0.10	100.00	50.05
50	11	1000	1011	1.10	100.00	50.55
60	55	1000	1055	5.50	100.00	52.75
70	172	1000	1172	17.20	100.00	58.60
80	377	999	1376	37.70	99.90	68.80
90	632	989	1612	62.30	98.90	80.60
100	828	945	1773	82.80	94.50	88.65
110	945	828	1773	94.50	82.80	88.65
120	989	623	1612	98.90	62.30	80.60
130	999	377	1376	99.90	37.70	68.80
140	1000	172	1172	100.00	17.20	58.60
150	1000	55	1055	100.00	5.50	52.75
160	1000	11	1011	100.00	1.10	50.55
170	1000	1	1001	100.00	0.10	50.05
180	—	—	—	—	—	—
190	—	—	—	—	—	—
200	—	—	—	—	—	—

Note: Notation is as follows (see Neter *et al.*, 1983, pp. 361–367): X_i = lower limits of class intervals along variable X; A_i = Group A frequencies along variable X; B_i = Group B frequencies along variable X; $p_i = B_i/(A_i + B_i)$; $p_i = \ln(p_i/(1 - p_i))$; $p' = \alpha + \beta_1 (X)$, where $\alpha =$ Y intercept, and β_1 = regression coefficient; $p(B) = 1/(1 + \exp(-(\alpha/\beta_1(X_i))))$; and $p(A) = 1 - p(B)$

variable X. From this we obtain the frequency distributions listed in Table 9.1, which are plotted as frequency polygons in Figure 9.2(a). Note that both distributions are normal, but each has a different central tendency along variable X.

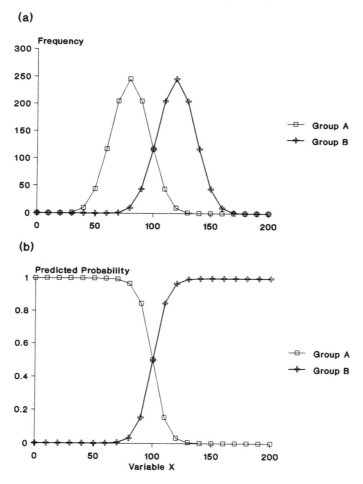

Figure 9.2. (a) Frequency polygons of two dichotomous groups (A and B) along a hypothetical variable X. (b) Plot of logistic regression functions along variable X, showing the predicted probability that cases are members of Groups A and B.

Our problem, once again, is to regress variable X on variable Y in such a way that we can predict the probability that any given location is a member of Group B (p[B]). The standard logistic regression formula for problems with a single independent variable is:

$$p(B) = \frac{\exp(\alpha + \beta X_i)}{1 + \exp(\alpha + \beta X_i)}$$

which reduces for the purpose of calibration to the following formula:

$$p(B) = \frac{1}{1 + \exp(-(\alpha + \beta X_i))}$$

where p(B) is the probability (p) that case i is a member of Group B, such that p(B) is equivalent to $p(Y_i = 1)$; exp is a function that raises the number e exponentially to a power of the value enclosed in parentheses, where the number e, Euler's number, is the irrational number whose natural logarithm is 1 ($\ln[1] = ca.2.71828$); α is a Y-intercept constant; β is a regression coefficient; and X_i is the value of the independent variable X for the ith case.

Binary dependent variables create problems in all regression models, including logistic regression. Two approaches have been developed for dealing with these problems. The first approach, which is widely used in sophisticated programs designed for use on mainframe computers, employs the binary codes (0 and 1) for the dependent variable Y and uses a maximum-likelihood fitting procedure to estimate the regression coefficients (α and β) for the independent variable. It is not possible to calibrate these coefficients with a direct algebraic solution, as the probabilities themselves are not observable. Instead, a complex series of iterations is performed in which trial coefficients are proposed, tested and refined until an optimum solution is obtained (see Aldrich and Nelson, 1984, pp. 49–54; Stopher and Meyburg, 1979, pp. 324–328).

The second, and simpler, of the two approaches estimates the regression coefficients (α and β) by calibrating a least-squares regression on an abbreviated version of the independent variable X versus a transformed version of the dependent variable Y. To do this it is necessary to consolidate the cases at arbitrary but regular intervals along the independent variable X, as we did earlier to obtain the frequency distributions presented in Table 9.1. Then we calculate aggregated proportions p_i of the Group B frequencies, and the proportions are treated as the observed values of the dependent variable Y. With this approach it is possible to obtain an algebraic solution to the problem. This is done by fitting a linear least-squares regression to the aggregate values of the independent variable X_i versus linear transforms of the values of the dependent variable p_i'. Although this approach is plagued by a number of statistical problems (see Aldrich and Nelson, 1984; Stopher and Meyburg, 1979), it is useful for the purpose of illustration.

Returning to our example, let us calibrate regression coefficients for our prediction problem using the least-squares approach to the dependent variable. We have already consolidated cases along the independent variable (Table 9.1), so we are ready to calculate the proportion of sites p_i in each interval of the independent variable using the formula:

$$p_i = B_i / (A_i + B_i)$$

These values are listed in column 4 of Table 9.1, where we see a monotonic but nonlinear increase in proportions along the independent variable X.

The next step is to transform the Group B proportions into a linear sequence of values p_i'. This is done by applying a logistic transformation to a truncated subset of the proportions. We include in this subset all fractional proportions that are greater than zero and less than one ($0 > p_i > 1$). The logistic transformation has the form:

$$p_i' = \ln(p_i / (1 - p_i))$$

where ln is a natural logarithm. The transformed values, which now have a linear relationship to variable X, are listed in column 5 of Table 9.1.

Although the logistic transformation linearizes the dependent variable p_i, it does not eliminate unequal variances of the error terms, which are heteroscedastic by default (Aldrich and Nelson, 1984, p.69). In serious applications it would be necessary to weight all independent and dependent observations (X_i and p_i') to ensure that the coefficients are not only unbiased, but also are efficient. Here, for the sake of economy, we ignore this procedure.

Once the Group B proportions have been linearized, we define the relationship between the independent variable X and the dependent variable p_i' with a linear least-squares regression. The regression is defined as:

$$p' = \alpha + \beta(X_i)$$

In our example we obtain an intercept constant α of -17.08 and a regression coefficient β of 0.1708. Column 6 in Table 9.1 lists the estimated values of the transformed dependent variable p'.

Finally, the predicted Group B probabilities p(B) are obtained by transforming the linear regression back into the nonlinear units of the logistic regression:

$$p(B) = \frac{1}{1 + \exp(-(-17.08 + 0.1708X_i))}$$

The predicted probabilities p(B) for each value of the independent variable X_i are listed in column 7 of Table 9.1. In our example, each prediction can be interpreted as the probability that a given location contains an archaeologoical site, as indicated by the distance of the location from the nearest stream (i.e. p[B] = p[Y] = 1 for a given value of X). Conversely, the probability that a given location does not contain a site is a simple function of the site-presence probability (p[A] = 1 − p[B]).

Both sets of probabilities are plotted against the independent variable X in Figure 9.2(b). Here we see that site presence probabilities p(B) increase monotonically at greater distances from streams, although the rate of increase is highly curvilinear. There is a corresponding decrease in the site-absence probabilities p(A). Note that the two curves intersect at a probability of 0.5, where p(B) = p(A). Also noteworthy is the fact that this poin t represents a critical threshold on the independent variable (ə X_i = 95), where the frequency polygons of Groups A and B also instersect, Figure 9.2(a).

The performance of our hypothetical logistic regression model can be assessed by comparing predictions of the model with the data from which it was derived. The most straightforward approach is to convert the site-presence and site-absence probabilities into predictions, and then measure the frequencies and percentages of locations that are classified correctly by the model as sites and nonsites. The results of this approach can be thought of as measures of the internal accuracy of a model.

It must be emphasized at the outset, however, that the number of possible 'solutions' to the question of accuracy is limited only by the number of distinct univariate or covariate patterns in or among the independent variables for which probabilities have been calculated. Thus, the number of possible solutions may be quite large in polynominal models that are derived from sizeable multivariate data sets. In our hypothetical univariate model there are only 15 distinct patterns. These correspond to the 15 values of the independent variable X for which at least one site or nonsite location was observed (X = 30, 40, 50, ... 170). Probabilities were calculated for all of these values, so we can measure the internal accuracy of our model at up to 15 meaningful reference points.

One of the most comprehensible ways to evaluate measures of internal accuracy is to plot them on axes of predicted probability. This is the scale shown in Figure 9.3, where the X axis is a gradient of site-presence probability (p[B]). The axis is labelled 'Cutpoint probability' to emphasize the fact that levels of accuracy vary at different cutpoints along the scale. The scale can also be referred to as the axis of discrimination, as it is along this axis that the logistic regression model seeks to maximize the discrimination between groups. This is evident in Figure 9.3(a), which shows frequency polygons of nonsites (Group A) and sites (Group B). The probability scale is nonlinear, and in comparison with Figure 9.2(a) we see how the logistic regression model has sought to push apart the distributions of the two groups.

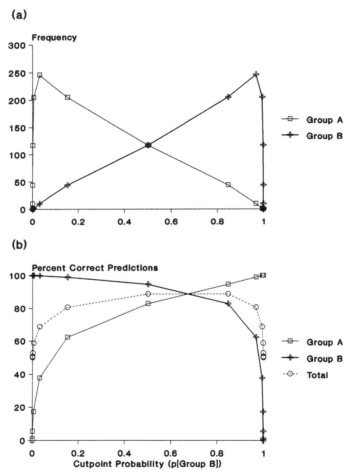

Figure 9.3. (a) Frequency polygons of two hypothetical groups along a gradient of cutpoint probability for Group B p(B). (b) Internal accuracy of the hypothetical logistic regression, expressed as percentage of correct predictions along the gradient of cutpoint probability p(B). The optimum cutpoint is between p(B) = 0.50 and p(B) = 0.85, where the Group A and Group B curves coincide and the total-percentage curve is at its maximum.

The frequencies and percentage of the correct predictions of our model are listed in the last six columns of Table 9.1, and the percentages are plotted in Figure 9.3(b). As is evident in these data, we can measure the internal accuracy of a model with reference to each of the dichotomous groups (A and B) and also in terms of the total number of predictions ($n = 2000$).

Less evident, perhaps, is the procedure used to determine the frequencies of correct predictions. Here it is useful to think of each of the 15 possible cutpoints along the probability scale p(B), or axis of discrimination, as a potential decision boundary. At each decision boundary we can arbitrarily decide that all locations at or above the boundary should be classified as sites. By the same token, all locations with lower Group B probabilities (or correspondingly higher Group A probabilities) should be classified as nonsites. Formal statements of these decision rules are as follows:

$$\text{If } p(B_j) \geq p(B_i), \text{ then } Y_{j...n} = 1$$
$$\text{If } p(B_h) < p(B_i), \text{ then } Y_{1...h} = 0$$

where the subscripts h and j respectively denote cutpoints just below and just above the decision boundary of interest (i); n denotes the highest meaningful cutpoint or decision boundary on the probability scale $pB_{1...n}$; and Y is the predicted binary code for sites ($Y = 1$) and nonsites ($Y = 0$).

For example, let us consider a decision boundary at the cutpoint probability of p(B) = 0.0318, which corresponds to a stream distance of 80 units (Table 9.1). Following our decision rules, all locations at or above this site-presence probability (p[B] ≥ 0.0318) are classified as sites, and all locations with lower probabilities (p[B] < 0.0318) are classified as nonsites. Cumulative frequencies indicate that at this decision boundary we classify 1622 locations as sites and 378 locations as nonsites (Table 9.1). Note, however, that many of these predictions are incorrect. At this level we misclassify 623 actual nonsites as sites and one actual site as a nonsite. Conversely, we correctly classify 377 of the 1000 actual nonsites (37.7 per cent), 999 of the 1000 actual sites (99.9 per cent) and 1376 of the 2000 total locations (68.8 per cent).

One important consideration in this approach to measuring accuracy is that the frequencies and percentages of correct and incorrect predictions change markedly at alternate decision boundaries. As is shown in Figure 9.3(b), the percentages of correct site predictions (Group B) progressively decrease from left to right along the cutpoint scale. In the same direction, however, there is a progressive increase in the percentage of correct nonsite predictions (Group A). Thus, when we move the decision boundary from left to right along the cutpoint scale we lose accuracy in predicting sites but gain accuracy in predicting nonsites. The two curves intersect at a probability of p(B) = p(A) = 0.673. The point of intersection is a critical juncture in terms of accuracy, and in this instance it coincides with the maximum percentage of total correct predictions. Here we find that 88.6 per cent of all locations are correctly classified.

In many situations, the optimum decision boundary for a predictive model will be the cutpoint probability at which the accuracy curves for Group A and Group B predictions coincide. It may also happen that the percentage of total correct predictions associated with this boundary can be viewed as an appropriate measure of the overall internal accuracy of the model. It should be emphasized, however, that measures and choices such as these are dependent not only on the data being analyzed, but also on the needs of the user.

Several data characteristics—including differences between groups in terms of sample size, means and variances—can affect the appropriateness of different decision boundaries and measures of internal accuracy.

(1) *Sample size.* As the sample size of one group increases in relation to the

other group, a series of changes ensues. For instance, if we increase by a factor of 10 the number of nonsites in our example (Group A) and hold constant the number of sites (Group B), then (a) the juncture of the Group A and Group B probability curves shifts to the right along variable X ($X_i = 108.6$; cf. Figure 9.2), (b) the juncture of the Group A and Group B curves of per cent correct predictions shifts to the left along the axis of discrimination ($p[B_i] = 0.265$; cf. Figure 9.3), (c) the apex of the curve of total correct predictions shifts to the right of its median location ($p[B_i] = 0.753$; cf. Figure 9.3), (d) the maximum total percentage of correct predictions increases (95.6 per cent; cf. Table 9.1), and (e) the shape of the curve of total per cent correct predictions becomes more asymmetrical, so that the per cent of correct predictions at $p(B_i) = 0$ is only one-tenth of the percentage at $p(B_i) = 1$ (cf. Figure 9.3).

(2) *Difference of Means.* As the frequency distributions of the two groups move closer together, such that there is a decrease in the difference between their means, we observe corresponding increases in the probability of error and decreases in the potential accuracy of logistic models. For instance, if the means of Group A and Group B in our example were only 20 units apart instead of 40, the maximum total percentage of correct predictions would decline from 88.6 per cent to 72.6 per cent. Conversely, accuracy increases as the means become further apart.

(3) *Difference of Variance.* If we hold the means of two groups constant and increase sample variance, there would be an increase in the probability of error and a decrease in the accuracy of logistic models. The converse is also true.

Sample-size problems are common in predictive models of site location, as sites generally are much less abundant on the landscape than nonsites. One way to minimize this problem is to extract a random sample of nonsites for analysis so that the frequencies of sites and nonsites are more nearly the same. A second method is to retain biased group sizes in the analysis and adjust the Y-intercept constant (α) after the model has been calibrated (see Kvamme, 1983b, pp. 18–19; Stopher and Meyburg, 1979, pp. 339). Group size has no effect on regression coefficients (β) in logistic regression problems, but it can bias intercept terms. This effect can be eliminated with the formula:

$$\alpha' = \alpha + \ln(n_2/n_1)$$

where α' is the adjusted Y-intercept constant; α is the raw constant as calibrated by a logistic regression program; ln is a natural logarithm; n_2 is the frequency of cases in the larger sample (e.g., nonsites in Group A); and n_1 is the frequency of cases in the smaller sample (e.g., sites in Group B).

Problems with proximate means and large variances may be more difficult to deal with, as they merely reflect patterns in data. On the other hand, it often happens that some variables are more powerful than others at discriminating groups. Also, multivariate models often are more powerful than univariate models. Thus, it may be possible to improve the accuracy of a model by testing a variety of variables.

The specific objectives of the user of a predictive model can also affect the appropriateness of different decision boundaries. If a user simply wishes to maximize the overall accuracy of predictions, the optimum boundary often will be the probability p(B) associated with the intersection of the Group A and Group B curves of per cent correct predictions, Figure 9.3(b). However, if the objective is to predict correctly the locations of sites at a higher rate, say 90 per cent, one would select a correspondingly lower cutpoint probability. The main tradeoff of this strategy would be a decrease in the accuracy of nonsite predictions.

The univariate example of logistic regression analysis described here provides a useful framework for interpreting the results of logistic models that are derived from more than one independent variable. Although calibration procedures differ substantially in bivariate and multivariate applications, such concepts as the axis of discrimination, the decision boundary and the measurement of internal accuracy apply equally well to complex problems. Moreover, the formula for multivariate logistic regression is simply an expansion of the univariate formula (Table 9.2). The multivariate formula differs only in that it provides regression coefficients for each independent variable included in a model.

Table 9.2. Multivariate logistic regression formula.

Standard formula:

$$p(B) = \frac{\exp(\alpha + \beta_1 X_{1i} + \beta_2 X_{2i} + \ldots + \beta_n X_{ni}}{1 + \exp(\alpha + \beta_1 X_{1i} + \beta_2 X_{2i} \ldots + \beta_n X_{ni})}$$

Calibration formula:

$$p(B) = \frac{1}{1 + \exp(-(\alpha + \beta_1 X_{1i} + \beta_2 X_{2i} + \ldots + \beta_n X_{ni}))}$$

Segregated calibration formula:

$$SCORE = \alpha + \beta_1 X_{1i} + \beta_2 X_{2i} + \ldots + \beta_n X_{ni}$$

$$p(B) = \frac{1}{1 + \exp(-(SCORE))}$$

Note: p(B) is the probability that case *i* is a member of Group B; exp is a function that raises the number *e* exponentially to a parenthetical value (the number *e*, Euler's number, is the number whose natural logarithm is 1[ln(1) = ca.2.71828]); α is a Y-intercept constant; $\beta_1 \ldots \beta_n$ are regression coefficients for variables $X_1 \ldots X_n$; $X_{1i} \ldots X_{ni}$ are values of variables $1 \ldots n$ for the *i*th case.

To help envision some of the multivariate analogues of univariate logistic regression, let us briefly consider a second hypothetical example with two independent variables. Figure 9.4 is a bivariate plot of the scores of two dichotomous groups of cases on two independent variables. As before, we can think of the two groups as representing the locations of nonsites (Group A) and sites (Group B) in a study area. Variable *X* can once again be a measure of distance to nearest stream. Let variable *Y* represent the ruggedness of local terrain, as measured by the amount of vertical relief near each location.

The plot of locations indicates, first of all, that there is a linear relationship between the two variables. As locations get closer to streams, there tends to be a decrease in local relief. Secondly, there is very little overlap in the scatters of the two groups, such that sites tend to occur in relatively rugged settings at greater distances from streams.

If we were to compute a logistic regression model for these data, the regression function would define an axis of discrimination that best separates the two groups (Figure 9.4). In our example this axis happens to coincide with the expected position of a least-squares regression line through the data set as a whole. However,

R.E. Warren

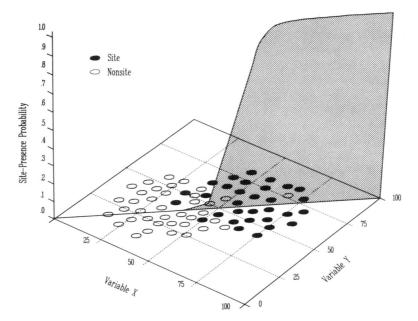

Figure 9.4. Idealized logistic regression of two groups of objects (sites and nonsites) across two independent variables (X and Y). The line running lengthwise through the horizontal scatter of points is the axis that best discriminates sites from nonsites. The vertical plane is defined by an S-shaped logistic regression line. This line shows an increase in site-presence probability from left to right along the axis of discrimination. Conversely, the percentage of correct site predictions decreases in the same direction.

there is no necessary relationship between these two lines. The position of the discriminant axis with respect to independent variables is defined by what we refer to in Table 9.2 as the SCORE component of the logistic regression formula. The probability component of the formula defines the predicted probabilities of Group A and Group B locations along the axis of discrimination. In our example, predicted site-presence probabilities p(B) increase from left to right along the discriminant axis, much as they do in Figure 9.2(b). As plotted in Figure 9.4, p(B) defines an S-shaped logistic ogive curve of site-presence probability projecting upward onto a Z axis.

Measures of internal accuracy for bivariate regressions also work in much the same manner as they did in our univariate example. For instance, the percentage of correct Group B (site) predictions would decrease from left to right along the axis of discrimination as it does in Figure 9.3(b). Note, however, that it would be necessary to rescale the discriminant axis in terms of cutpoint probabilities p(B) if we were to plot percentages of correct predictions as in Figure 9.3. The optimum decision boundary along the probability scale in our bivariate example would be a line running perpendicular to the discriminant axis at a point that best segregates sites from nonsites. This is where the per cent-correct curves of Group A and Group B predictions would intersect, and the percentage curve of total correct predictions would reach its apex.

Most of the salient features of univariate logistic regression models are also found in models that make use of more than two independent variables. However, in multivariate models the axis of discrimination becomes more complex as variables are added to the model, and decision boundaries become hyperplanes rather than points, lines or simple planes. Also, the calibration of a logistic model becomes unwieldy with two or more variables. This is especially true when the data used to develop a model consist of one subject per case. Here it is necessary to use binary codes (0 and 1) for the dependent variable instead of the group proportions (p_i) of tabulated data we used in our univariate example. Binary codes require a maximum-likelihood fitting procedure to estimate the regression coefficients of independent variables. This is accomplished through a complex series of iterations in which trial coefficients are tested and refined until an optimum multivariate separation between groups is obtained. In practice, this procedure requires access to sophisticated computer programs that are readily available only on mainframe computers. To my knowledge, the only commercial logistic regression program that operates on untabulated data (data with one subject per case) is the BMD-PLR program in the BMDP package of statistical software (Engelman, 1985, pp.330–344).

As noted earlier, once a logistic regression model has been developed it is necessary to test the true accuracy of the model with independent data. Technically speaking, this step is not requisite to the development of a logistic model. However, testing is essential if a model is to be applied in an appropriate manner for the purposes of management, planning or research.

The statistical rationale for independent testing stems from the fact that the internal tests of accuracy described above are inherently biased (Gong, 1986; Kvamme, 1983b). The sample used to develop a logistic model is the *training sample*. When one uses a training sample to determine the proportion of correct predictions of a model, the resulting accuracy rating, which we call its 'apparent accuracy,' will almost always overestimate the 'true accuracy' of the model. This is because prediction rules are heavily dependent on the data from which they are derived. The true accuracy of a model can only be determined by applying it to an independent set of observations called a *test sample*. A test sample is composed of cases that were excluded from the process of model development and, thereby, have no effect on the prediction formula. Only by applying a model to an independent test sample is it possible to evaluate the performance of the model in a realistic manner.

Several methods have been developed for assessing the true accuracy of predictive models, including the cross-validation, jackknife and bootstrap procedures (see Gong, 1986). All are resampling plans that isolate sets of observations from the calibration of prediction rules. In cross-validation, the most widely used approach, an investigator simply omits a random sample of cases from the training sample and then uses these cases as a test sample to reassess the model. The resulting difference between the 'apparent' and 'true' rates of accuracy provide a gauge of both the adequacy of the training sample and the validity of the model derived from it.

Conclusions

With the growing availability of GIS and the increased statistical capabilities of many archaeologists, prediction is bound to play a bigger role in future archaeological research. This is especially true with regard to predictive models of archaeological site location, which are proving to be valuable tools for both cultural resource management and topical research. The integrated approach to predictive modelling outlined here promises to be a useful procedure for accomplishing these goals in years to come.

Acknowledgements

I would like to thank Kathleen Allen for commenting on the manuscript and Jacqueline Ferguson for assisting with the references and illustrations. This work was supported by contracts with the USDA-Forest Service and the Illinois Department of Energy and Natural Resources

References

Aldrich, J. H. and Nelson, F. D., 1984, *Linear Probability, Logit, and Probit Models*. Sage University Papers on Quantitative Applications in the Social Sciences, No. 07–045. (Beverly Hills, CA: Sage)

Anderberg, M. R., 1973, *Cluster Analysis for Applications* (New York: Academic Press)

Berkson, J., 1944, Application of the logistic function to bio-assay. *Journal of the American Statistical Association*, **39**, 357–365

Blalock, H. M., 1979, *Social Statistics,* 2nd edition, (New York: McGraw-Hill)

Campbell, A. N., Hollister, V. F., Duda, R. O. and Hart, P. E., 1982, Recognition of a hidden mineral deposit by an artificial intelligence program. *Science*, **217**, 927–929

Carr, C., 1985, Introductory remarks on regional analysis. In *For Concordance in Archaeological Analysis: Bridging Data Structure, Quantitative Technique, and Theory*, edited by C. Carr, (Kansas City, MO: Westport Publishers), pp. 114–127

Conner, E. F. and Simberloff, D., 1986, Competition, scientific method, and null models in ecology. *American Scientist*, 155–162

Cox, D. R., 1970, *The Analysis of Binary Data.* (London: Methuen)

Engelman, L., 1985, Stepwise logistic regression. In *BMDP Statistical Software Manual,* edited by W. J. Dixon, (Berkeley: University of California Press), pp 330–344

Gong, G., 1986, Cross-validation, the jackknife, and the bootstrap: Excess error estimation in forward logistic regression. *Journal of the American Statistical Association*, **81**, 108–113

James, S. E., Knudson, R., Kane, A. E. and Breternitz, D. A., 1983, Predicting significance: A management application of high-resolution modeling. Paper presented at the *48th Annual Meeting of the Society for American Archaeology*, Pittsburgh, PA

Johnston, J., 1984, *Econometric Methods*, 3rd edition, (New York: McGraw-Hill)

Kellogg, D. C., 1987, Statistical relevance and site locational Data. *American Antiquity* **52**, 143–150

Klecka, W. R., 1980, *Discriminant Analysis*. Sage University Papers on Quantitative Applications in the Social Sciences, No. 07–019. (Beverly Hills, CA: Sage)

Kohler, T. A. and Parker, S. C., 1986, Predictive models for archaeological resource location. In *Advances in Archaeological Method and Theory*, Volume 9, edited by M. B. Schiffer, (New York: Academic Press), pp 397–452

Kvamme, K. L., 1983a, Computer processing techniques for regional modeling of archaeological locations. *Advances in Computer Archaeology*, **1**, 26–52

Kvamme, K. L., 1983b, A Manual for Predictive Site Location Models: Examples from the Grand Junction District, Colorado. Submitted to Bureau of Land

Management, Grand Junction District, Colorado.

Kvamme, K. L., 1984, Models of prehistoric site location near Pinyon Canyon, Colorado. In *Papers of the Philmont Conference on the Archaeology of Northeastern New Mexico,* edited by C. J. Condie, (New Mexico Archaeological Council Proceedings 6), pp. 349–370

Kvamme, K. L., 1985, Determining empirical relationships between the natural environment and prehistoric site locations: A hunter-gatherer example. In *For Concordance in Archaeological Analysis: Bridging Data Structure, Quantitative Technique, and Theory,* edited by C. Carr, (Kansas City, MO: Westport Publishers), pp. 208–238

Kvamme, K. L., 1989, Geographic information systems in regional archaeological research and data management. In *Archaeological Method and Theory*, Volume 1, edited by M. B. Schiffer, (Tucson: University of Arizona Press)

Lewis-Beck, M. S., 1980, *Applied Regression: An Introduction*, Sage University Papers on Quantitative Applications in the Social Sciences, No. 07–022. (Beverly Hills, CA: Sage)

Limp, W. F. and Carr, C., 1985, The analysis of decision making: alternative applications in archaeology. In *For Concordance in Archeological Analysis: Bridging Data Structure, Quantitative Technique, and Theory,* edited by C. Carr, (Kansas City, MO: Westport Publishers), pp 128–172

Maynard, P. F. and Strahler, A. H., 1981, The logit classifier: A general maximum likelihood discriminant for remote sensing applications. Paper presented at the *Fifteenth International Symposium on Remote Sensing of Environment,* Ann Arbor, MI

Neter, J., Wasserman, W., and Kutner, M. H., 1983, *Applied Linear Regression Models.* (Homewood, IL: R. D. Irwin)

Parker, S., 1985, Predictive modeling of site settlement systems using multivariate logistics. In *For Concordance in Archaeological Analysis: Bridging Data Structure, Quantitative Technique, and Theory,* edited by C. Carr, (Kansas City, MO: Westport Publishers), pp 173–207

Prentice, R. L., 1976, A generalization of the probit and logit methods for dose response curves. *Biometrics,* **32,** 761–768

Press, S. J. and Wilson, S., 1978, Choosing between logistic regression and discriminant analysis. *Journal of the American Statistical Association,* **73,** 699–705

Scholtz, S. C., 1981, Location choice models in sparta. In *Settlement Predictions in Sparta*, edited by R. Lafferty, J. Otinger, S. Scholtz, W. F. Limp, B. Watkins and R. Jones. Research Series No. 14. (Fayetteville: Arkansas Archeological Survey, University of Arkansas)

Stopher, P. R. and Meyburg, A. H., 1979, *Survey Sampling and Multivariate Analysis for Social Scientists and Engineers.* (Lexington, MA: Lexington Books)

Warren, R. E., Oliver, S. G., Ferguson, J. A. and Druhot, R. E., 1987, *A Predictive Model of Archaeological Site Location in the Western Shawnee National Forest.* Technical Report No. 86–262–17. (Springfield: Quaternary Studies Program, Illinois State Museum)

10

GIS algorithms and their effects on regional archaeological analysis

Kenneth L. Kvamme

Introduction

In recent years archaeologists have made increasing use of geographic information systems (GIS) and related technologies for the management and analysis of regional data sets. The unique ability of GIS to incorporate and manage vast bodies of spatially distributed archaeological and environmental data—with the spatial structure of the data intact—has proved to be a major boon to regional studies. Archaeological and various forms of environmental data from large regions can be encoded and spatially co-registered within GIS allowing rapid investigation of relationships between the locations of sites or other archaeological remains and any number of environmental parameters. It is a relatively simple matter to address archaeological locational questions whether using a strategy that tests specific hypotheses about environmental pattern or using a data dredging approach that merely attempts to find non-random patterns in the locations of sites with respect to the regionally encoded environmental data. Furthermore, the development and application of models of regional archaeological distributions is greatly facilitated through GIS. Model decision rules based on various environmental conditions and constraints can be computed and plotted across vast areas allowing ready portrayal of results and assessments of performance (see Wansleeben, 1988, and Kvamme, 1989, for recent overviews of GIS applications in archaeology).

When analyzing distributions of archaeological phenomena with respect to environment, archaeologists traditionally have measured such variables as slope, aspect, elevation, or distance to water either (1) directly in the field, or (2) on paper maps (e.g., Judge, 1973; Parker, 1985; Kellogg, 1987). Data obtained from maps, of course, represent a level of abstraction from the real world. In GIS contexts regional analyses are necessarily based on environmental data produced and generated through computer means, which represent a second level of abstraction because the data input to the computer normally are map data and the end product is an abstraction from this source. In other words, to create an elevation data theme in a GIS setting, usually referred to as a digital elevation model (DEM), contour lines on maps must be electronically digitized and input to the system. One of a variety of available interpolation methods is then employed by the computer systematically to estimate an elevation at locations between the contour lines (Kvamme and Kohler, 1988). Slope, aspect, local relief, drainage locations and other data of relevance to archaeological analyses then may be derived from the DEM based on local interrelationships among stored elevation values, and there are various algorithms used to generate each of these data types as well (e.g., Collins and Moon, 1981; Monmonier, 1982; Burrough, 1986).

It seems prudent, and of some importance, to investigate the effects of different computational procedures used in GIS or terrain analysis packages on the nature of the environmental data actually generated. Additionally, given that alternative computational methods might yield different environmental outcomes, it is worthwhile to examine how the conclusions of archaeological analyses based on such data might be affected. This chapter attempts to offer a few insights into these issues. A number of interpolation methods for estimating *elevation*, used in DEM construction, are examined as are several algorithms used to generate *slope* (gradient or ground steepness) data. In each case differences are emphasized through comparison. Finally, the effects of these differences are examined on the results of an archaeological locational model developed for an east-central Arizona study region.

Digital elevation models

Regional distributional analyses of archaeological phenomena frequently focus on such environmental variables as elevation, slope, aspect, local relief, and proximity to drainages and landform features like mesa or canyon rims (Judge, 1973; Thomas and Bettinger, 1976; Roper, 1979). It is noteworthy that many of these variables relate to terrain or landform shape characteristics. In GIS settings the DEM is the representation of landform. Terrain variables, like slope, aspect, local relief, drainage locations and others, are derived from the DEM through various software techniques. The accuracy of the DEM, therefore, should be an important issue and concern because it controls not only the quality of the overall landform representation, but also the nature of all the secondary data types derived from it. If archaeologists are interested in analyzing prehistoric or historic regional distributions with respect to terrain form variables using GIS, one would think that a general goal would be DEMs of fairly high accuracy, portraying ridges where ridges are located, knolls and hilltops where knolls and hilltops are located, drainages where drainages are located, and the like. In practice, this often is not the case.

High quality DEMs generally are hard to obtain for a particular region of study. The US Geological Survey (USGS) currently is producing, on tape for general distribution at low cost, DEMs of their 7.5 minute series, 1:24,000 scale maps. Only a small portion of the available quadrangles in this series have been converted to digital form thus far, however (National Cartographic Information Center, 1988). Chances are that a DEM of a particular map of interest will not be available. One alternative is to obtain the services of one of the growing number of commercial enterprises that are able to create DEM data for a specified region, but this can be expensive. A second alternative is to create a DEM in-house, but this requires special hardware, usually a digitizing tablet, vast amounts of labour, to trace manually with the digitizer all the contour lines on the map(s) of the region, and special software to convert the digitized contours to a DEM of the region (Kvamme and Kohler, 1988). It should be noted that state-of-the-art technology utilizes optical means to computer encode map contour data, greatly reducing the manual labour requirement, but this means of DEM creation is only beginning to be available at low cost (e.g., Sircar and Ragan, 1984).

Because of the difficulty of obtaining a high-quality DEM of a region an alternative, and compromise, is typically adopted. The US Army's Defense Mapping

Agency digitized, more than two decades ago, the elevation contours on the
1:250,000 scale maps which are available for the 48 contiguous United States. DEMs
were created from these contours (which can be purchased through the USGS), but
it must be realized that at this scale, and with the 200 ft contour interval that is used
in many of the western maps, whole ridges, drainages, hills and other landscape
features are underrepresented or even missing (Stow and Estes, 1981). To illustrate,
Figure 10.1a portrays a portion of the USGS 7.5 minute series (1:24,000 scale)
Dotsero, Colorado quadrangle. In Figure 10.1b is a computer-contoured map of the
Army-produced DEM data of approximately the same area. Although there is a
general correspondence, in the latter (1) many small ridges and hills are absent, (2)
minor drainages are missing, (3) large features are greatly smoothed, and (4) there
is a major error in the form of a 120 ft high cliff face which would surely make a
spectacular waterfall on the Colorado River (which flows down the central valley)
if it really existed!

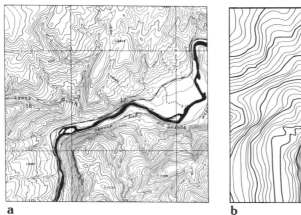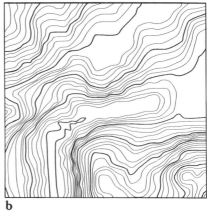

a   b

*Figure 10.1. (a) Portion of the USGS 7.5 minute, 1:24,000 scale, Dotsero, Colorado
quadrangle. (b) Computer-contoured map of the Army-produced digital elevation data of the
same region derived from 1:250,000 scale maps. The contour interval is 40 ft in both maps.*

From an archaeological standpoint it is exactly the kinds of data that are
missing—small ridges, hills, drainages—that usually are most important in regional
locational studies. Analyzing the distributions of the small lithic scatter sites that
occur in the depicted area with respect to such variables as elevation, slope, aspect,
local relief and topographic setting, when the small ridges and other features upon
which they usually occur do not exist in the computer data set, is an error-prone
exercise of questionable worth. The absence of strong archaeological patterning on
these kinds of variables in several GIS-based regional analyses (e.g., Marozas and
Zack, Chapter 14; Warren *et al.*, 1987) has been blamed, in part, on the lack of
accuracy in the DEM representation of the study region.

Of course, an alternative is to construct a DEM utilizing data obtained through
digitization of elevation contour lines on maps of better quality. Even with
reasonably small-scale maps and a low contour interval, however, the DEMs that
result can vary, and yield differential quality, depending on the nature of the
computer algorithms used to create them.

The basic problem in DEM creation from digitized contour lines is to estimate
elevation values between the contour lines of known elevation. Literally hundreds

of interpolation algorithms exist (Lan, 1983; Hodgson, 1989) and each, of course, offers a slightly different result. Elevation values normally are interpolated to yield a grid or matrix of elevations which forms the DEM (Kvamme and Kohler, 1988).

To illustrate some of the foregoing a contour map of a hill with a saddle running to the northeast and connecting with a second hill is shown in Figure 10.2a. The map portrays a 300 × 300 m region with a contour interval of 10 m. The contour lines were digitized and three elevation interpolation algorithms were investigated that offer alternative and slightly different DEMs. In each case elevation values were estimated every 10 m yielding a 30 × 30 matrix.

The first algorithm, called here the 'steepest ascent' method, searches from a locus in eight directions through the digitized elevation contour data until known elevations are encountered in each search. An elevation is then linearly interpolated at the locus based on the two opposite elevations that offer the steepest slope across the locus (Leberl and Olson, 1982).

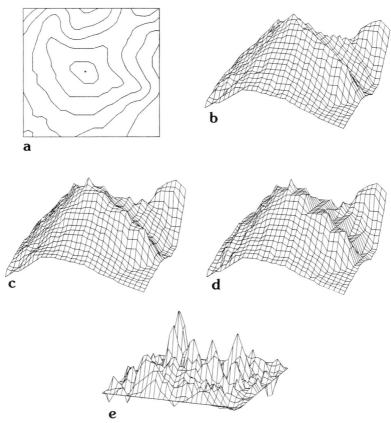

Figure 10.2. (a) Contour map of the test region; (b) DEM produced by the steepest ascent algorithm; (c) DEM produced by the weighted average algorithm; (d) DEM produced by the vertical scan algorithm; (e) difference map of b–c.

This procedure is illustrated in Figure 10.3 where the eight search lines are indicated radiating outward from the central locus of unknown elevation (point 0). The line of steepest slope or ascent across the locus is clearly that between points 4 and 8 (Figure 10.3). Calculations for the steepest ascent interpolation algorithm are shown in Table 10.1. It should be realized that the steepest ascent algorithm closely follows the traditional manual method for determining the height of any point on a contour map. Application of it to the digitized contour data in Figure 10.2a yields a DEM that appears to offer a natural representation of the terrain form (Figure 10.2b)

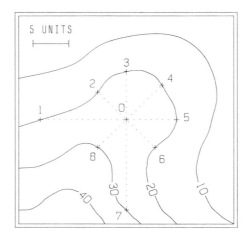

Figure 10.3. Search pattern for known elevations used to interpolate an elevation at locus 0.

The second interpolation method, referred to as the 'weighted average' algorithm, also searches in eight directions from a locus until elevation contour data are encountered (Figure 10.3). Following the logic that the elevation at any point is more closely related to nearby known elevations than ones farther away, this procedure simply takes a weighted average of the eight surrounding elevations, where each is weighted by the reciprocal of its distance from the central locus. This is a common spatial interpolation method (Monmonier, 1982; Lan, 1983), but regions within horseshoe-shaped contours tend to receive too much weight from that contour. This situation is illustrated in Figure 10.3 and Table 10.1 where the weighted average algorithm yields a very low elevation estimate owing to the large influence of the nearby elevations in the surrounding horseshoe-shaped contour. In application, this shortcoming results in a 'bench-like' effect along ridges and drainages as shown in Figure 10.2c.

The final, and simplest, method of interpolation is termed the 'vertical scan' technique, which searches only vertically, in opposite directions, from any locus through the digitized contour data until known elevations are encountered (the line between point 3 and 7 in Figure 10.3). An elevation is then linearly interpolated at the locus based on these elevations (Yoeli, 1975). As might be expected, this procedure can yield a poor result if the principal direction of elevation change is orthogonal to the scan line because the algorithm ignores data to the left or right. This situation is illustrated in Figure 10.3 and Table 10.1 where the vertical search lines between points 3 and 7 clearly are not coincident with the principal direction of elevation change causing an elevation estimate that is too low. In application, regions within horseshoe-shaped contours can even receive identical elevations causing a 'stair-step' effect in the resulting DEM (Figure 10.2d). The vertical scan algorithm was introduced here only to show the extreme nature of results that can be obtained.

Seriously considering only the steepest ascent and weighted average results

Table 10.1. Elevation interpolation calculations for locus 0 in Figure 10.3.

DATA (see Figure 10.3)

Elevations (e_i)

$e_1 = e_2 = e_3 = e_4 = e_5 = e_6 = 20$; $e_7 = e_8 = 30$

Euclidean distances (d_{ij}) between opposing elevation pairs

$d_{1,5} = 19$; $d_{2,6} = 11.31$; $d_{3,7} = 20$; $d_{4,8} = 12.73$

Slopes (s_{ij}) between opposing pairs (* = slope by vertical scan; ** = steepest slope)

$s_{1,5} = (e_5 - e_1)/d_{1,5} = (20 - 20)/19 = 0$

$s_{2,6} = (e_6 - e_2)/d_{2,6} = (20 - 20)/11.31 = 0$

$s_{3,7} = (e_7 - e_3)/d_{3,7} = (30 - 20)/20 = 0.5$*

$s_{4,8} = (e_8 - e_4)/d_{4,8} = (30 - 20)/12.73 = 0.79$**

Weights (reciprocals of Euclidean distances) between each known elevation (e_i, $i = 1,8$) and the unknown locus (e_0):

$w_i = 1/d_{0,i}$

$w_1 = 1/d_{0,1} = 1/12 = 0.083$; $w_5 = 1/d_{0,5} = 1/7 = 0.143$

$w_2 = 1/d_{0,2} = 1/5.66 = 0.177$; $w_6 = 1/d_{0,6} = 1/5.66 = 0.177$

$w_3 = 1/d_{0,3} = 1/7 = 0.143$; $w_7 = 1/d_{0,7} = 1/13 = 0.077$

$w_4 = 1/d_{0,4} = 1/7.07 = 0.141$; $w_8 = 1/d_{0,8} = 1/5.66 = 0.177$

INTERPOLATION ALGORITHMS

Steepest ascent (steepest slope is between points 4 and 8)

$e_0 = s_{4,8}d_{0,4} + e_4 = (0.79)(7.07) + 20 = 25.59$

Weighted average

$$e_0 = \frac{\sum_{i=1}^{8} w_i e_i}{\sum_{i=1}^{8} w_i}$$

$= [(0.083)(20) + (0.177)(20) + (0.143)(20) + (0.141)(20) + (0.143)(20) + (0.177)(20)$
$+ (0.077)(30) + (0.177)(30)]/(0.083 + 0.177 + 0.143 + 0.141 + 0.143 + 0.177 + 0.077$
$+ 0.177) = 22.27$

Vertical scan (vertical line is between points 3 and 7)

$e_0 = s_{3,7}d_{0,3} + e_3 = (0.5)(7) + 20 = 23.5$

(Figures 10.2b,c), a number of important differences in the outcomes seem to occur. These differences can be made more apparent by resorting to a popular technique used in some GIS known as 'map subtraction' (Berry, 1987), where one data theme is subtracted from another. Here, the matrix of 900 elevations obtained from the weighted average method is subtracted from the 900 elevations generated by the steepest ascent algorithm to produce a data set of differences, which is plotted three-dimensionally in Figure 10.2e. In Figure 10.2e the bisecting plane represents no difference between the two DEMs; the numerous peaks and valleys elsewhere indicate that some major differences exist. Statistically, the most extreme difference is 5 m in elevation, about nine per cent of the range in the original elevation data (in both DEMs the range in elevation is 56 m). The mean difference is 0.23 m indicating that the steepest ascent method tends to yield higher values.

Slope data

To explore one step further how alternative computer algorithms for the same data type can yield different results, several computer procedures for deriving slope (ground steepness) data from a DEM are investigated. Like elevation interpolation methods, a wide variety of algorithms have been used to yield slope estimates (Evens, 1972; Monmonier, 1982; Jensen, 1984). The two examined here utilize interrelationships among elevations in a 3 × 3 moving 'window' which is centred in turn around each elevation in the DEM matrix (except those in the first and last row and first and last column where a full 3 × 3 matrix is not possible; alternative methods are used here). The first method determines two orthogonal slope vectors, one along the current row and a second along the current column, and returns the square-root of the vector summation as a slope value. Calculations for a hypothetical 3 × 3 matrix are shown in Table 10.2. The second method fits a least-square plane to the nine elevations and computes the maximum gradient (slope) on the plane. Note that in this simple context of nine elevations, each spaced an equal distance or cell width apart, the least-squares arithmetic greatly simplifies (Table 10.2). Because this method considers all elements of the 3 × 3, including the diagonals, it is more sensitive to the extreme slope lying across the principal diagonal of the hypothetical data in Table 10.2, yielding a higher slope value. The drawback, of course, is the increased computational effort.

Both slope estimation techniques were applied to the same steepest ascent DEM data (Figure 10.2b) to yield two slope surfaces, portrayed as grey level plots in Figures 10.4a,b (with dark regions indicating steeper ground). Although each slope map yields generally parallel results, which is to be expected, subtle differences exist. These differences again are made more apparent through map subtraction and plotting of the differences (Figure 10.4c). Simple descriptive statistics (Table 10.3), further illustrate the nature of the dissimilarities between these slope algorithms where it is apparent that the univariate distributions are different, as revealed by the moments and other statistics. For example, the least-squares procedure tends to yield lower and less variable slope values than the root sum of vectors method (Table 10.3).

Compounding of differences

Each of the slope data sets portrayed in Figures 10.4a,b were derived from the same DEM obtained by the steepest ascent algorithm. It is instructive to examine how differences between final end-products can be compounded through GIS. This can be accomplished by ascertaining the degree of dissimilarity between data obtained from one slope procedure obtained on a DEM created by one interpolation method, and a second slope procedure derived from DEM data generated by a different elevation interpretation algorithm. The least-squares-produced slope data obtained from the steepest ascent DEM (Figure 10.4a) constitutes one data set and a second slope data set (not illustrated) was generated by application of the root vector summation

Table 10.2. *Illustration of slope calculations by the root sum of vectors and least-squares algorithms. The example data exhibit a strong slope on a northwest-southeast axis (upper left to lower right).*

DATA: ELEVATIONS IN A 3×3 MATRIX

COLUMN

		$j-1$	j	$j+1$
	$i+1$	47	35	31
ROW	i	38	30	26
	$i-1$	33	27	21

COMPUTATION OF SLOPE AS PER CENT GRADE (s_{ij}) AT POSITION I, J
Root sum of orthogonal vectors algorithm

let:

$$b_1 = \frac{(e_{i,\,j+1}-e_{i,\,j-1})}{2r} \quad ; \quad b_2 = \frac{(e_{i+1,\,j}-e_{i-1,\,j})}{2r}$$

then:

$$s_{i,\,j} = (100(b_1^2 + b_2^2)^{0.5}$$

example:

$$b_1 = \frac{(26-38)}{2(10)} = -0.6 \quad ; \quad b_2 = \frac{(35-27)}{2(10)} = 0.4$$

$$s_{ij} = 100(-0.6^2 + 0.4^2)^{0.5} = 72.11\%$$

Least-squares algorithm

let:

$$b_1 = \frac{(e_{i-1,\,j+1}-e_{i-1,\,j-1})+(e_{i,\,j+1}-e_{i,\,j-1})+(e_{i+1,\,j+1}-e_{i+1,\,j-1})}{6r}$$

$$b_2 = \frac{(e_{i+1,\,j-1}-e_{i-1,\,j-1})+(e_{i+1,\,j}-e_{i-1,\,j})+(e_{i+1,\,j+1}-e_{i-1,\,j+1})}{6r}$$

s_{ij} is defined as above.

example:

$$b_i = \frac{(21-33)+(26-38)+(31-47)}{6(10)} = -0.667$$

$$b_2 = \frac{(47-33)+(35-27)+(31-21)}{6(10)} = 0.533$$

$$s_{ij} = 100(-0.667^2 + 0.533^2)^{1/2} = 85.37\%$$

Table 10.3. *Descriptive statistics for slope data obtained by alternative computational procedures applied to the steepest ascent DEM.*

	ALGORITHM	
	root vector summation	least-squares
mean	32.899	31.906
variance	149.831	122.555
skewness	0.665	0.554
kurtosis	0.951	0.676
minimum	2.231	5.116
maximum	75.000	71.144
range	72.769	66.028

K.L. Kvamme

procedure applied to the weighted average DEM. The least-squares-produced slopes were then subtracted from the root vector summation slopes to yield a data set of compounded differences, illustrated in Figure 10.4d.

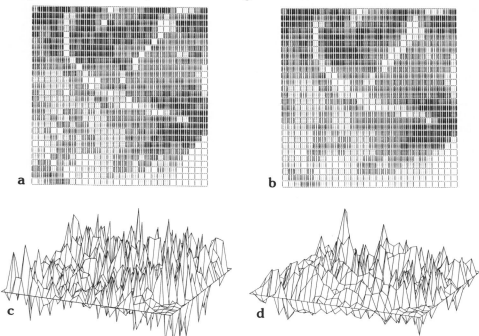

Figure 10.4. (a) Slope data produced by the least-squares algorithm; (b) slope data generated by the root sum of vectors algorithm; (c) difference map of b−a where both were computed on the same DEM; (d) difference map of b−a where each was computed on a different DEM.

Although comparison of Figure 10.4d with Figure 10.4c does not strongly indicate the greater dissimilarities obtained by the compounding of estimation differences (owing to the nature of the plots), simple descriptive statistics amply demonstrate this to be the case. Given in the first column of Table 10.4 are statistics describing the differences between the slope procedures when each was computed on a different DEM (i.e. compounded differences). The second column summarizes the differences obtained by computing both slope estimates on the same DEM (i.e. the difference between the data summarized in Table 10.3). The variance of the compounded differences is nearly three times greater than the variance of the difference between slopes computed on the same DEM and the range of the differences is nearly twice as great, for example.

Table 10.4. Descriptive statistics for differences between the root vector summation and least-squares slope algorithms.

	Slope data sets derived DIFFERENT DEMs	Slope data sets derived SAME DEM
mean	−0.678	0.993
variance	59.244	22.669
minimum	−28.561	−12.905
maximum	30.215	16.667
range	58.776	29.572

Effects of different environmental algorithms on archaeological analyses

The effects of the potentially different values of basic environmental data that can be obtained by computers on archaeological analysis results should be obvious. Different means, variances, coefficients of skewness, ranges, and distributional form in general, have been demonstrated for just two environmental data types when alternative computational methods are employed. As a result, in regional archaeological analyses that employ such data, different outcomes, and possibly significantly different outcomes, could conceivably be obtained depending on whichGIS package is used and the nature of the particular environmental estimation procedures present in each. This raises the question of the extent to which the conclusions reached in an analysis are the result of real characteristics of the data or of the particular computer procedures used to generate the data. A second issue raised is whether different conclusions would be reached if a different GIS package (with different algorithms) was used. Because the conclusions of any analysis are only as good as the data upon which they are based, a question that apparently should be examined in GIS contexts pertains to the basic quality and nature of the computer-produced environmental data.

To illustrate the foregoing a geographic data set established by the students of my 'Regional Analysis with GIS' class at the University of Arizona is employed. As part of a class project the students digitized from 1:24,000 scale maps such data as elevation contours, soil class polygons, and river and stream locations in a 9×8 km region around the Grasshopper Pueblo in east-central Arizona. This area was selected because more than 25 seasons of survey and excavation have been conducted in the region by the University of Arizona Field School and numerous sites dating to the 13th and 14th centuries are known (Reid, 1974; Longacre *et al.*, 1982). GIS techniques were used to generate a DEM from the elevation contours as well as slope, aspect, relief, soil type, distance to agricultural soil, distance to water, and various other terrain form indices variables which were expected by prior work to have some bearing on the prehistoric site distributions. The DEM was produced using the steepest ascent algorithm and the slope data were obtained by the least-squares method. The data for each variable were held in a 90×80 matrix (7,200 observations) with a spacing between each observation or cell width of 100 m. The DEM of the region is illustrated in Figure 10.5a and portrays a central valley, and several smaller valleys, flanked by gently rising hills. The class used these data to perform univariate and multivariate analyses of the known archaeological spatial distributions (more than 200 sites), and to develop models of various prehistoric patterns of settlement and land use.

One model that I developed was a multiple logistic regression equation for the locations of multi-room habitation sites. This model was a function of slope, local relief, a terrain form index that describes a location relative to ridge-like or drainage-like settings (see Kvamme, 1989), horizontal distance to nearest drainage, vertical distance to nearest drainage, and distance to the nearest agricultural soil in the study region. The slope, relief, terrain index and vertical distance to nearest drainage data all were derived from the DEM (Figure 10.5a). The regression coefficients were estimated from a sample of 30 known habitation sites and a control sample of 198 locations taken randomly from the background environment of the study region (see Kvamme, 1988, for details of this procedure). They are given in the first column of

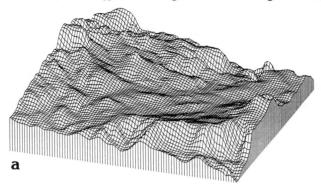

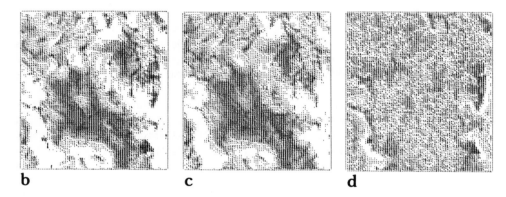

Figure 10.5. (a) DEM of the east-central Arizona study region; (b) habitation site model 1; (c) habitation site model 2; (d) absolute difference of b−c.

Table 10.5. Logistic regression coefficients for the habitation site models. Coefficients significant at the 10 per cent level are indicated with an asterisk.

	FIRST MODEL (Steepest ascent DEM; least-squares slope)	SECOND MODEL (Weighted average DEM; root vector sum slope)
Intercept	0.0372	1.4439
Slope	−0.2278*	−0.1817*
Terrain form index	0.0059*	0.0090*
Relief	−0.0169*	−0.0111*
Horizontal distance to drainage	−0.0041*	−0.0019
Vertical distance to drainage	0.0212*	0.0024
Horizontal distance to agricultural soil	0.0010*	0.0007*

Table 10.5. The mapping of the model, obtained by applying the regression coefficients to the GIS environmental data point-by-point (i.e., to all 7,200 locations), is portrayed in Figure 10.5b, and generally suggests habitation site locational tendencies in certain valley or gently-sloping uplands contexts (compare Figures 10.5a and b).

In order to illustrate how environmental data generated by alternative computational methods can yield different archaeological results, a second DEM of the same region was generated using the weighted average interpolation algorithm. The DEM-derived variables (slope, relief, the terrain form index and vertical distance to nearest drainage) were then recomputed from this data source. Additionally, the slope algorithm was changed from the least-squares method to the root sum of vectors procedure. The distance to soils and horizontal distance to drainage data remained unchanged.

Using the same sample habitation site and background control locations, but with the new environmental data, a second logistic regression function was established. The coefficients of this model are given in the second column of Table 10.5 and its mapping is shown in Figure 10.5c.

Comparison of the rival models suggests broad similarities although some major differences exist. For example, the coefficients of both models (Table 10.5) possess the same sign. Thus, we can infer from both models that habitation sites tend to be located on level ground (low slope), in locally uniform terrain (low relief), close to drainages (low horizontal distance to nearest drainage), but on local high points (high terrain form index and high vertical distance to nearest drainage) and on non-agricultural soil (high distance to agricultural soil). Although the signs of the coefficients are the same, their magnitudes differ markedly and some do not achieve statistical significance in the second model (Table 10.5). The terrain form index, for example, has about one-and-a-half times more weight in the second model than in the first, while horizontal distance to the nearest drainage has about half as much weight, and vertical distance to the nearest drainage has about a tenth as much weight. The magnitude of these differences, by themselves, must yield different modelling outcomes when mapped, but since each model is applied to a different digital data set obtained by alternative computational procedures, outcomes with major dissimilarities must occur (compare Figures 10.5b and c).

The lack of agreement between the two models is made more apparent by subtracting one model from the other and mapping the absolute differences (Figure 10.5d). In Figure 10.5d it is clear that large sub-regions exhibit major differences between the rival models and data sets thereby emphasizing that different outcomes can be achieved in archaeological analyses solely on the basis of the computer algorithms that generate the environmental data upon which the analyses are based. We must therefore question the degree to which a particular analysis outcome is due to real pattern in the archaeological data, or to the computer methods used to produce the data in GIS contexts. More research is clearly needed in this area.

Conclusions

Archaeologists are usually concerned only with the quality of archaeological data, not the quality of data obtained by computer means. In GIS environments archaeologists are only too happy to obtain vast amounts of elevation, slope, aspect and other data, with relatively little effort. Seldom questioned are the nature or validity of these data. Moreover, the documentation that accompanies most GIS often does not reveal exactly how such data are calculated, making informed assessment of the issues presented here difficult. Rather, there seems to be a general tendency to accept computer-generated data as 'truth'. Hopefully, this paper has interjected a note of caution in this regard.

Acknowledgements

Much of the above material was performed using a computer program called *TERRAIN PAC* written by the author. The 3-D graphics were obtained using the

SAS/GRAPH program. I am indebted to the students of 'Regional Analysis with GIS' (1988) for their efforts in establishing the archaeological application data set. The archaeological data analyzed in this paper were originally collected by the University of Arizona Field School at Grasshopper with funds provided by Historic Preservation Survey and Planning Grants-in-Aid from the State Historic Preservation Officer and with the encouragement of the White Mountain Apache Tribal Council. This paper was originally presented at the annual meeting of the Society for American Archaeology in 1988.

References

Berry, J.K., 1987, A mathematical structure for analyzing maps. *Environmental Management*, **11**, 317–325

Burrough, P.A., 1986, *Principles of Geographical Information Systems for Land Resources Assessment*. (Oxford: Clarendon Press)

Collins, S.H. and Moon, G.C., 1981, Algorithms for dense digital terrain models. *Photogrammetric Engineering and Remote Sensing*, **47**, 71–76

Evens, I.S., 1972, General geomorphometry, derivatives of altitude, and descriptive statistics. In *Spatial Analysis in Geomorphology*, edited by R.J. Chorley, (London: Methuen), pp 17–90

Hodgson, M.E., 1989, Searching methods for rapid grid interpolation. *Professional Geographer*, **41**, 51–61

Jensen, S.K., 1984, Automated derivation of hydrological basin characteristics from digital elevation model data. *Proceedings, Auto Carto 7*, pp. 301–310

Judge, W.J., 1973, *Paleoindian Occupation of the Central Rio Grande Valley, New Mexico* (Albuquerque, University of New Mexico Press)

Kellogg, D.C., 1987, Statistical relevance and site locational data. *American Antiquity*, **52**, 143–150

Kvamme, K.L., 1988, Development and testing of quantitative models. In *Quantifying the Present and Predicting the Past: Theory, Method, and Application of Archaeological Predictive Modeling*, edited by W.J. Judge and L. Sebastian, (Washington D.C.: US Government Printing Office), pp. 325–428

Kvamme, K.L., 1989, Geographic information systems in regional archaeological research and data management. In *Archaeological Method and Theory*, volume 1, edited by M.B. Schiffer, (Tucson: University of Arizona Press), pp. 139–203

Kvamme, K.L. and Kohler, T.A., 1988, Geographic information systems: technical aids for data collection, analysis, and display. In *Quantifying the Present and Predicting the Past: Theory, Method, and Application of Archaeological Predictive Modeling*, edited By W.J. Judge and L. Sebastian, (Washington, D.C.: US Government Printing Office), pp. 493–547

Lan, N.S.N., 1983, Spatial interpolation methods: a review. *The American Cartographer*, **10**, 129–149

Leberl, F.W. and Olson, D., 1982, Raster scanning for operational digitizing of geographical data. *Photogrammetric Engineering and Remote Sensing*, **48**, 615–627

Longacre, W.A., Graves, M.W. and Holbrook, S.J., (editors), 1982, Multidisciplinary Research at Grasshopper Pueblo, Arizona. *Anthropological Papers of the University of Arizona*, **40**.

Monmonier, M.S., 1982, *Computer Assisted Cartography: Principles and Prospects*

(Englewood Cliffs, NJ: Prentice-Hall)

National Cartographic Information Center, 1988, *Index to DLG 1:24,000 Quadrangles and DEM* (Reston, VA: US Department of the Interior, Geological Survey).

Parker, S., 1985, Predictive modeling of site settlement systems using multivariate logistics. In *For Concordance in Archaeological Analysis: Bridging Data Structure, Quantitative Technique, and Theory*, edited by C. Carr, (Kansas City Westport Publishers), pp. 173–207

Reid, J.J., (editor), 1974, Behavioral Archaeology at the Grasshopper Ruin (special issue). *The Kiva*, **40.**

Roper, D.C., 1979, Archaeological Survey and Settlement Pattern Models in Central Illinois. *Illinois State Museum Scientific Papers*, **16.**

Sircar, J.K. and Ragan, R.M., 1984, An interactive technique to generate elevation data using a vidicon camera. In *Spatial Information Technologies for Remote Sensing Today and Tomorrow, Proceedings of Pecora 9* (Silver Springs, MD: IEEE), pp. 411–418

Stow, D.A. and Estes, J.E., 1981, Landsat and digital terrain data for county-level resource management. *Photogrammetric Engineering and Remote Sensing*, **47,** 215–222

Thomas, D.H. and Bettinger, R.L., 1976, Prehistoric pinon ecotone settlements of the Upper Reese River Valley, Central Nevada. *Anthropological Papers of the American Museum of Natural History*, **53.**

Wansleeben, M., 1988, Applications of geographical information systems in archaeological research. In *Computer and Quantitative Methods in Archaeology, 1988*, edited by S.P.Q. Rahtz, BAR International Series **446**, volume 2, pp 435–451

Warren, R.E., Oliver, S.G., Ferguson, J.A. and Druhot, R.E., 1987, A Predictive Model of Archaeological Site Location in the Western Shawnee National Forest, Report submitted to Shawnee National Forest by the Illinois State Museum Contract Archaeological Program, Springfield.

Yoeli, P., 1975, Compilation of data for computer-assisted relief cartography. In *Display and Analysis of Spatial Data*, edited by J.C. Davis and M.J. McCullugh, (New York: John Wiley) pp. 352–367

PART III

Data sources, hardware and software

11

Coping with space: commentary on GIS data sources, hardware and software

Ezra B. W. Zubrow and Stanton W. Green

Introduction

Perhaps the best way to begin a discussion on data sources is to discuss how data are created and presented. Methodologies for creating data sources range from ground survey through computer digitizing and from scanning to remote sensing. Some systems such as ARC/INFO include programs which allow one to take standard surveying readings (i.e. ground trail readings) and transform them into a coordinate map system which creates coverages for a GIS. Coding, a second method for creating a GIS data source, is accomplished by gridding a map and coding the cells according to values of particular variables which are then used primarily in raster-based GIS systems. Digitizing is also a map-based data source. It is accomplished using a system of manual tracing. This is usually performed with an electronic stylus, pointer, wand or mouse on a digitizing tablet or table. Digitizing creates a computer-readable file. Scanning is the next level of technological abstraction for database creation. Here an optical scanner reads a map and creates a computer readable file. Even more recently, there has been the development of video digital camera scanning. This has developed rapidly as part of the ongoing video-computer interactive market. We predict that in a few years you may be able to go down to your local video store and rent digital map data to put in your local GIS for trip planning. Furthest from the ground is remote sensing. These techniques use the reflectance of different land covers read via aircraft or satellite to create raster-based data matrices.

Sources

Federal and private sources use these techniques to create data sources to sell and trade. Together they provide researchers with access to GIS data. A sample of these agencies includes USGS, National Cartographic Information Center, Soil Conservation Service, National Geophysical Data Center, Defense Mapping Agency, and the Department of Defense, to name only a few.

These databases come in a variety of forms. Two of the most common are digital elevation models (DEM)—raster-based (i.e. grid-based) elevation data that represent topography and digital line graphs (DLG), vector-based data that often include roads, political boundaries and hydrology. Many of the major GIS have the capability of combining data from multiple sources using open-ended hooks. For example, even student-oriented image processing packages such as Dragon have hooks which allow one to send and receive data to and from ARC/INFO.

There are several issues regarding governmental sources which will become of increasing importance in the next decade. For example, it appears there is an increasing privatization of public data. Is it a good trend based upon efficiency or does it provide unfair monopolistic advantages to particular companies?

Output

At the other end of the data continuum is output or data presentation. To begin, there are a variety of output media. The most traditional is paper output which can be produced using a variety of printers and plotters including: dot matrix, laser printers, pen plotters, ink-jet printers and plotters, and thermal transfer plotters (Plates 3 and 11).

Photographic output is also possible. One may photograph from the monitor with a variety of equipment and techniques. We use the old fashioned 'shoot-off-the-screen-with-your-home-Nikon' technique. This works best by covering your monitor with a cardboard box with a hole for the lens to reduce background reflection (Plates 7 and 8). There is, of course, professional photographic equipment which attaches to the computer and or monitor to produce high quality photographs. Video output is also possible and is particularly useful for dynamic and animation aspects of GIS.

Temporary output can be sent to monitors which come in a variety of flavours and resolution qualities (colour, black and white, high resolution vga). For most of these systems one needs a graphics card or graphics adapter for the computer. Some systems use separate specialized graphics terminal such as TEKTRONIC 4207 or a VISUAL 600. These will frequently provide useful hardware and software capabilities. There are also software emulators which imitate graphics adapters and graphics terminals (TGRAF, SOFTGRAF).

Finally, it is also possible to output to such computer media as files, tapes, CD-ROM and disks.

Data themes

Databases are similar to symphonies. They have themes, motifs and a variety of instruments giving similar information different emphases. Data themes transfer data sources into useful cartographic and geographic information. Among these are the traditional cartographic themes that many maps would usually contain including hydrology, elevations, roads, vegetation, land use and ownership, soils, political boundaries and population centres.

The archaeological section of the symphonic database is analogous to a single instrumental section such as the cellos. They play their own themes which are not identical to the other sections (violins), but are related harmonically. Archaeologists use such thematic coverages as the areas which have been field surveyed, surface counts for artifacts, site classifications, chronological attributes, agricultural field walls, canals and prehistoric subsistence, to name only a few. There are specialized themes or coverages which the individual researchers create for specific needs. These might include plant species for ethno-botantists and palynologists, chert outcroppings for lithic specialists, and stratigraphy for geo-archaeologists.

Integrating data sources

Frequently there are problems in integrating data sources. New GIS users are normally unaware of this problem. Let us consider a brief example in which prehistoric sites are located near streams. Futhermore, the hydrology system is created from two different data sources. For example, two archaeologists, one from Columbia and one from Buffalo, have studied the area, each digitizing his own data. A third archaeologist from Tulane wishes to use both sets of data. However, the hydrological coverages were created by different digitizers (people) using different digitizers (equipment) and using different GIS software. The Cajun archaeologist must consider the following questions. Do the two mouse-pushers digitize streams in the same direction, i.e. in the act of digitizing does one proceed from the northeast to the southwest corner of map or vice versa? Did the digitizers proceed equivalently from the source of the stream to the outlet or vice versa? At what level of specificity was the hydrology digitized? Were river tributaries included; were streams; and were creeks?

A second problem relates to archaeological excavation. Problems in this area are based upon the implications of digitizing a site as a point or as an area. This is not simply a question of resolution. It has implications for the entire method by which one excavates, analyzes artifactual data and interprets a site. For example, how does one correlate point-provenanced data with grid cell data? How do you treat point-provenanced depths when they must be combined with the more traditional artifactual locations within a stratum? Finally, what are the implications of digitized arbitrary stratum versus digitized natural strata for analysis?

These problems may be solved if one has both grid and point-provenanced information for every artifact in the database. However, this is usually unrealistic. Therefore, archaeologists are frequently faced with the problem of integrating grid and point-provenanced data. This is by no means impossible, but it requires systematic methods—many of which are provided by GIS. A final problem to note is rectification. This commonly results when a field archaeologist needs to regrid a site after a winter's abandonment. Despite the hypothetical precision of the datum and the high accuracy of many of today's surveying instruments, the new grid must be rectified with the previous one, both on the ground and in the GIS.

The papers

Stine and Decker (Chapter 12) integrate a variety of data sources using ERDAS. The paper is a technical demonstration in rectification and reconstruction. They create palaeo-landscapes of the Florida coast with the use of satellite, bathymetric and topographic data. Their product is high quality, three-dimensional colour output which models the prehistoric coast. It is a 'view' which includes the sub-surface oceanic floor. In a real sense, it is a 'new view' of the past topography. One area, not emphasized but important, is the way this 'new view' provides a 'new insight' into possible relationships of environment to human activity. One may literally view the prehistoric shoreline from the sea. It supplies a basis for modelling the relationship of sea level change to prehistoric settlement patterns. In fact, if one wished to be a post-processual archaeologist, one might claim that viewing these reconstructed landscapes from the perspective of the prehistoric native in a canoe could actually provide the basis for a 'reconstructed cognition'.

The paper by Farley, Limp and Lockhart (Chapter 13) is an illustration of how a 'tool box' approach to GIS may be applied to large archaeological problems. As part of the Arkansas Archaeological Survey they frequently must control large amounts of data in an efficient manner. The tool box approach has both advantages and disadvantages. The advantages are the ability to apply specific tools to particular problems. The disadvantage of the tool box approach is that there is a tendency to 'band aid' one's analysis. By this we mean that instead of designing an entire analysis based upon theoretical considerations, one tries each tool in turn to see what one gets. It is analogous to 'throwing one's data against the wall to see what sticks'. It is the standard argument against over-empiricism. If one is not careful, one can accept second best analyses.

Two tools which this paper use are exceptionally useful. One is exploratory data analysis (EDA). Exploratory data analysis attempts to show the characteristics of the data as well as prove relationships. Thus, there is an interactive approach to finding out relationships or outliers among the data and then rediscovering new relationships after taking the previous findings into account. This interactive approach is very fruitful in combination with GIS. The complicated spatial relationships can be segmented into their individual aspects. The second tool which we find particularly valuable is 'Chernov faces'. These faces are used to show visually the relationships of multivariate data. This tool takes advantage of human ability to recognize complicated relationships among human faces far more easily and rapidly than in any other statistical or graphical form.

Marozas and Zack's paper (Chapter 14) on prediction of site location develops a graphic representation of a probability surface of site density zones. Since this analysis is based upon interpolation procedures, one of the critical variables is the degree of error which is created by the particular interpolation. When combined with the errors created by problems in spatial resolution, spatial measurement and general imprecision, as well as the errors created by data capture algorithms and manipulation functions of a GIS, the accumulated error could be sufficient to discredit the results. However, this paper shows that the actual accumulated error is relatively small and controllable in adequate GIS analyses. Their suggestions for control are threefold. First, one carefully scrutinizes the scale and resolution of input data so that they match the intended uses of the GIS. Second, there should be methodical definition of the research problem so that the problem, data and GIS match. Finally, one integrates disparate data sets using methods of scale matching.

Madry's paper (Chapter 15) provides a broad survey of the types of hardware which are available. There is a large amount of information available. The potential user should be aware of a variety of different types of CPUs, storage media, input and output devices such as digitizers, and CRTs which Madry describes in an organized manner. One of his most valuable contributions is to provide the new user with some reasonable sample systems. Finally, Madry delineates the major trends in hardware. In particular, he points out the unification of diverse systems into multi-purpose systems so that with the new developments in such software as GIS/CAD and GIS/Remote Sensing systems, the hardware has to become more generalized and able to do a broader range of tasks. In addition, the editors believe that there is an ongoing movement towards distributed mini-computers of the SUN, VAX or IBM workstation type which provide the increased power which is necessary for new types of software. Finally, we expect that GIS hardware will ultimately be attached to a variety of laboratory equipment which does image processing.

Zubrow's article (Chapter 16) provides a paper on how to evaluate software.

He wants the user to maintain a sceptical attitude. He initiates the discussion by reminding the reader to determine whether they really need a GIS. Next, he provides a set of evaluative criteria. A new user will find these criteria worthwhile in thinking about which software to purchase. They are particularly useful in that rather than making positive suggestions about types of functions to obtain, they suggest areas of consideration where many users are unaware of the pitfalls and traps. He provides two tables of useful data. The first is a survey of software, the second a list of governmental agencies and people (as well as their telephone numbers) who are using particular software. Finally, he comments on the 'dream system'.

Conclusions

Archaeologists will be able to cope with space in a far more sophisticated, efficient and perspicacious manner by using GIS. The nature of the data, the quality of the data and their definition will ultimately effect the results. However, it is important to remember that there is a degree to which this entire section is very ephemeral. Data sources, hardware and software are moving targets. The questions that one may ask one day may be obsolete the next. The rate of change is faster than in any other area of scientific and technical endeavour of which the editors are aware.

Ultimately, there is no problem independent of theory; nor when using GIS is there any solution independent of the quality of the data sources or the techniques which are built into the software and hardware.

12

Archaeology, data integration and GIS

Roy S. Stine and T. Drew Decker

Introduction

'Integration of diverse data is the key GIS capability...' (Chrisman *et al.*, 1989). For the archaeologist data can range from actual excavation data, or ground truth, to satellite imagery. The variety and quantity of anthropological data make geographic information system (GIS) technology an appealing option for the researcher.

A GIS should not be thought of as a 'quick fix' to one's spatial data problems (Berry, 1987). In today's market GIS come with different data structures, capabilities, platforms and, very importantly, cost. It is necessary, therefore, that archaeologists formulate their information needs within the spatial framework of a GIS (Stine and Lanter, 1989; see Chapter 8). This information includes traditionally drafted site maps, purchased maps (eg. 7.5 minute USGS quadrangles), aerial photographs, digital imagery, digital information (eg. census materials) and all ancillary data (eg. site and artifact descriptions). After this initial step one needs to review and understand the abilities of different GIS, including aspects of both hardware and software. The Federal Government published a pamphlet entitled *A Process For Evaluating Geographic Information Systems* (Guptill, 1988) which should help in this endeavour.

This paper will review theories and methods incorporated in other projects that have utilized a GIS. Next, some of the standard digital sources of spatial data available to archaeologists will be discussed. Finally we model the landscape of Florida during the Paleo-Indian period to demonstrate how digital image processing and GIS technology can be applied to archaeology.

GIS and the integration of data sources

Analysis of ancient landscapes has been stimulated by GIS technological development. Overstreet *et al.* (1986) use a GIS with the goal of a comprehensive environmental reconstruction for the interpretation of human prehistory. They feel that '...knowledge of a three-dimensional matrix in which archaeological sites occur, and an understanding of which landscape components have been removed by erosional forces' will greatly aid archaeological interpretations. This environmental matrix approach has been utilized earlier by Butzer (1982). Overstreet *et al.*, however, modelled the environmental and cultural factors within a GIS framework. This allows for more rapid update, storage and modelling capabilities than would be possible with manual cartographic methods.

'The landscape is the spatial manifestation of the relations between humans and their environment' claim Crumley and Marquardt (1987), who take a multiscalar, interdisciplinary approach to landscape interpretation. They are interested in understanding the interrelationship between various physical and cultural features at particular times and in particular places and demonstrate how these relationships change over time and space. Their work incorporates the disciplines of anthropology, geography, geology, history, art history and ecology. Most pertinent to this particular discussion is the variety of data sources utilized by Crumley and Marquardt, ranging from archaeological field survey and excavation to digital MSS remotely sensed data. Madry (1987) discusses the multidimensionality of this data. He refers to the four resolutions inherent in remotely sensed data; spatial, spectral, temporal and radiometric (for a full discussion of each of these see Jensen, 1986). Integrating these data sources is made possible through GIS.

Johnson *et al.* (1988) utilize the ERDAS software driven by an IBM-AT PC to design large scale archaeological survey in the north Mississippi platform. This software is primarily an image processing package with GIS capabilities. In this study they find that the multidimensional nature of remotely sensed data combined with statistical analysis allows development of predictive site models. The remotely sensed data generate a 100 per cent coverage of the survey area. This coverage is then coupled with soils information and the results of on-site testing. This information was placed onto USGS 7.5 minute quadrangle sheets to rectify the images. Investigation with this wide variety of data, integrated within a GIS, thereby gives the researchers the ability to combine different landscape features in various ways—that is, it allows them to model landscape.

Digital sources of spatial data

Archaeologists collect spatial information from a wide variety of sources. Some of the traditional sources include historic maps, state and county road maps, property and tax maps, deed plats, USGS 1:24,000 and 1:100,000 topographic maps, soil maps and geologic maps. In addition archaeologists generate their own maps. These include site maps, unit and feature maps, distribution maps and soil stratigraphy maps. In most of these cases ancillary data are also generated to accompany the maps. Several issues must be faced when one integrates these data in a GIS.

First, when a GIS is incorporated into the project all of this information must be converted to digital form. The four major methods for doing this are manual digitizing, automatic scanning, entry of coordinates using Coordinate Geometry (COGO), and conversion of previously automated information (Jenks, 1981; Dangermond, 1988). Each of these input methods has its strengths and weaknesses.

Second, regardless of the source of spatial information, error is always present (Goodchild, 1982). The amount and traceability of the error depends upon the data source. In an automated environment this error will produce problems in digitizing, edgematching, polygonization, labelling, reformatting digital data between different systems, and database creation (Guptill, 1988). These conflicts between maps can cause great difficulties within a GIS. Understanding and utilizing preventive and corrective measures will save processing time and prevent the production of 'elegant trash.' Both of these problems can be integrated through data standardization.

Recently, *The American Cartographer* (1988) published an issue entitled 'The Proposed Standard For Digital Cartographic Data'. This volume covers standards, definitions and references, spatial data transfer specification, digital cartographic data quality and cartographic features. In it was proposed the digital cartographic standard endorsed by the Federal Government mapping agencies (Guptill, 1988). Since the Federal Government is one of the leading producers of digital data, conforming to the quality standards set forth in *The American Cartographer* will eliminate many problems in the collection and generation of data.

The standards also include accuracy ratings for maps. This is a particular problem when historical maps, unrectified aerial photographs (eg. many county soil maps), and other less reliable sources are used. Referencing maps as to how and where these maps were derived will help one in understanding what maps can be overlaid with one another and which maps may only be compared side by side. Archaeologists should take the same care in recording the sources, lineage and error rates of their maps as they do in recording each aspect of the site.

In an environment where cartographic data are digitally input, revised, reorganized and generated, keeping track of map lineages is essential. Overlaying maps of exactly the same geographic region that have conflicts in representing the same features will produce polygons and distinctions that do not exist on either map (Dangermond, 1988). Placing an image that is not geo-referenced into a layer will mean that the resulting image does not correspond to an accurate map of the same geographic area. If the two maps are overlaid, they produce erroneous information. Fully documenting graphic products will greatly reduce the possibility of misusing them.

Of the various sources of digital spatial information available at present, the Federal Government is the largest supplier. The US Geological Survey (USGS) presently has the 1:100,000 Digital Line Graphs (DLG) entered into the *National Digital Cartographic Data Base* (NDCDB). Aerial photographs are also available through the USGS. This agency is currently working on Digital Elevation Models (DEM), and on digitizing the 1:24,000 topographic quadrangles for the entire United States.

Other federal sources of digital data include: the Soil Conservation Service's *Soil Geographic Data Base*; the US Fish and Wildlife Service's *National Wetlands Inventory Data Base* and the National Ocean Service, (including shore line data, aides to navigation data and nautical charts). One of the best sources of digital data promises to be the US Bureau of the Census *Topological Integrated Geographic Encoding and Referencing* (TIGER) files for the 1990 census. This will contain census block information that is accurately represented by a topologically-structured DLG.

Private sources of satellite remote sensing data include the EOSAT Corporation which handles US multi-spectral scanner (MSS) data and thematic mapper (TM) data. The French, Satellite Pour l'Observation de la Terre (SPOT), Image Corporation also produces high quality digital imagery. This brief listing does not include the many private companies which fly independent missions. More complete listings of digital data sources are available from the listed federal agencies or the private companies.

A View of Paleo-Indian Period Florida

This study models the Paleo-Indian landscape of Florida by combining various sources within a GIS. The system used is similar to the one utilized by Johnson *et al.* (1988). The lowered sea level of this period would have extended the land area around John Pennekamp Coral Reef State Park, Florida, the site of the present coral reefs. The creation of a modelled image along the Florida coast during the Paleo-Indian period is detailed.

We begin by creating a three dimensional (3-D) model of the ocean floor near the reef. The image is formed through the integration of satellite data and bathymetric data of the southern coast of Key Largo, Florida in John Pennekamp State Park. This study area covers the island of Key Largo, Florida and the reef lying south of the island from Rodriguez Key northeast to the Elbow Reef. Data for the area include: (1) a SPOT three-band multispectral satellite scene with a ground resolution of 20 × 20 m; (2) three USGS 7.5 minute (1:24,000 scale) quadrangle maps; and (3) bathymetric charts of water depths from the Florida Department of Natural Resources. The University of South Carolina Remote Sensing Laboratory and RPI Corporation, Columbia, South Carolina supplied the maps. The ERDAS image processing computer system at the Remote Sensing Laboratory was used to create the 3-D image.

Three USGS quadrangles (Rock Harbor, Blackwater Sound and Garden Cove) were used to remove geometric distortion from the satellite image. The process involves matching easily distinguishable points found on the image to the same points found on the maps at known coordinate locations. Approximately 60 of these ground control points were used to correct the image. Error in matching the ground control points between the maps and the image resulted in each point in the corrected image being within 1.49 pixels of its real location. Since each pixel is 20 m square, the image has an accuracy rate of under 30 m.

The bathymetric map was placed on an optical enlargement and reduction machine to match the scale of the park map to the scale of the USGS maps. Water depths on the bathymetric map were then transferred to the quadrangles. This allows the depth data to be placed on the same accepted base map that is used to correct the image; each value thus has a Universal Transverse Mercator (UTM) coordinate location. This procedure introduces some error as the two maps do not perfectly register. The approximate error does not exceed 30 m and is attributed to the lack of land reference points in the water-dominated study area.

Elevation points for the area are then digitized from the USGS quadrangles that contain the bathymetric data. Since the computer software does not allow negative numbers, sea level is given the arbitrary value of 100 as no depths exceed 100 feet. The completed elevation file is called DEPTH.1.

The next step is to input topographic and bathymetric data into the ERDAS 3-D generation program. To accomplish this both the satellite image file (KEYLARGO.LAN) and elevation file (DEPTH.DIG) had to be reduced to 512 × 512 pixels. A site within the study area is then selected to provide the observer with the optimum view of the subsurface landscape. For this we choose an area combining the present-day island, the coral reef and the channel between them. First the satellite image is trimmed and its exact corner UTM coordinates noted. Then the coordinates are used to trim the corresponding portion of the depth file so that the two areas will match exactly. Plate I shows the new satellite image with its UTM coordinates.

Several programs are accessed to create a contour surface of the area from the elevation data and satellite image (Plate II). Each contour shading represents an approximate 5 foot contour interval. Next we establish the viewing position. A viewing position above the reef itself and looking back toward shore would convey the most relief. An observer's point-of-view position was centred over the reef. This requires specifying the width of the field of view, observer height, direction of view and vertical exaggeration. This procedure allows for 'quick looks' of the topography based on the viewing geometry before the 3-D image is created.

The final 3-D image is created by overlaying the contour surface with the satellite image based on final viewing geometry specifications. The 3-D image (Plate III) looks northwest from the reef toward Key Largo Island. The image uses the same colours assigned to the original SPOT image. Thus, various blue and green hues represent the subsurface terrain while the distant vegetated coast appears false-colour infrared. The result is a new view of a very old perspective of a portion of the Florida landscape.

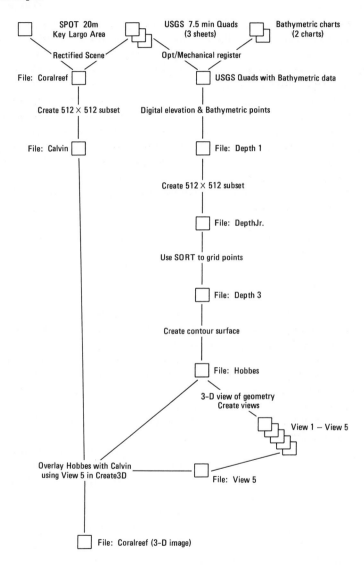

Figure 12.1. Outline of three-dimensional generation process.

Conclusion

One of the key contributions of a GIS to archaeology is that it gives researchers powerful methods to test theories and produce models. Several rules of thumb will aid archaeologists in making the best use of this powerful methodology. Cartographic standards and digital cartographic standards must be developed and followed. A proper 'lineage' of how each map layer is produced should be kept (Figure 12.1). The assumptions, statistics and functions utilized on the different layers must also be recorded. These procedures help the operator to keep track of what they have accomplished and the limitations of their data set. Most significantly, it allows other researchers to recreate the work.

One of the strengths of a GIS is its ability to integrate diverse data sets. In the present experiment, 'traditional' paper maps were utilized in combination with satellite data to help generate a new image. Manual cartography, manual digitizing, digital image processing and various GIS functions were combined to create a theoretical Paleo-Indian landscape. By keeping careful records of each step in the process, a lineage can be charted for this final image. The stated commands were intentionally kept general so as to indicate their function rather than using the exact terminology of a particular software package. The accuracy rate of the final product is equivalent to the largest amount of error at any stage in the process, in this case 30 m. As a result, the project can be replicated by the same or other systems with the same or better accuracy rates.

GIS and remote sensing technologies are rapidly becoming powerful tools for archaeologists who, along with their fellow researchers in related disciplines, must come to know the strengths and weaknesses of their databases as well as the GIS technology. Only then will the potential of both be fully exploited.

References

Berry, J. K., 1987, Introduction to Geographic Information Systems: Management, Mapping and Analysis for Geographic Information Systems (GIS). Special Workshop for South Florida Water Management District

Burrough, P. A., 1986, *Principles of Geographic Information Systems for Land Resource Assessment* (Oxford: Clarendon Press)

Butzer, K., 1982, *Archaeology as Human Ecology* (Cambridge: Cambridge University Press)

Chrisman, N. R., Cowen, D. J., Fisher, P. F., Goodchild, M. F. and Mark, D. M., 1989, GIS: Past and Future. In *Geography in America*, edited by G. L. Gaile and C. J. Willmott, (Columbus, OH: Merrill Publishing Co.)

Cowen, D.J., Hodgson, M., Shirley, W. T., Wallace, T. and White, T., 1985, Alternative approaches to display of USGS land use/land cover digital data. *Auto-Carto VII*, pp. 116–125

Crumley, C. and Marquardt, W. (eds.)., 1987, *Regional Dynamics: Burgundian Landscapes in Historical Perspective*. (New York: Academic Press)

Dangermond, J., 1988, A Review of Digital Data Commonly Available and Some of the Practical Problems of Entering Them Into A GIS, *ACSM Proceedings Annual Conference, Volume I*, pp 1–10

Goodchild, M., 1982, Accuracy and Spatial Resolution: Critical Dimensions For Geoprocessing. In *Computer Assisted Cartography and Geographic Information Processing: Hope and Realism*, edited by D. H. Douglas *et al.* (Ottawa: Cartographic Association)

Guptill, S. C., (ed.)., 1988, A Process For Evaluating Geographic Information Systems. US Geological Survey Open-File Report 88–105

Jenks, G. F., 1981, Lines, Computers, and Human Frailties. *Annals of the Association of American Geographers*, **71**, (1)

Jensen, J. R., 1986, *Introductory Digital Image Processing*, (Englewood Cliffs,NJ: Prentice-Hall)

Johnson, J. R., Sever, T. L., Madry, S. L. H. and Hoff, H. T., 1988, Remote Sensing and GIS Analysis in Large Scale Survey Design in North Mississippi. *Southeastern Archaeology*, **7**.

Kvamme, K. L., 1989, Geographic Information Systems In Regional Archaeological Research And Data Management. In *Method and Theory in Archaeology*, Volume I, edited by M. B. Schiffer, (Tucson, AZ: University of Arizona Press)

Madry, S. L. H., 1987, A multiscalar approach to remote sensing in a temperate regional archaeological survey. In *Regional Dynamics: Burgundian Landscapes in Historical Perspective*, edited by C. L. Crumley and W. H. Marquadt (San Diego, CA: Academic Press)

McHarg, I. L., 1971, *Design With Nature* (New York: Doubleday and Company)

O'Brien, M. J., (ed.)., 1984, *Grassland, Forest, and Historical Settlement: an analysis of dynamics in northeast Missouri* (Lincoln, NB: University of Nebraska Press)

Overstreet, D. F., Smith, C. R. and Bruzewicz, A. J., 1986, The Archaeology of Lost Landscapes: Geographic Information Systems at Coralville Lake, Iowa. In *Geographic Information Systems in Government*, Volume I, edited by B.K. Opitz (Hampton, VA: A. Deepak Publishing), pp. 313–378

Stine, R. S. and Lanter, D. P., 1989, Application of Geographic Techniques in Archaeology. Paper presented at the *Society for Historical Archaeology Conference*, Baltimore, MD

Wagstaff, J. M., (ed.)., 1987, *Landscape and Culture: Geographical and Archaeological Perspectives*, (Oxford: Basil Blackwell Ltd.)

13

The archaeologist's workbench: integrating GIS, remote sensing, EDA and database management

James A. Farley, W. Fredrick Limp and Jami Lockhart

Introduction

Technological advances in many areas of computer science are providing archaeologists with a powerful set of tools to pursue their investigations. Specifically, the availability of high performance graphic work stations and sophisticated software for database management (DBM), geographic information systems (GIS), remotely sensed image processing (RS) and exploratory data analysis (EDA) combine to provide the archaeologist with the means to identify and explain cultural and biophysical patterning more effectively. These tools have been effectively used in archaeological applications for a number of years. However, in most instances these applications have focused on the utility of a single tool to resolve a specific problem.

This paper examines the results which can be obtained through the integration of these techniques in a 'tool box' for automated data analysis and explores the rationale behind such an approach. We will illustrate the use of this tool box with specific examples at three different spatial scales; multi-state, regional and local. In addition, we will discuss the range of data which is appropriate for a GIS at each of these various scales and the types of information they can be expected to produce.

The approach presented here is a direct outgrowth of the organizational orientation of the Arkansas Archeological Survey. This orientation is driven by our long term responsibility to acquire and preserve information pertinent to state cultural resources and the desire to use that information to conduct a broad range of archaeological research. Because we are in the position of having to function as a repository for data while simultaneously using these data in ongoing research we must develop general purpose tools to accommodate many different applications and interests. However, we also have the fiscal and staffing resources afforded to a statewide institution. This combination of charter, research interests and resource availability have both directed us and permitted us to pursue these research directions.

Our statewide archaeological databases for sites, projects, artifacts and citations illustrate our first attempt at developing integrated systems for management applications and research. These databases are multipurpose and must serve as instruments for research and meet management needs on many levels. The elements of each database are interlinked allowing a comprehensive view of the full range of existing data. It is important to emphasize that these automated systems could not have been developed without the substantial historical foundation provided by the existing manual record system. Even with a high quality records system in place, system development and data acquisition involved many person-years. The rewards of such a database are not immediate and the effort and time required to develop a system of this sort may be unacceptable to those who require more immediate results. However, like a long term investment programme, the benefits increase with time and eventually come to fruition, overwhelming the initial, though substantial, effort.

Cowen succinctly summarizes this fact in a discussion of GIS system development:

> '...successful applications of GIS must occur in an institutional setting...[and]...operation of such systems must be conducted with a long-term perspective' (Cowen, 1987; pp.53–54; see also Marble and Peuquet, 1983)

The benefits derived from our general purpose databases have multiplied as we have expanded our applications and incorporated additional hardware and software. This is primarily due to strategic decisions made during the initial stages of database design and development which insured that 'hooks' were preserved in each database to allow for future database expansion and subsequent integration with peripheral software packages. As we became involved with GIS and EDA applications the existing databases and these 'hooks' were used to provide useful input on the range, distribution, preservation and overall character of the archaeological sites throughout the state (e.g. Limp, 1986, 1989; Parker *et al.*, 1986; Farley, 1988; Limp *et al.*, 1989). In time, although each isolated application area remained important, this individual importance was substantially enhanced because data could easily be moved from one application to another and prototype programs developed for a specific application could be quickly generalized and reused in other applications. Our site, project artifact and citation databases have been developed using INFORMIX, a commercial database system which is based on the relational model popularized by E. F. Codd in the late 1960's and supports the structured query language (SQL) standard. Our GIS applications are run under the Geographic Resources Analysis Support System (GRASS), a raster-based GIS which was developed by the Construction Engineering Research Laboratories (Westervelt *et al.*, 1987; Goran *et al.*, 1987). Remotely sensed image processing is conducted using the ELAS system developed by NASA-ERL (Graham *et al.*, 1987), and EDA applications are conducted using 'S' an interactive environment for data analysis and graphics developed at Bell Laboratories (Becker and Chambers, 1984). This software was all developed in, or in the case of ELAS adapted to, the UNIX operating system environment. It exhibits the modularity and the flexibility associated with modern software design and because of this our applications can be integrated and effectively portrayed as an automated 'tool kit' for analysis (Figure 13.1).

Tools for automated data processing

The notion of the tool kit is ingrained in archaeology and in fact the efforts of many archaeologists have been directed exclusively towards identifying and explaining the relevance of these groups of associated artifacts. The concept is a simple one. A single tool, a saw for instance, provides the ability to cut wood. This function in itself is useful, but if a hammer and nails are added to the saw you create a set of

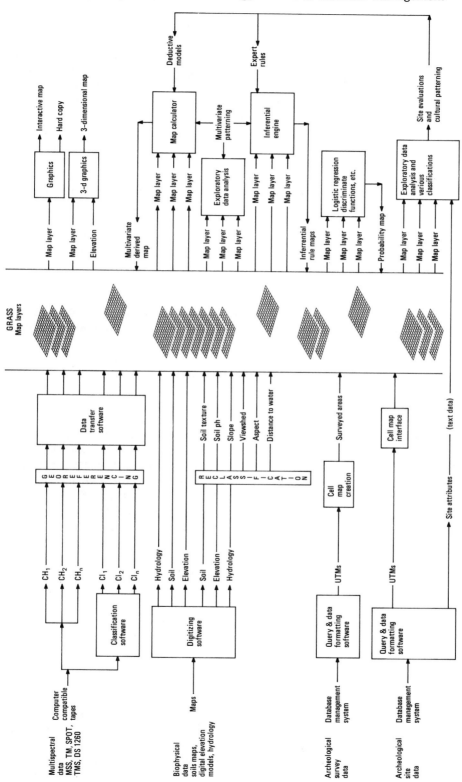

Figure 13.1. Components of the automated tool kit.

tools capable of building a structure. Thus by making an additional investment in the hammer and the nails the value of the saw is enhanced. In addition, the manner in which the saw is perceived has been changed. It is no longer simply a wood-cutting device, it is now perceived as a house-building tool. These benefits and this change in perception would not occur if an ice pick, a useful but unrelated tool, was acquired in place of the hammer and nails.

There are a number of practical lessons to be learned from this simple example. First, a group of tools used in an integrated framework is more powerful than the same tools used individually in isolated or unrelated applications. Second, the acquisition of new tools requires an additional investment. Third, the use of tools in such an integrated fashion requires a change in perception on the part of the tool user. Finally, the acquisition of new tools cannot be *ad hoc*, rather it must be directed by a well defined objective. Implicit within all of this is the assumption that the product or range of products to which any tool can contribute will ultimately determine its value and by extension justify the investment required to obtain it.

There are direct corollaries in the preceding discussion which are relevant to the application of GIS, RS, DBM and EDA techniques to analytical problems in archaeology. First, the cost of obtaining these tools is substantial. The tools in this case are not simply the requisite pieces of hardware and software that combine to form a platform for analysis. These are just the building blocks or the foundation for analysis. The real cost is incurred in identifying and acquiring the data which these tools will manipulate to produce useful information. For instance, the construction of the AMASDA database (Hilliard and Riggs, 1986), which contains observations on as many as 130 variables for more than 23,000 archaeological sites, required continual refinement of the database schema and took almost 10 years to complete. This investment far outweighs the actual cost of the hardware on which the AMASDA system is run and does not take into consideration the aggregate cost associated with the production of the manual records from which AMASDA was derived. GIS data development presents a similar problem. For example, development of the numerous data themes which together comprise the Fort Hood GIS has taken almost 8 years to complete (Williams *et al.*, 1990). Similarly, the construction of a single statewide data theme representing the surficial geology of the state of Arkansas required more than three months and cost in excess of $10,000.

Because of the significant costs, measured in both time and capital, required by these types of applications the range of problems to which they are applied must be maximized. In addition, the number of managers and research specialists having access to the data, and the tools required to manipulate it effectively, must also be maximized. DBM, GIS, RS and EDA can be utilized in such an environment, but these integrated applications must be carefully planned and data must be collected and stored in a way that preserves maximum flexibility in the exchange of data and information between these tools. In the absence of such planning, or modelling, of data flow, large amounts of both time and money may be invested only to create compartmentalized analytical modules which never realize their potential.

In order to achieve this level of integration some commonly held perceptions need to be modified. First, the techniques themselves must be perceived as components within a system. When a database (spatial or non-spatial) is designed, its structure and the data categories selected must reflect not only the immediate needs but future applications as well. A new database cannot be developed for each project that comes along. In addition, 'hooks' must be created which permit one module to process data and provide this processed data as input or information to another module. For instance, the non-spatial database should be capable of generating spatially referenced point information (e.g. UTM coordinates, site number, site type), and introducing it as a data theme to the GIS. The results of image processing using remotely sensed data should also be available for export to the GIS, thus simultaneously taking advantage of the robust classificatory algorithms common to remote sensing software and the data manipulation and display strengths of the GIS. There should also be two way communication between the GIS and the EDA modules. This will permit primary or derived GIS data layers to be exported for processing using EDA and will insure that the results of this analysis can be reintroduced to the GIS as a new data layer. For instance, point information from the GIS along with its associated environmental data can be exported to the EDA where a variety of univariate, bivariate and multivariate analyses are conducted. The results of these analyses can then be exported back to the GIS creating a statistical surface for display and further comparative analyses (Figure 13.2).

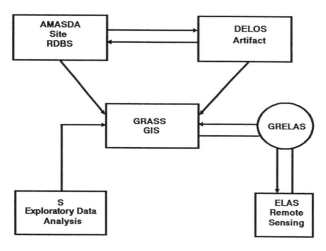

Figure 13.2. Information exchange between integrated modules.

To avoid creating a series of similar but isolated applications the principles of information management must be applied. These principles are simply a framework for managing data in which the cost of the data is acknowledged and a top-down approach for its management is emphasized while specific applications are developed from the bottom up. The goal in this case is to insure that a maximum amount of information is obtained from data and that this information is applied in as many contexts as possible. This approach stresses the linkages which exist between applications and requires a modelling process in the design of integrated systems to insure that these logical connections are identified and supported. James Martin (1977, 1983) refers to this as an organizational approach while others have labelled it as a corporate strategy for data management. As Raines (1987, p. 112) has noted, the focus of GIS must necessarily switch from technology, emphasizing software and hardware, to an information focus which concentrates on the overall management of information. In other words information and its management must drive the technology.

There is one additional change in perspective or perception which must accompany the integrated use of these tools. This pertains to the conceptual approach taken in problem-solving and the manner in which we view the tools themselves. The tendency exists to adopt a new tool or technology and simply overlay it on an existing conceptual model. While this may be warranted, it is more frequently the case that the new tool may provide different information or information in a new format which does not conform or directly correlate with existing models. As one example we can point to the fact that many complain that GIS systems (particularly cell-based systems) do not present their results in forms that correspond to our traditional view of the way a 'map' should appear. But as Robinove (1981) has clearly argued, maps are historical constructs which themselves simply attempt to represent reality within the constraints of paper and ink. 'Reality' is in fact much more irregular, multivariate and complex than paper maps portray. This complexity is graphically portrayed in our GIS products! Thus what initially appears to be a limitation in the technology may instead be a limitation in our conceptual models which were themselves formed by an earlier technology.

Integrated systems at work

In the remainder of this paper we illustrate these concepts in an applied environment, reviewing a number of projects which involved the analysis of spatial data and the integration of these tools. These projects have ranged in scope from a continental multi-state overview, to a statewide study, to small geographically circumscribed projects within the state of Arkansas. They are different not only in geographic scale but in the types and resolution of information they were intended to produce. In this paper we will present only an overview of these projects, more complete elaboration of the substantive results of these studies can be found elsewhere.

Continental-scale applications

A regional overview of the archaeology within the Southwestern Division of the Corps of Engineers was initiated in 1983 and was completed late in the summer of 1989. This overview encompassed more than one-fifth of the continental United States (Figure 13.3), and required the participation of many regional specialists. The project was designed to compile and assess the range of existing archaeological and bioarchaeological data across this eight state area (Sabo *et al.*, 1988; Limp, 1989; Jeter *et al.*, 1989; Limp *et al.*, 1989; Simmons *et al.*, 1989; Hofman *et al.*, 1989; Story *et al.*, 1989; Hester *et al.*, 1989).

GIS applications at this scale are useful for portraying broad spatial patterns in the data. The primary analytical unit for much of the data was the county. In the Southwestern Division there are more than 600 counties each with a wealth of demographic and archaeological data consistently collected over a broad time frame. These data were compiled and recorded in a relational database. We also developed GIS data themes reflecting geopolitical boundaries to provide the base data layer against which all these other data would be analyzed and displayed. These map layers were digitized from an Alberts equal area projection to minimize the distortion which can occur over broad distances. In addition, digital data layers were constructed for such diverse themes as historic tribal distributions, linguistic patterns and the distribution of natural vegetation and AVHRR imagery. Data from within the relational database were used to construct a number of derived data layers measuring the frequency of archaeological sites, sites with human burials and

National Register sites etc. (Table 13.1). These data layers were also compared with census data to identify rapidly developing areas where threats to the cultural resource base have existed in the past or where similar threats might be forecast for the future.

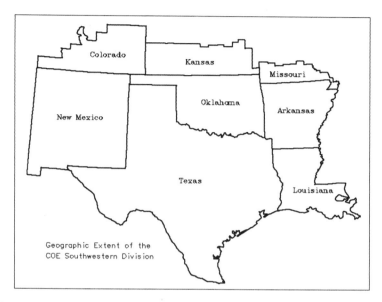

Figure 13.3. Geographic extent of the Southwestern Division project

Table 13.1. *Selected components of the Southwestern Division database.*

Data layer	Data source
Archaeological site density Burial site density Ratio of burials to burial sites Ratio of burial sites to total sites Density of National Register sites Density of National Register eligible sites Population change 1960–1980 Population change 1960–1970 Population change 1970–1980 Population Density/Square Mile	Derived from SWD relational DB
State boundaries County boundaries	Digitized from Alberts Equal Area Projection
COE district boundaries	Digitized from COE Base Map
Ecological research areas Major forest types Early Indian tribes Culture area and linguistic stocks Physiographic regions Physical divisions Surficial geology Potential natural vegetation	Digitized from National Atlas of USA, USGS

Cartographic communication was an important element of this application. As a result it was necessary to consider the cartographic issues and principles associated with the effective communication of graphic data. While computerized mapping is derived from cartographic origins many recent GIS applications have tended to ignore the contribution of effective visual communication which may be achieved using basic cartographic principles. This illustrates another important issue. Use of new technologies is often enhanced through the incorporation of 'old' or established methods.

One traditional method used to portray quantitative data which is also widely used in GIS involves the production of choropleth maps. The primary purpose of these thematic maps is to compare and evaluate spatial variation and the relationships between geographic phenomena (Robinson *et al.*, 1984). Simple choropleth maps are quantitative thematic maps designed to symbolize the magnitude of ordinal level data within the boundaries of unit areas (e.g. population density in Washington county in 1920). They are generally used to provide an overall impression of the patterns and trends of a geographic distribution.

Choropleth mapping requires consideration of cartographic conventions which are often overlooked. For instance, raw numbers should be standardized so that data take the form of either ratios involving area (e.g. densities such as archaeological sites per square mile, Figure 13.4) or ratios independent of area (e.g. percentages or proportions such as population change between 1960 and 1980, Figure 13.5).

Normally, choropleth mapping involves the grouping and display of similar data elements into classes using a range-graded symbolization scheme (Dent, 1985). The purpose of range grading is to categorize data values creating a generalized surface which facilitates the recognition of mapped patterns.

In this context effective use of choropleth maps is a form of exploratory data analysis where the spatial component is an important element of the patterning. In order to maximize classing accuracy, areas which are quantitatively similar should be grouped together and represented by the same map symbol or colour. However, due to the existence of a variety of classing algorithms, it is possible to generate several different map distributions using a single data set. The choice of a classing procedure therefore should be made with care.

In addition, the accuracy of certain classing methods appears to be dependent on the distribution of the data set being mapped (Smith, 1986). The classing methods most likely to produce accurate and reliable results with any distribution are optimization procedures which follow a mini–max strategy. First proposed by Jenks and Caspall (1971), optimization classing is an iterative process which establishes class boundaries by minimizing variation within classes and maximizing variation between classes. This process results in a series of data categories which can be displayed and visually investigated for the existence of embedded patterns in the data.

State-level GIS applications

A statewide GIS can function as both a management tool and a vehicle for broad research. In addition, a large proportion of the data preserved within a statewide GIS (e.g. elevation, soils, hydrography and transportation networks), can be used by groups which cross cut the traditional boundaries of academic disciplines and Federal and State agencies with land management responsibilities. This point was emphasized at the recent URISA meetings when an entire week was devoted to issues

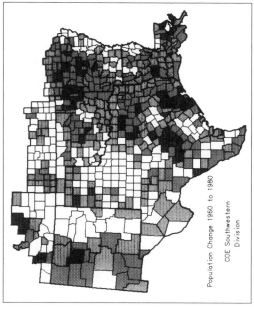

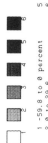

Figure 13.5. Population change between 1960 and 1980, derived from SWD relational database.

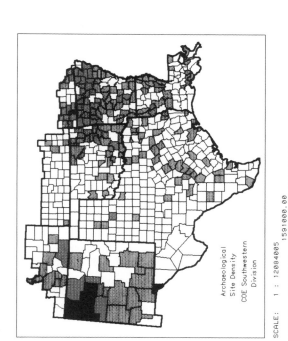

Figure 13.4. Densities of archaeological sites, derived from SWD relational database.

associated with the development of statewide geographic databases and topics pertaining to data sharing among disparate groups. Clearly this indicates a recognition of the importance of a corporate strategy towards data in the development of these broad-based systems.

At the state level the resolution of the GIS is fine enough to develop map layers which can contribute to our understanding of the mechanisms responsible for the distribution of archaeological sites. It is possible for us, using a locally developed interface between AMASDA and GRASS, to assess the distribution of many compliance and preservation-related variables such as National Register Status and the degree and cause of archaeological site destruction. In addition it is possible to compare and contrast the distributions of archaeological sites using broad chronological parameters and a suite of generalized environmental data themes (e.g. soils, surficial geology, elevation and hydrology).

For instance, to produce site density maps the site database is queried from within GRASS for all site locations meeting a specific criterion such as all Archaic sites. A report is then produced from within the AMASDA database providing the UTM coordinates for all sites meeting the specified criterion. GRASS uses this report as input and the UTM coordinates for archaeological sites are converted into cell frequencies where the cell resolution is calculated from the current resolution being used by the graphics monitor. This allows us to calculate density maps quickly, at various scales of resolution, for any of the more than 135 variables which are recorded for archaeological sites within AMASDA and registered to the UTM coordinate system (Table 13.2). Using an alternative function in GRASS it is also possible to create data sets which consist of the point data corresponding to the UTM coordinates. This enables us to conduct broad-scale comparative analysis of statewide site distributions. Because of the range of data retained by AMASDA the possible permutations of criteria used to generate point information for introduction into GRASS are virtually endless.

Table 13.2 Selected relations from the AMASDA database

Table name	Description of contents
Site management	Primary site data table includes all geographic locational information, National Register status, general site characteristic data and information on site disturbances and subsurface investigations. Each site can have only one entry.
Cultural affiliation	Includes all cultural assignment information for archaeological sites in Arkansas. Each site can have multiple entries.
Archaeological material	Includes general data on material classes recovered from each site.
Historic site function	Includes general information on historic site function (e.g. domestic, industrial, military, etc.)
Nonstructural features	Includes general presence/absence data for burials, midden, etc.
Structural features	Includes general presence/absence data for both prehistoric and historic structural features.

More than 19,000 sites in the database have UTM coordinates. Figure 13.6 illustrates the distribution of these 19,000 sites with each pixel representing a single site. These point data were converted to a cell map (Figure 13.7), utilizing 10×10 km cells. A total of 1840 cells are now represented. Site frequency ranges from zero up to 198 sites per cell. There are 12 cells with more than 100 sites and 68 cells with frequencies ranging between 50 and 100. These data have been reclassified for the

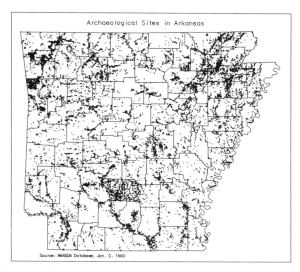

Figure 13.6. Distribution of archaeological sites in Arkansas, derived from the AMASDA relational database.

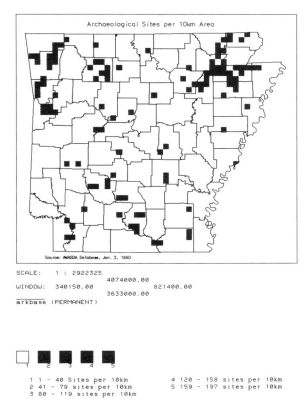

Figure 13.7. Density of archaeological sites in Arkansas, using 10 × 10 km resolution.

purposes of black and white display. A more generalized version of the same data set is presented using a cell resolution of 25 × 25 km (Figure 13.8). If we compare

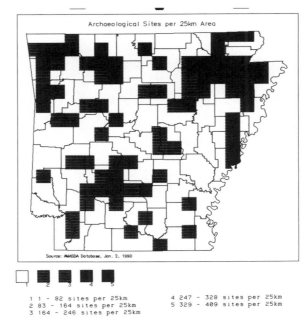

Figure 13.8. Density of archaeological sites in Arkansas using 25×25 km resolution.

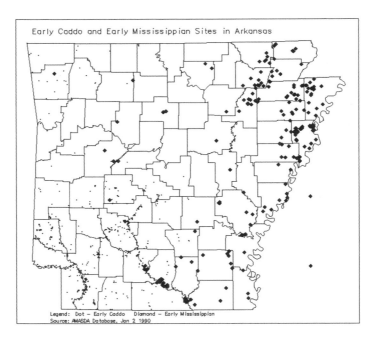

Figure 13.9. Distribution of Early Caddo (n = 643) and Early Mississippian sites (n = 251).

Figures 13.6 to 13.8 it is clear that each illustrates different aspects of the statewide site patterning, and as we have suggested this process can be seen as the spatial equivalent of EDA. Among other patterns these maps graphically display the gaps as well as the errors in our site records. The varying site densities across the state are a function of both surveying intensity and prehistoric and historic occupational densities.

We can move beyond these general pictures and focus on a variety of more interesting issues. The regional distribution of cultural units is a matter of concern to many archaeologists. Figure 13.9 compares the distribution of two roughly contemporaneous, ca A.D. 1000, archaeological groups, the Early Caddo and the Early Mississippian as recorded in the AMASDA database. In this figure the actual site locations are shown. There are 643 Early Caddo sites and 251 Early Mississippian sites. In general these two groups are spatially isolated except for the area in the south central portion of the state where a significant amount of overlap exists.

Truncated or 'flat top' mounds are a clear indicator of a major socio-political centre in this part of Arkansas. Such mounds were present in both the Caddo and the Mississippian areas. An examination of the spatial correlation between the presence of these mounds and the densities of both Early Caddo and Early Mississippian sites highlights an interesting phenomenon (Figures 13.10 and 13.11). Several factors are clear from these figures. First, there is a much closer relationship between the location of truncated mounds and cells containing Early Caddo sites, located predominately in the southwestern portion of the state (Figure 13.10), than exists between this same type of mound and the Early Mississippian sites (Figure 13.11). Secondly, there is a marked differentiation in the density of recorded occupation between these two groups. Site densities for the Early Caddo range up to 24 sites per 10 km cell and these sites are concentrated along the lower Red River and the upper reaches of the Ouachita River which are both major drainages in the southern portion of the state. Conversely, the Early Mississippian occupation appears to be less dense with a maximum value of 14 sites per 10 km cell, and the spatial distribution of these sites is much more dispersed and not confined to areas immediately adjacent to major drainages. This type of information presents an intriguing hypothetical basis for a number of avenues of research which is of interest to archaeologists throughout the state.

We can also compare site distribution changes through time. To do this we have superimposed the locations of Middle Caddo sites on our original Early Caddo distribution map (Figure 13.12). There are only 65 occurrences of Middle Caddo sites represented as black boxes on this figure. As we can see the distribution of these sites is quite similar to that of the Early Caddo. A new concentration appears, however, along the Ouachita River in the extreme south-central portion of the state.

The spatial continuity of the Caddo can be contrasted with the dramatic change in the location of Mississippi populations through time. Figure 13.13 indicates the distribution of the later Mississippi sites which can be compared with Figure 13.11. As we can see site density has increased (the maximum is now 51 sites per 10 km cell as opposed to the earlier value of 14) as has the extent of the regional distribution.

It is apparent that large volumes of comparative data can be generated when using a comprehensive relational database and a flexible GIS. It is equally apparent that these data are capable of highlighting some interesting spatial patterning which must subsequently be examined in detail and explained. The data which we have presented here have been used to emphasize these points. They do not represent an exhaustive examination of either the data in the database or the capabilities of the GIS system, rather they have been presented as a proof of concept in the hope that the general approach will be applicable in other settings.

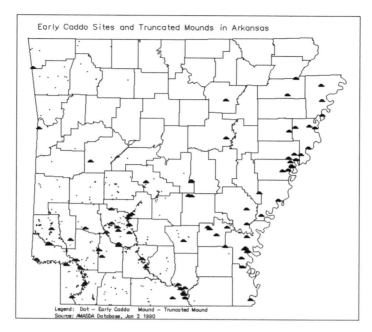

Figure 13.10. Spatial correlation between the distribution of truncated mounds and Early Caddo sites

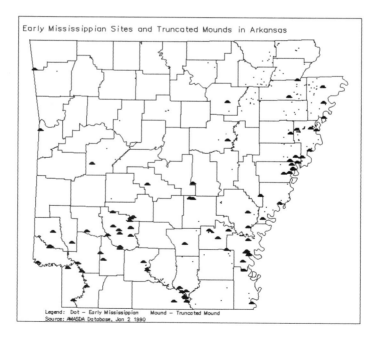

Figure 13.11. Spatial correlation between the distribution of truncated mounds and Early Mississippian sites.

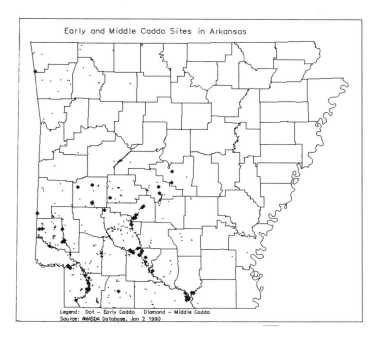

Figure 13.12. Geographic continuity in Early and Middle Caddo settlements distribution.

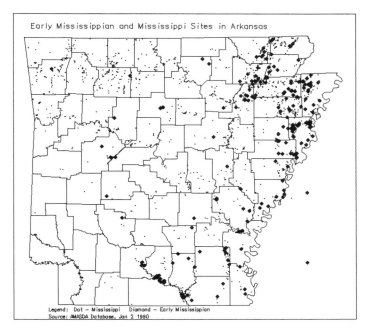

Figure 13.13. Distribution of Early Mississippian and Late Mississippian sites.

Project-level GIS applications

Project-level activities more closely approximate the environment most people associate with GIS applications than do either of the two preceding examples. These applications are confined to smaller, spatially circumscribed areas and in consequence tend to have a more complete array of data themes developed using a finer resolution. Because of this a greater range of techniques can be applied in a closer examination of the relationship between the environment and human settlement location behaviour. For this reason we will present two different examples of the use of integrated systems in project-level applications. The first will focus on the integration of AMASDA, GRASS and the ELAS remote sensing software in an application designed to investigate diachronic changes in settlement location selection through the use of satellite imagery. The second will examine the application of exploratory data analysis techniques in a comparative analysis of prehistoric and historic settlement patterns using data drawn from both AMASDA and GRASS.

Lee Creek project—remote sensing applications

Multispectral digital imagery is an important data source for GIS applications and will become increasingly valuable as sensor systems and analysis methods improve. In this example we will illustrate the use of imagery acquired from the SPOT sensor. SPOT data are recorded in both multispectral and panchromatic formats, are captured using three spectral bands and have a resolution of 20 m on the ground. The panchromatic scenes are single band data; however the ground resolution improves to 10 m. A single SPOT scene is approximately 60×60 km and is recorded as the satellite traverses the earth's surface along vectors which run from southeast to northwest. The application of robust classificatory algorithms to a data set of this size requires a hardware platform which is designed for intensive numeric applications. A more complete discussion of digital imagery, sensor types and the basics of digital classification is found in Limp (1989).

The Lee Creek study area includes an upland forested area north of the Arkansas River between Arkansas and Oklahoma. This area has been subjected to intensive archaeological survey in preparation for the construction of a projected reservoir to service the community of Fort Smith, Arkansas which is located south and east of the project area.

A full scene SPOT multispectral image was obtained for this area. A subset of this image focusing on the project area was created and subjected to preliminary image analysis using each of the three available spectral bands. Using the results of this process a fourth 'band' was derived using a ratio measurement of bands 1 and 3. An unsupervised classification was performed first using bands 1–3 and subsequently incorporating all 4 bands. This final classification identified 23 different spectral signatures in the project area. However, the fabric of the environment in the project area is such that over 62 per cent of the project locale is encompassed by just 4 spectral classes.

In an unsupervised classification the algorithm is used to identify the spectral attributes which will be used to discriminate between categories of phenomena on the ground, thus eliminating the possibility of introducing bias into the process by classifying the spectral data using predefined categories. The principle here is that the algorithm is more sensitive to subtle changes in the multivariate make-up of the environment than the human eye.

Following the classification a false colour infrared (CIR) colour table was created to display the image using a more real-world colour scheme. The image was then georeferenced. This is necessary to orient the SPOT scene to true north from the southeast to northwest orientation it is captured in. The georeferenced image was transferred to GRASS using the GRELAS software (Farley *et al.*, 1987) creating a cell file of the classified image in the GIS. Stream courses and landforms were then digitized from USGS 7.5 minute quads to complete the data themes needed. The landform data constituted 6 discrete classes of readily observable features associated with the Lee Creek drainage and the surrounding uplands (e.g. floodplain, first terrace, etc).

A total of 56 archaeological sites were found in the AMASDA database from within the project area. These sites ranged from Paleo-Indian lithic scatters to historic standing structures. In addition to the attributes of cultural affiliation and site function the presence or absence of midden was noted, as was the area of the site, measured in hectares, and the general surface condition of the site area (e.g. pasture, woods, ridge slope). The actual site boundaries were digitized from USGS quadrangle maps and made a data layer in the GIS.

We then performed a series of correlation analyses, first comparing landforms with spectral classes and subsequently comparing spectral classes with site locations and site distributions. Spectral classes were also compared to sites with middens, Mississippian sites and historic sites. This process revealed a number of interesting patterns in the data. Spectral class 13 represented only 5 per cent of the total area but included 15 per cent of the total sites. More significantly, this class accounted for 2 per cent of the Mississippian sites and 23 per cent of the historic site locations. Spectral classes 4, 9, 13 and 21 together covered only 14 per cent of the area, but they were able to 'predict' 75 per cent of the Mississippian site locations and 61 per cent of the historic sites (Figure 13.14).

We must emphasize that the Lee Creek project represents a pilot study designed to explore the utility of these integrated techniques. The results are not being presented as conclusive evidence based on exhaustive research. However, the results of this study do highlight some significant findings. First, with the use of multispectral digital imagery we were able to partition the environment effectively into units which apparently have some cultural relevance, at least where Mississippian and historic sites are concerned. Second, with this technique we were able to delimit the effective environment for Mississippian and historic sites. Simply, if you are interested in these types of sites then you will be able to locate 75 per cent of them by investigating only 14 per cent of the area in this project location. Identified through unsupervised classification of the SPOT image, this finding should be of some relevance to land managers and those operating under stringent fiscal constraints. Finally, there is the apparent relationship between Mississippian and historic site location selection strategies which presents an interesting hypothetical basis for future research. For a more in-depth discussion of the logic of the potential relationships between site characteristics and spectral classes see Limp (1989).

Ozark project—model building and EDA applications

A simple GIS database was developed for a 270 square km area in the rugged uplands of the Ozark mountains. Map layers reflecting archaeological site location (exported from AMASDA), locations of previous archaeological surveys, drainage patterns, the network of existing roads and property ownership boundaries were created. These data layers were developed using a cell resolution of 100 × 100 m. A

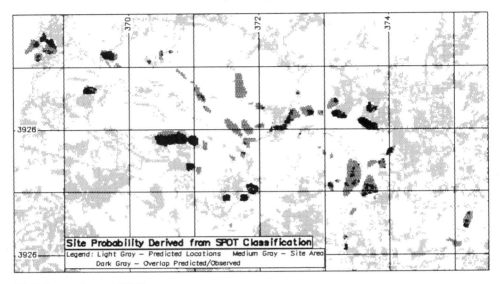

```
TITLE:        (lcreek.model)
LOCATION:     Lee Creek Valley Study Area
```

SCALE: 1 : 44328

Figure 13.14. High probability locations for Mississippin and historic settlements.

map layer for elevation was constructed using DMA format DEM data. The elevation data were then used to produce the derived map layers of slope and aspect.

Using these data in the GRASS GIS, proximity analysis and coincidence tables
were conducted to create a profile of the environment within the project area. Data
collected for each cell included elevation, aspect, slope, proximity to available water
and the relative location to the existing road network. Using the UTM registered
point data for site location imported from AMASDA we then compiled a similar
body of data reflecting the environmental and locational attributes associated with
archaeological site locations. This data set was then stratified by both time (e.g.
prehistoric, historic, Archaic, Woodland, Mississippian) and function (e.g.
domestic,historic, bluffshelters). Using this data set as a model we then evaluated
all cells within the project area to identify those candidate locations which displayed
environmental properties similar to those associated with archaeological sites.

To expand our understanding of the nature of the distribution of historic sites
and to assess the degree of homogeneity inherent in this sample we relied on the
techniques associated with EDA. Exploratory data analysis (EDA) is a body of
univariate, bivariate and multivariate statistical techniques which relies on the
graphic presentation of data. These techniques emphasize the investigation of data
structures leading to the identification of patterns or trends which may exist in the
data.

Using these techniques we compared the distribution of a number of GIS data
with the distribution of historic sites. This was accomplished by exporting the
environmental and site location data from GRASS and AMASDA to the 'S'

statistics package. Basic EDA techniques such as histograms, boxplots and quantile plots indicated that the distribution of historic sites was bimodal. Using Chernov faces (Figure 13.15), Tukey stars (Figure 13.16) and a hierarchical clustering algorithm (Figure 13.17), historic sites were grouped on the basis of values associated with five dimensions; distance to major stream, slope, elevation, distance to road and distance to minor stream. The resulting phenogram clearly discriminated the early Pioneer settlement locations from the later Developed settlement locations. This technique was also sensitive enough to identify subsets within the later Developed settlement period sample creating nodes on the phenogram which represented rural and city locations.

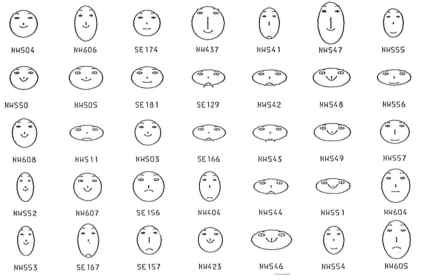

Figure 13.15. Chernov faces showing multidimensional variations in historic period occupations (Columns 1–3, Developed settlement: Columns 4–7, Pioneer settlement).

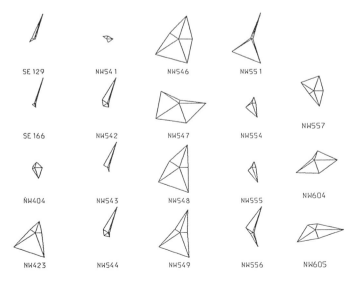

Figure 13.16. Tukey stars showing range of variation in selected historic period occupations.

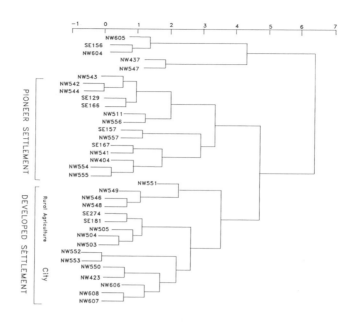

Figure 13.17. Phenogram of selected historic period occupations.

Using this information and similar comparative data measuring cultural and functional attributes associated with prehistoric site locations, we were able to construct three different models of archaeological site location. These models assess the selection criteria associated with the location of prehistoric base camps and bluff-shelters and the general pattern of historic site location. Each of these models is additive-linear in type assuming that the process of settlement location selection involves the iterative weighing of a number of environmental factors. Clearly the total distribution of archaeological sites in an area results from a mix of different processes. For example, we would expect large macro-band base camps to be selected on the basis of criteria which are different from those applied in the selection of special function localities (lithic extraction quarries). Similarly, the selection of bluff-shelter sites would be constrained by specific geologic conditions. At this point we are concerned exclusively with the environmental variables associated with the site selection process. However, it should be understood that we are in no way dismissing or diminishing the importance of the less tangible cultural variables (i.e. proximity to kin, political boundaries and the overall infrastructure, etc), which certainly played a significant role in the ultimate location of any site.

By weighing a combination of slope, proximity to water and proximity to large streams we were able to produce an effective model for predicting the location of large base camps. The second model, for bluffshelter locations, was developed weighing factors such as steep valley walls, appropriate geological strata interfaces, south-facing slopes and large streams, and was also effective (Figure 13.18).

The third model focused on the distribution of historic sites. In this area historic settlement history can be separated into two primary temporal categories, the Pioneer settlement and the later Developed settlement (Sabo, 1989). The existing data suggested that initial Pioneer settlements were selected using criteria that closely mirrored the operative criteria applied in the selection of late prehistoric base camps,

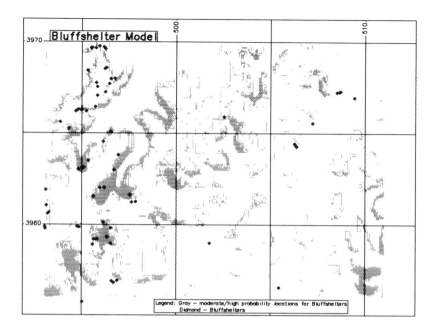

Figure 13.18. Moderate to high potential locations for bluffshelter occupations.

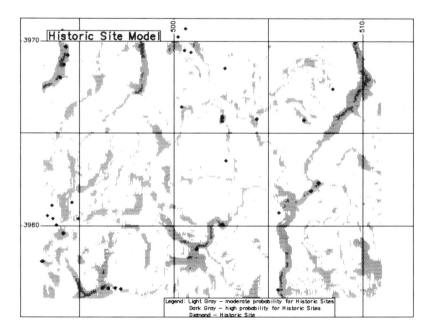

Figure 13.19. Moderate to high probability locations for historic occupation locations.

and for similar reasons. Over time, however, the demographics and the development of an infrastructure (comprised of historic road networks and defined communities complete with municipal and social services) introduced new, more complex associations. It was these new, more intricate associations that were responsible for the observed pattern of the later Developed settlement sites. The model developed to account for these complex relationships correctly identified 64 per cent of the known historic resources within a moderate to high probability zone which encompasses only 16.5 per cent of the total project area (Figure 13.19).

Summary

This paper has presented the organizational approach used by our group to consider a wide range of information relevant to the analysis and management of archaeological data. In this necessarily brief presentation a number of important issues were not addressed and the whole range of site-specific applications has not been considered. For a discussion of archaeological site-specific applications of GIS see Williams *et al.* (1990). Our key concern has been with presenting a comprehensive and integrated approach to data management and its use.

We have presented both GIS and database management applications at a variety of scales ranging from the continental scope of the Southwestern Division Overview to the more manageable project locale at Lee Creek. We have also illustrated the application of a number of analytical techniques which are appropriate at these varying scales and resolutions of information. These techniques include choropleth mapping strategies, the use of multispectral digital imagery and the application of exploratory data analysis. Each of these techniques has its individual strengths and weaknesses and a range of suitable applications. However, each of these tools has been presented using a common underlying theme. They have all been used in an effort to elicit and identify hidden patterns and internal relationships within the data. In addition, they have all been presented as components within an integrated system.

From our perspective the significant issues do not revolve around the technology, nor even the individual tools themselves. Rather, it is the possibility of arranging the technology and the tools in such a way that a symbiotic environment for research and management is created. In this environment large amounts of disparate, but interrelated, information may be assembled and compared in a relatively rapid fashion to address a multitude of different types of problems. In such an environment the research scientist is relieved of some of the more mundane tasks associated with conducting comparative analysis. In addition, many of the traditional constraints associated with the comprehensive investigation of multivariate data are either relieved or removed. These factors in turn alleviate the need to restrict research to the investigation of a small set of predetermined relationships. Instead, an atmosphere is created where virtually any perceived relationship may be examined and quickly evaluated.

Acknowledgements

Research discussed here has been supported by a number of sources. These include; US Army Construction Engineering Research Laboratories (CERL), US Army Corps of Engineers Southwestern Division, US Forest Service, National Park Service, Arkansas Historic Preservation Program, INFORMIX Inc. and Concurrent Computer Corporation. Without the support of these groups much of our research would have been impossible. We acknowledge their contributions with sincere appreciation.

References

Becker, R. A. and Chambers, J.M., 1984, *S: An Interactive Environment for Data Analysis and Graphics* (Belmont, CA: The Wadsworth Statistics and Probability Series)

Cowen, D. J., 1987, GIS vs. CAD vs. DBMS: What are the differences? In *Proceedings from the Second Annual International Conference on Geographic Information Systems* (Falls Church, VA: American Society for Photogrammetry and Remote Sensing), pp. 46–57

Dent, B.D., 1985, *Principles of Thematic Mapping,* (Reading, MA: Addison-Wesley Publishing Company)

Farley, J.A., 1988, The Land Manager's Dilemma: Monitoring multidimensional resource associations in a three dimensional world. In *Proceedings of the U. S. Army Corps of Engineers Sixth Remote Sensing Symposium* (Ft. Belvoir, VA: US Army Corps of Engineers Water Resources Support Center, Institute for Water Resources), pp. 198–222

Farley, J. A., 1989, Integrating relational database capabilities with the GRASS Geographic Information Managment System, In *Proceedings of the 4th Annual GRASS Users Meeting* (Champaign, IL: US Army Construction Engineering Research Laboratories)

Farley, J.A., Limp, W. F., Johnson, I. and Powell, B., 1987, The GRELAS Software. Submitted to the Space Remote Sensing Center, NSTL. Manuscript on file.

Goran, B.J., Westervelt, J., Shapiro, M. and Johnson, M., 1987, *GRASS Users Manual Version 2.0* (Champaign, IL: US Army Construction Engineering Research Laboratories)

Graham, M., Junkin,B., Kalcic, M., Pearson, R. and Seyfarth, B., 1987, *ELAS Users Guide* (NASA Earth Resources Laboratory)

Hester, T.R., Black, S.L., Steele, D.G., Olive, B.W., Fox, A.A., Reinhard, K.J. and Bement, L.C., 1989, *From the Gulf to the Rio Grande: Human Adaptation in Central, South and Lower Pecos Texas* (Fayetteville, AR: Arkansas Archeological Survey) Research Series No. 33

Hilliard, J. and Riggs, J., 1986, *AMASDA Site Encoding Manual* (Fayetteville AR: Arkansas Archeological Survey Technical Series)

Hofman, J. L., Brooks, R. L., Hays, J. S., Owsley, D. W., Jantz, R. L., Marks, M.K. and Manhein, M. H.,1989, *From Clovis to Comanchero: Archeological Overview of the Southern Great Plains* (Fayetteville, AR: Arkansas Archeological Survey) Research Series No. 35

Jenks, G.F. and Caspall, F.C., 1971, Error on choropleth maps: definition, measurement, reduction. *Annals of the Association of American Geographers*, **61** (2), 217–244

Jeter, M. D., Rose, J. C., Williams, G. I. and Harmon, A. M., 1989, *Archeology and Bioarcheology of the Lower Mississipi Valley and the Trans-Mississippi South in Arkansas and Louisiana*, (Fayetteville, AR: Arkansas Archeological Survey) Research Series No. 37

Limp, W. F., (editor) 1986, *Guidelines for Historic Properties Management, Southwestern Division Management Plan* (Fayetteville, AR: Arkansas Archeological Survey)

Limp, W. F., 1989, *The Use of Multispectral Digital Imagery in Archeological Investigations* (Fayetteville, AR: Arkansas Archeological Survey) Research Series No. 34

Limp, W. F., Parker, S., Farley, J. A., Waddell, D. and Johnson, I., 1989, *Integrated tools for Cultural Resource Management* (Champaign, IL: US Army Construction Engineering Research Laboratory)

Marble, D. F. and Peuquet, D. J., 1983, Geographic Information Systems and Remote Sensing: In *The Manual of Remote Sensing* (Falls Church,VA: American Society of Photogrammetry and Remote Sensing), 1, pp. 923–958

Martin, J., 1977, *Computer Database Organization,* 2nd edition, (Englewood Cliffs, NJ: Prentice Hall)

Martin, J., 1983, *Managing the Database Environment,* (New York: Prentice Hall)

Parker, S., Limp, W., Farley, J. and Johnson, I., 1986, *Integrated Approach to Database Management and Geographic Information Systems for the U.S. Forest Service* (Fayetteville, AR: Arkansas Archeological Survey)

Raines, M. T., 1987, The role of GIS in Spatial Resource Information Management: A Forest Service Perspective. In *Proceedings from the Second Annual International Conference on Geographic Information Systems,* (Falls Church, VA: American Society for Photogrammetry and Remote Sensing), pp. 111–122

Robinove, C. J., 1981, The Logic of Multispectral Classification and Mapping of Land. *Remote Sensing of the Environment,* 11, 231–244

Robinson, A.H., Sale, R.D., Morrison, J.L. and Muehrcke, P.C., 1984, *Elements of Cartography* (New York: John Wiley and Sons)

Sabo, G., III, Early, A.M., Rose, J., Burnett, B., Vogel, L. and Harcourt, J.,1988, *Human Adaptation in the Ozark and Ouachita Mountains* (Fayetteville, AR: Arkansas Archeological Survey Research Series)

Sabo, G., III, 1989, Images of the Past in Ozark Cultural Landscapes. Paper presented at the *Annual Meeting of the Southern Anthropological Society* Memphis, TN

Simmons, A.H., Stodder, A., Dykeman, D.D. and Hicks, P.A.,1989, *Human Adaptations and Cultural Change in the Greater Southwest* (Fayetteville, AR: Arkansas Archeological Survey) Research Series No. 32

Smith, R.M., 1986, Comparing Traditional Methods for Selecting Class Intervals on Choropleth Maps, *Professional Cartographer,* **38,** 62–67

Story, D.A., Guy, J.A., Burnett, B.A., Rose, J.C., Steele, G. and Olive, B., 1989, *Human Adaptations in the Gulf Coast Plain* (Fayetteville, AR: Arkansas Archeological Survey)

Tukey, J., 1977, *Exploratory data analysis* (Reading, MA: Addison-Wesley)

Westervelt, J., Goran, W. and Shapiro, M., 1987, Development and applications of GRASS: The geographic resources analysis support system. In *Geographic Information Systems in Government,* Volume 2, edited by B.K. Opitz (Hampton VA: A. Deepak Publishing), pp. 605–624

Williams, G. I., 1989, Application of Geographic Information Systems in Archeological Intra-Site Spatial Analysis. Paper presented at the *46th Annual Southeastern Archaeological Conference,* Tampa, FL

Williams, I., Limp, W.F. and Briuer, F.L., 1990, Using geographic information systems and exploratory data analysis for archaeological site classification and analysis. In: *Interpreting Space: GIS and archaeology,* edited by K.M.S. Allen, S.W. Green and E.B.W. Zubrow (London: Taylor & Francis). [Chapter 21 of this volume]

14

GIS and archaeological site location

Bryan A. Marozas and James A. Zack

Requirement for a GIS

A popular methodology by which archaeologists can develop empirical predictions of the probability of site occurrence is through quantitative site location studies. To many archaeologists, these site location studies are based upon the assumption that non-cultural aspects of the environment (independent variables) will correlate with, and predict, site locations (Ebert *et al.*, 1984). Although many archaeologists refer to such quantitative site location studies as predictive modelling, these efforts might be seen as 'projections' since they depend solely upon empirical observations which are then used to project probable site locations (Ebert and Kohler, 1988). In other words, they are simple correlational models.

Some of the more common environmental (or independent) variables for which measurements are collected include slope, elevation, aspect, relief and distance to water. Parameters for all of these variables can be manually observed in the field or measured in the office using topographic data. However, to collect the number of manual measurements necessary to develop a database for a large study area is exhausting, time consuming and potentially error-prone work.

Therefore, to avoid these problems, it is advantageous to apply the abilities of a geographic information system (GIS) to quantitative site location studies. The archaeologist can use a GIS to save time and eliminate human error in the measurement of independent variables which will be input into a correlational model. However, he/she must be aware that the measurements, although expediently obtained, are likely to possess a low level of accuracy as a result of error in source data and error created by GIS manipulation routines (Walsh *et al.*, 1987a). This error may manifest itself through the identification of misleading correlations.

Surface model error

To illustrate the nature and origin of some possible errors in source data and GIS functions, several data collection methods developed to support a site location study in north-central Montana will be examined. These methodologies are significant as they were developed and used to collect approximately 47,500 environmental measurements within an 11,738,240 acre study area in about two months.

Simplistically, the goal of this site location study was to develop a probability surface for the study area which would represent graphically the projected distribution of high and low site density zones. This was accomplished by training a model to discriminate sites from non-sites based upon the values calculated for a number of noncultural independent variables. Crucial to the integration of this model or any model is the knowledge that a degree of error exists between the probability model and reality.

With this in mind, note that chi-square and other goodness-of-fit statistics are available to help establish the confidence limits of a model. However, a different type of error is present which may affect a model's confidence, especially when environmental or proximity measurements are used to obtain values for the independent variables in a model. This is the error embedded in the source data and in the methodology used to generate a digital model of the study area topography. Hence, error also exists between a surface model and reality. This is not a value which can be looked up in statistical tables.

One source of error, called inherent error (Walsh *et al.*, 1987a), can result from the error present in analogue or digital source data. Inherent error may include the spatial resolution at which source data are measured, systematic data measurement error (i.e., accuracy of the measurement device) and random measurement 'mistakes' or imprecision. The inherent error may also be compounded by operational errors 'produced through the data capture and manipulation functions of a GIS' (Walsh, 1988, p. 343) (i.e., GIS hardware and/or software imprecision). The next section of this paper will briefly describe the steps utilized to create a surface model with Triangulated Irregular Network (TIN) commands. The remainder of this paper will then explain how inherent and operational errors are incorporated into a surface model, as well as other map layers in a GIS.

Developing the surface model

Although a variety of independent variables can be incorporated into a site location study, this section will only focus on the methodology used to calculate slope, elevation and aspect. The ability of the GIS to calculate automatically or derive values for these independent variables is dependent upon the generation of a digital model of the topography. Both the surface model and calculation of independent variables were produced with various components of the TIN modelling software package in the ARC/INFO toolbox.

Digital Elevation Model

Central to the ability of the TIN software to generate a surface model is the fundamental data source, a Digital Elevation Model. Digital Elevation Models (DEM) are composed of a digital array of regularly spaced points with X, Y and Z values. 'They are produced from map contour overlays that have been digitized from automated or manual scanning of aerial photographs' (Elassel and Caruso, 1983, p. 2). The only source of Digital Elevation Model data available for the Montana study area was obtained from digitized 1:250,000 US Geological Survey (USGS) topographic sheets with the sampling interval modified to 50 m, compared to the three arc-seconds (approximately 60 m at 50 degrees latitude) of the standard USGS DEMs.

Data structures

One vital component of digital surface model generation is the identification of a structure with which to store and manipulate the DEM. Due to the types of measurements which must be made on the independent variables, the array of points had to be converted into structures suitable for automated interpolation routines. These routines estimate the values of a new point, with respect to the position of the known points representative of the surface topography. The two data structures employed in this project were the lattice and TIN structures.

Past experience has indicated that the data structure most representative of irregular topography would be the TIN, which is a structure in which a surface is represented as a series of non-overlapping contiguous triangular facets, of irregular size and shape (Chen and Guevara, 1987). Since each triangle contains information on slope, aspect and area for its interior, as well as the topological relations to its neighbouring triangles, it was decided that this data structure would be most suitable for obtaining the slope and aspect measurements.

Elevation measurements can be obtained from the TIN surface, which calculates elevation through bivariate interpolation of the three known elevations of each triangle's vertices. However, this operation was unnecessary for this project since elevation measurements could be obtained directly from the lattice structure. The lattice structure is an array of sample points containing locational coordinates and elevation data. Since elevation measurements can be expediently obtained from the lattice, this structure seemed most appropriate for obtaining elevation measurements.

DEM to TIN conversion

Prior to the calculation of slope, aspect and elevation, it was necessary to convert the DEM-type data into the functional TIN format.

Initially, ESRI run-encoded Single Variable Files (SVF) of the DEM-type data file were formed using a simple FORTRAN read-write program. To create a TIN coverage from the SVFs in this study, several TIN procedures were invoked. First, each SVF was converted to an ESRI lattice structure, which is simply a mesh of points (rather than a grid of areas) containing X (easting), Y (northing) and Z (elevation) data for each point.

The lattice, however, contained more than 600,000 points which, due to hardware limitations, is far too many to create a TIN coverage at this level of analysis. Moreover, the TIN generation program requires a point, line or point/line coverage as input. Therefore, the lattices had to be both reduced in information content and converted into point coverages.

The reduction of information content appears at first to be a necessary evil. The complete lattice, however, is redundant, and to use every lattice point to generate a TIN would be overkill. In addition, two advantages of the TIN data structure would not be exploited; simplification and generalization (Chen and Guevara, 1987). The problem of choosing the most important lattice points for input to the TIN generation program is approached by calculating each lattice point's 'significance'. Chen and Guevara (1987) define 'significance' as 'how great a contribution the lattice point can make to the representation of the surface' (p. 50). They continue 'Our goal is to construct a triangulated irregular network (TIN) to represent the original surface by using the least amount of points' (p. 50). The selection of significant points is achieved by a program which selects only a predetermined percentage (20 per cent) of the total lattice points and creates an ARC point coverage out of these points. These point coverages were then input into the TIN generation program and TIN coverages were created.

Inherent and operational error affecting the data structures

The preceding delineation of methodological steps has been included to convey the simplicity and speed with which a DEM can be re-structured and manipulated to derive topographic measurements. These procedures were developed not only to provide a rapid turn-around for database compilation, but to reduce the error associated with measurements obtained manually. Since the archaeologist is concerned with accuracy, he/she should be aware that the TIN model of surface topography is only an approximation of the actual surface. The levels of generalization which contribute to the approximation of the surface are inherent in the primary source data (the DEM structure, Walsh *et al.*, 1987a) and operationally within the routines which generate the TIN data structure.

The first source of inherent error which exists in the TIN model of surface topography comes from the resolution of the input source (in this case the 1:250,000 scale DEM). Inherent error begins with the 1:250,000 scale USGS basemaps. With regard to the horizontal accuracy of a 1:250,000 scale USGS map, the National Map Accuracy Standards (NMAS) specify that not more than 10 per cent of the test points can be in error by more than 0.02 inch. In other words, the 1:250,000 scale USGS basemaps have a horizontal accuracy of ±127 m. The NMAS also states that vertical accuracy for these maps can be within half of the contour interval if the test point is within the horizontal accuracy permitted. This means that for a 1:250,000 scale map with a contour interval of 200 ft. (61 m) the elevation can be off by 100 ft. (31 m) if the test point is within 127 m of its true position. According to these statements on accuracy, an error in contour placement can be regarded as error in elevation (Goodchild and Dubuc, 1987). Hence, any elevation error inherent in the 1:250,000 USGS basemap is carried over into the 1:250,000 scale DEM. In addition, generalization of the 1:250,000 scale DEM occurs when the 1:250,000 scale basemaps were sampled in a grid-like manner every 50 m. This level of generalization translates directly to the lattice structure from which the elevation measurements were derived. At this stage, operational error is introduced when elevation values for the independent variables are obtained through an interpolation process between lattice points. These values, of course, are not the actual elevations, but those mathematically most likely, based on the known values of points around them (ESRI, 1988, p. 24).

A second level of operational error is introduced in the form of generalization which results from algorithms which generalize the model surface by choosing only the most significant lattice points to triangulate. The accuracy of the TIN structure, from which slope and aspect are derived, therefore, depends upon the quality and quantity of the triangle vertices chosen by the point selection procedure.

Before the archaeologist can make any judgement as to the inaccuracy of measurements from the TIN structure, he/she must realize that the intent behind the algorithms is that they generate the most accurate surface possible, using the least number of triangles. This results in reduced storage space and ultimately more rapid data retrieval and manipulation. Hence, generalization of surface data is unavoidable if one wishes to derive measurements in an automated and efficient manner. Such factors should be taken into consideration if the archaeologist is attempting to develop correlational models with measurements from representative topographic models. This is an important consideration since correlational models are actually developed by measuring the proximity of archaeological sites to environmental indicators. If the levels of generalization from (1) the 1:250,000 scale USGS source map, (2) the 1:250,000 scale DEM, and (3) the TIN data structure are great enough, these measurements may introduce a level of error which could bias any ability to project site location accurately. It should be noted that there would be inherent error introduced (of the same magnitude) if the archaeologist were to interpolate elevation data from a topographic map by assuming constant slope between contour lines.

Error in feature location

The type of correlational modelling which relies upon measures of association is dependent upon the installation of a variety of layers into the GIS. Thus far only one layer, the model of surface topography and its generalized nature, has been discussed. Therefore, it is also important to consider the type of inherent and operational errors which may be present in the development of other GIS data layers.

As previously mentioned, one source of error in a GIS data layer is the error inherent in the basemaps used to create the layer. Further consideration of basemap error is necessary because feature location error can have an impact upon the accuracy by which features are thought to be associated in a correlational model. For example in the worst case, according to National Map Accuracy Standards (NMAS), a point on a 1:250,000 scale quad may be in error by 127 m. Thus, the relationship between two points, in the worst case, has the potential of being in error by 254 m.

It is interesting to note that for the 1:250,000 scale USGS maps, only 10 per cent of the total points can be in error by as much as 127 m, and these points must be 'well defined' points such as road intersections. One begins to speculate on the degree of error of the other 90 per cent of the tested points, and what the error in model-related feature locations really might be. Of further concern is the NMAS fact that test points chosen on a map to be compared with points in the field have to have locations identifiable on the ground, thus excluding timber lines, soil boundaries, etc. The fact that such boundaries are excluded from any accuracy statement is important for correlational models since they often include the boundary location of such zones in their measurements.

To illustrate feature location error we can examine the error found in a layer containing archaeological site locations. Since archaeological site location coordinates are not derived photogrammetrically or with engineering survey precision, their locations can be the first source of error in developing associations. Quite often the location of a site is mapped as a dot on a 1:24,000 scale USGS topographic sheet and the Universal Transverse Mercator (UTM) coordinates are identified and recorded. Since human judgement and biases are involved in this operation, it can be expected that a degree of operational error is included. Next, the UTM coordinates are put into a file and a point layer is generated which can be displayed at any desired scale (i.e., 1:250,000). Due to the inherent error in a 1:250,000 scale map, if the archaeological site point layer was registered to this map, the site locations would now have a probability of having a horizontal accuracy of 127 m. In addition, this error could be compounded by the 12 m error from the 1:24,000 scale source map and any operational error encountered in delineating the site's position.

If in a single layer we can identify a compounded source of error, what happens then when all of these layers with their inaccuracies are overlaid in a GIS for measurement purposes? The answer is that the errors would be additive; that is, the horizontal accuracy of the final product or overlay would be equal to the sum of the errors of each of the GIS layers. For example, using only two layers from a GIS compiled at a 1:250,000 scale, archaeological sites and perennial streams, a measurement of distance to water from the archaeological site could be in error by as much as 254 m, based upon a possible horizontal error in feature location of 127 m for any point on each layer. In a more complex situation there might be a GIS with an archaeological site layer, vegetation layer, surface water layer, soils layer and a terrain surface model layer all compiled or registered at a scale of 1:250,000. If we were to develop a composite predictor for site location from these layers, the variables might include archaeological sites, distance to water, distance to vantage point, type of soil zone and vegetation zone. Consider then, that in the worst case, a feature location in each layer can have an error of 127 m. Since the error is additive, when these layers are overlaid to obtain measurements, the mapped data could now contain a possible linear lack of correspondence to actual feature location of 635 m. In other words, the overlay contains an additive feature location error of 635 m which could affect any possible association produced by the composite predictor.

As stated above, the horizontal errors used in the examples are the worst possible cases, and actual horizontal error is unlikely to be as high as the maximum cited here. However, the use of the maximum horizontal error in feature location was used to make archaeologists aware of actual error in feature location on USGS topographic maps. The actual error in feature location can be checked for each layer by making a comparison between map position and actual field locations for a sample of points. Discrepancies in horizontal accuracy would properly be referred to as the root mean square error (RMS), better known as the standard deviation of the discrepancies (Merchant, 1987). To obtain the RMS for an overlay composed of several layers, the variance from each layer would be summed (Watkins, 1983) and then the square root taken to produce the root mean square error of the overlay.

Merely from the description of the methodology used to determine error of an overlay, it should be apparent that the composite overlay product is less accurate than any of the layers utilized (Walsh *et al.*, 1987b). Such knowledge should alert archaeologists to the accuracy they can expect from multiple layer overlays, and perhaps inspire a rethinking of the analytical methodology being employed.

Since most of what has been discussed in this section deals with inherent error, it seems worthwhile to mention operational error in GIS operations. The archaeologist should also be aware that operational error, although difficult to measure, does contribute some error to the GIS product. Several common types of operational error are: digitizing operator error, digitizing from paper rather than stable-based Mylar, reduction of digitizing tolerance, raster-to-vector conversion, raster sampling (smoothing), vector-to-raster conversion, data aggregation, inaccurate measurement, erroneous boundary delineation and totally unidentifiable generalization related to and resulting from GIS algorithms (i.e., allowable feature movement known as fuzzy creep) and actual computer limitations (i.e., single versus double precision).

Conclusion

When considering the amount of area covered by a 1:250,000 scale map, the levels of generalization in the surface terrain model and the degree of horizontal error in a GIS layer may seem insignificant. For instance, if the horizontal error in feature location for an overlay equalled between 300 m and 500 m, planimetrically this distance is only about 1 to 2 mm on a 1:250,000 scale map sheet. However, this horizontal error could actually be significant in determining the difference between high-site density and low-site density for correlational models which propose that the material remains of early people can be found in association with commonly occurring 'geographic features' (*Photogrammetric Engineering and Remote Sensing*, 1987).

Since surface model error and feature location error are consequences of source data and the use of a GIS, consideration should be given to how the archaeologist may account for and minimize this error. The following is a list of suggestions which the archaeologist may wish to take into consideration:

(1) The resolution of the output product is determined by the resolution of the input data. In other words, 'the accuracy of any GIS output product can only be as accurate as the least accurate data plane of information involved in the analysis' (Walsh *et al.*, 1987a, p. 26).

A surface model resulting from highly generalized input data cannot be expected to be compatible with a fine-grained analysis. Hence, the results of a correlational model based upon generalized input data are likely to be useful only for making generalized statements on trends rather than site-specific statements (Walsh *et al.*, 1987b). This is acceptable, provided that the results meet the requirements of the user. However, one has to wonder whether or not generalized data on site probability has any value as a cultural resource management tool.

(2) The most effective method of dealing with surface model error and feature location error is to design the research problem methodically based upon a defined resolution. Are you conducting a coarse-grained or fine-grained analysis? What is the accuracy of the archaeological data that have been collected? With this knowledge and some consideration of the resolution and limitations of mapped data and DEMs at various scales, it is possible to determine whether or not the independent variable measurements calculated from the digital database will be accurate enough to help solve problems and meet the requirements of the user as a cultural resource management tool. If the methodical process of defining a problem and identifying data specifications is followed, the collection and manipulation of unnecessary data will be prevented.

(3) The problem design phase must also consider that although a GIS can handle disparate data sets, either digital or analogue, the data must be appropriate and compatible for problem-solving rather than merely entered into the GIS database because they were available. Simply entering sets of readily available data together into a GIS does not substitute for accuracy nor does it solve problems. In fact, it will not utilize the full potential of a GIS as a tool to help manage cultural resources.

Since GIS applications are a fairly recent addition to archaeological analysis, it is important to lay out and define the problem carefully and to provide accurate input and output data specifications in an effort to minimize problems, and to promote the effective use of a GIS to future users.

Acknowledgements

The authors wish to acknowledge Dr James I. Ebert for his review of this paper and the valuable insights he provided.

References

Chen, Z. and Guevara, J., 1987, Systematic selection of Very Important Points (VIP) from digital terrain model for constructing Triangulated Irregular Networks. In *AutoCarto 8 Proceedings, Eighth International Symposium on Computer-Assisted Cartography,* pp. 50–56

Ebert, J.I. and Kohler, T.A., 1988, The theoretical and methodological basis of archaeological predictive modeling and a consideration of appropriate data-collection methods. In *Quantifying the Present and Predicting the Past,* (Denver, CO: Bureau of Land Management), pp. 97–171

Ebert, J.I., Larralde, S. and Wandsnider, L., 1984, Predictive Modelling: Current Abuses of the Archaeological Record and Prospects for Explanation. Paper presented at the *49th Annual American Meeting of the Society for American Archaeology.* Portland, Oregon, April 1984

Elassel, A.A. and Caruso, V., 1983, Digital Elevation Models. In *USGS Digital Cartographic Data Standards,* edited by R. B. McEwen, R. E. Witmer and B. S. Kamey. Geological Survey Circular 895-B, USGS, Reston, Virginia

ESRI, 1988, *TIN Rev. 4.0 User's Manual.* (Redlands,CA: Environmental Systems Research Institute, Inc.)

Goodchild, M.F. and Dubuc, O., 1987, A model of error for choropleth maps, with applications to geographic information systems. In *Auto-Carto 8 Proceedings, Eighth International Symposium on Computer-Assisted Cartography,* pp. 165–174

Merchant, D.C., 1987, Spatial accuracy specification for large scale topographic maps. *Photogrammetric Engineering and Remote Sensing,* 53(7), 958–961

Photogrammetric Engineering and Remote Sensing, 1987, GIS News: Developing GIS, Arkansas Archaeological Survey: GRASS Integrators. *Photogrammetric Engineering and Remote Sensing,* 54(3), 405–406

Walsh, S. J., 1988, Author's reply in Letters to the Editor. *Photogrammetric Engineering and Remote Sensing,* 54(3), 343–344

Walsh, S.J., Lightfoot, D.R. and Butler, D.R., 1987a, Assessment of and operational errors in geographic information systems. In *Technical Papers, 1987 ASPRS-ACSM Annual Convention, Volume 5, GIS/LIS,* pp. 24–35

Walsh, S. J., Lightfoot, D.R. and Butler, D.R., 1987b, Recognition and assessment of error in geographic information systems. *Photogrammetric Engineering and Remote Sensing,* 53(10), 1423–1430

Watkins, T., 1983, The archaeological application of high precision radiocarbon dating. In *Archaeology, Dendrochronology and the Radiocarbon Calibration Curve,* edited by B.S. Ottaway, University of Edinburgh Department of Archaeology, Occasional Paper No. 9, pp. 74–82

15

The realities of hardware

Scott L. H. Madry

Introduction

Any GIS consists of a combination of software, hardware and one or more digital databases. Software options and the characteristics and problems associated with digital data are addressed elsewhere in this volume. This chapter will attempt to shed some light on the many issues relating to current hardware components for GIS, and the future directions of hardware and systems developments. Any mention of brand name products in this chapter is included for information purposes only, and is not intended as an endorsement of any particular product or vendor by the author, editors or publisher. The state-of-the-art in computer hardware is changing so rapidly that any such discussion is bound to be outdated by the time it is published, but it is important to attempt to get such information into the hands of potential users of GIS and related technologies. At any rate, such a discussion may be an interesting historical footnote for future researchers as to the state of hardware capabilities and costs in our time.

Obviously, the three components of hardware, software and databases are closely interwoven, and the requirements of one significantly affect the others. The appropriate decisions in choosing a total GIS system will have enormous impact on the ability to address the management or research issues required successfully. GIS hardware configurations range from the smallest microcomputers, to mini workstations, to the largest mainframe systems.

A GIS can be considered to be a computerized system that can collect, manage, display, manipulate, analyze and produce hardcopy of spatially referenced information (Figure 15.1). The hardware required to develop, operate and generate such information includes some combination of the following (Figure 15.2):

  –computer CPU (central processing unit) and operating system
  –storage media (hard disks)
  –colour graphics terminal(s)
  –tape drive (cassette 1/4 inch cartridge or 9 track 1/2 inch tape)
  –digitizer table (or digital scanner)
  –line printer
  –colour printer
  –pen plotter

The digitizing table, digital scanner and tape drive are used in the acquisition of data into the system (Figure 15.3). This is done either by creating data by manually digitizing existing maps or by reading in data already in digital format such

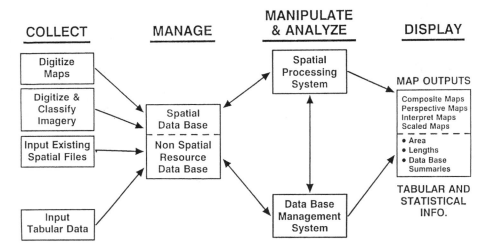

Figure 15.1. A GIS schematic.

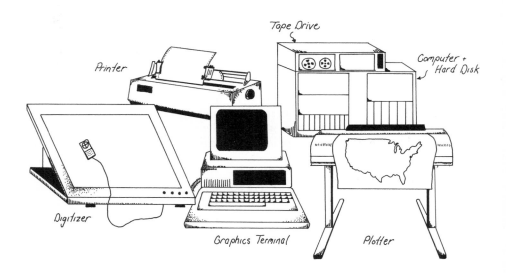

Figure 15.2. Typical GIS hardware.

as digital elevation models (DEMs). The computer's CPU, operating system and the system software manage, manipulate and analyze the data. The GIS software (working within the computer's operating system) often works with other software such as relational databases, statistical packages and image processing packages to do the manipulation and analysis work. Line printers, colour printers, pen plotters and digital film recorders produce hard copy data, either in the form of maps or as tabular information. Tape systems, modems and local area networks allow the transfer of data between users.

Data input

Data must be put into the GIS in a format that the specific software system can manage. This is done in one of three general ways; manual digitizing of maps using a digitizing table, digital scanning using a scanner, or by the reading of computer compatible tapes (CCTs) that have data on them already in digital format. These CCTs can contain either vector data (such as USGS DLG roads, streams or political boundaries) or raster data, such as digital elevation models (DEMs) or remote sensing data derived from satellites or airborne scanners such as Landsat, SPOT or Daedelus scanners.

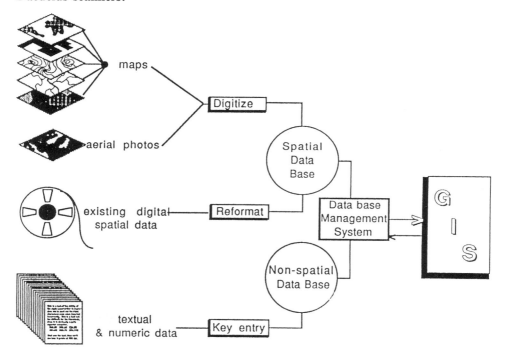

Figure 15.3. GIS database management system.

Digitizing tables

Digitizing tables are the most common input devices, and are available from a variety of vendors. Major current US manufactures include Altek, Calcomp and Kurta. There many others available. Digitizing tables use a mouse-like device, called a cursor or puck, to trace the outline of a map which is laid out on the table's surface. The table itself has an electronic grid within it that captures the X, Y position of the cursor. Tables come in a variety of sizes, ranging from 12×12 inches up to 60×90 inches. Standard sizes are 24×36, 36×48, and 42×60 inches. Accuracies are available from ± 0.010 to 0.005 inches and even 0.003 inches. Several models are available with backlit surfaces, power tilt and height accessories, and other options. Four button cursors are standard, but sixteen button cursors permit some automation of digitizing functions. Prices range from around US $3,000 to well over $9,000. The accuracy of the table you choose is a most important factor that must match your requirements (and pocketbook).

Digital scanners

Digital scanners currently fall into three general types, the raster scanning digitizer,
drum scanners, and line-following scanners. Several systems are available that scan
data digitally. These devices are essentially analogue-to-digital conversion systems that
use some sort of a charged couple device (CCD) to turn a map or photograph into
a two dimensional digital array. These systems are commonly used to scan aerial
photographs and to enter them into a digital database or manipulate them using image
processing software. Maps can also be scanned, but the raster to vector conversion
process that is required to enter such data as soil maps or roads is complex and not,
at this time, a mature, inexpensive technology. Typical of systems that are currently
available is the Eikonix scanner. These systems can scan in 512, 1,024, 2,048 or 4,096
arrays. Colour photographs or maps can be scanned three times (with a red, green
and blue filter) to produce true colour output. Such systems can cost around US $18,000
and require a tape drive to read the digital data into a computer system, or a specific
software interface to read the data directly into your system.

  Much recent research has been done in the area of automatic digitization using
scanners, due to the cost and time required for manual digitizing. One example of
this approach is LIDES, or Local Interactive Digitizing and Editing System,
developed by the US Forest Service. It is operational at several US National Forests,
running on microcomputers.

  Drum scanners use a different approach, in that the map to be scanned is placed
on a rolling drum which is rotated before a digitizing device. These devices produce
very high quality output, but again are very expensive. One example is the Tangent
Engineering ColorScan system that accepts maps up to 44 × 66 inches in size. It can
scan in up to 256 colours in real time and can separate features by colour (roads in
red, streams in blue, etc.), all at up to 1000 points per inch. Eighteen companies in
the US advertised their services in the 1989 *GIS Sourcebook*.

  A third approach is the line-following technique developed by LaserScan
Laboratories of Great Britain. This system uses a laser device actually to follow lines
on a paper map. It also produces very high quality output, but is also very expensive. For high quality, affordable digitizing, manual digitizing with a table is still
the most commonly used method, but this is likely to change at some point in the
future.

Computer systems for GIS

The computer system, consisting of the CPU, operating system, storage, display, and
input and output devices, is the heart of any GIS system. The choice of computer
hardware platform will define a great deal of what kind of GIS capability you will
have. Software packages work within a specific operating system, and your decision
on hardware will largely define your options in terms of the software you can use.
From the other perspective, if you decide on a specific software package, that decision will, to some extent, define the hardware platforms you can use. For example,
if you wish to run the GRASS system, you will require a computer that can use the
UNIX operating system.

  One could say that there are three general classes of GIS user. The low-end
system configuration would consist of a single modified PC or low-end workstation,

a single user at a time, work with small databases, do little remote sensing or image analysis, and have few if any turn-around requirements to conduct production level work.

A middle range system could consist of two or more computers, frequently linked together, with more memory and peripherals, larger or multiple databases, and requirements to do some image analysis work. A high-end configuration would consist of several mini and workstation systems linked together in a network, have several concurrent users on-line at any one time, use multiple databases and software systems on-line at any one time, have multiple peripherals, large memory needs, and frequently conduct significant amounts of remote sensing image analysis. Such a system tends to be production-oriented, with required turn-around of data and analysis, significant database development and digitizing activity, and frequently supports governmental or regulatory activities.

To support these user requirements, there are basically four kinds of computer platforms at the present time; IBM DOS and compatible personal computers, Apple Macintoshes, UNIX workstations, and proprietary mini and mainframe systems (like VAX/VMS and IBM mainframe systems). The most popular computers (in terms of numbers) are the IBM DOS based 286 systems and 'compatibles'. They comprise some 90 per cent of the personal computers in the world at the present time. This is reflected in the fact that just over 50 per cent of the GIS software packages listed in the most recent survey of GIS systems operate in the DOS environment (*GIS Sourcebook,* 1989). DOS does have inherent array limitations which limits its applicability for large, power users. The introduction of PS-2 resolves these problems, but there are few software packages currently available that run under PS-2.

Apple Macintosh systems are growing in popularity, mostly due to their simple, graphic user interface and easy learning curve, but relatively few software options are available in GIS at this point (although this is changing).

Proprietary mainframe systems such as IBM mainframes are not usually considered to be GIS platforms, although many universities still run SYMAP and POLYVERT on their mainframes. The problems involved in batch processing, lack of interactive graphics, and remote log-ins make these systems impractical for GIS use, particularly with the increased capability and lowered cost of today's mini workstation systems. Proprietary operating system mini systems such as VAX/VMS systems are located in a variety of facilities that do remote sensing or GIS work.

The major growth area in GIS is in the standards-based mini-computers and workstations using the UNIX operating system. UNIX is an extremely robust operating system developed by AT&T Bell Labs, and is very popular in the academic and government sectors. The primary benefit of UNIX is the portability of software between platforms. Software can be easily ported from one platform to another as computer hardware designs rapidly improve or new systems become available to users for whatever reason. Most governmental agencies in the US have standardized on UNIX systems for their procurements. The reduced cost and increased capability of such mini systems has permitted a major increase in the use of these systems. One problem with UNIX as it currently exists is that such systems generally require a systems administrator to manage the system. Adoption of the X-windows graphics standard will help to make UNIX more 'user friendly' as well as making ports to new hardware platforms even easier than it is currently by reducing the problem of writing individual graphics drivers for each vendor's machine. Current major hardware manufacturers in the GIS minicomputer market include Prime, Intergraph,

DEC and SUN Microsystems. Other manufacturers include IBM, Concurrent, Tektronix, Apollo and Silicon Graphics.

Storage requirements

Depending on the needs of the GIS user, varying amounts of data storage will be required. Magnetic media come in a variety of forms, with most computer systems including a floppy disk and some amount of hard disk. Most mini systems also include some sort of tape cartridge back-up system. Bernoulli removable disks are becoming more common, and allow individuals to remove specific data sets as needed (this can be an important security benefit, if such a capability is required). Nine track, or 1/2 inch tape in 1600 or 6250 bits per inch (bpi) readable formats are generally required for reading in remote sensing data such as Landsat or SPOT, and for reading in digital elevation or planimetric data supplied from governmental agencies such as the USGS. These tape drives are relatively expensive but are required for users who intend to conduct remote sensing activities as part of their GIS work. They also provide a quick means of backing up your data frequently, which is an absolute must.

Remote sensing also requires significant amounts of hard disk storage. One single Landsat scene, which covers a 185×185 km area of the Earth's surface, comes on four 6250 bpi tapes and takes over 300 Mb of storage just for the raw data. These data must be stored and then analyzed before integration into a GIS. Many GIS projects use multiple scenes, or data from more than one remote sensing source, so storage requirements can be a key issue in terms of defining your needs for a system. The future of CD-ROM seems to be bright, and could be an answer to providing a convenient and inexpensive medium for the distribution of remote sensing and GIS raw data. Erasable CD-ROM will become more common, and less expensive, but WORM disks (write once, read many) will still have a role in raw data distribution, where you do not want ever to be able to erase the raw data. It would be a significant step if data suppliers such as SPOT, EOSAT, DMA, USGS and others would begin to offer data on CD-ROM. This would allow smaller GIS systems efficiently to use the vast amounts of remote sensing and other digital data that are already available, without the cost of expensive tape drives.

Output devices

GIS output is an essential aspect of the methodology as it often lays the basis for interpretation. Output can be channelled to display devices, printers, plotters and photographic devices.

Colour display devices

Colour devices are required for GIS systems, and are available in a variety of configurations, ranging from 640×480 pixels (picture elements, or individual colour elements on the screen) up to new 2048×2048 systems recently introduced. The current standard is 1024×1024, but many 512×512 or 640×480 devices are still in use and are quite serviceable for many GIS requirements.

Another important issue for colour display devices is the ability to display true colour. For many remote sensing purposes, display devices need to be able to display in 'true colour', or a minimum of 24 planes per pixel (8 bits each for red, green and blue), with additional planes for graphics overlays. This is generally not required for general GIS work, and adds a significant amount to the total cost of the system. Be certain that your software runs with the specific configuration of graphics cards, etc., that you have. In the PC domain, the abundance of graphics boards can be very confusing and can cause problems in terms of compatibility.

Printers

Printer output devices cover a wide range of costs, capabilities and technologies. Some of the key issues involved in deciding what technology you should choose include the cost, your software's ability to drive that device, the size of the output product, the resolution (usually referred to in dots per inch, or dpi), the types of medium the printer supports and the speed of output production.

Line printers

Standard dot matrix line printers are used to produce tabular data, statistical tables and text reports. These are normally a part of any computer system.

Colour printers

Colour printers are generally required by GIS for producing hardcopy maps. These maps can include raster, vector, point, label or text data, or any combination of these. There are a variety of technologies available at different resolution and cost. The most common (and least expensive) systems are the colour ribbon dot matrix printers such as the Epson LQ-2500 or Genicom 3310. These use an impact, dot matrix technology using 3 or 4 colour ribbon systems. They can use serial or parallel connections, and can produce output ranging from 36 to 90 dots per inch (dpi), depending on the settings. They are relatively slow and require frequent changing of the ribbons (every month or so) with normal use. Costs range from US $1,600 to $2,400. The ink tends to fade significantly over time unless the map is protected with mylar and kept out of the sunlight.

Another type of colour printer is the colour ink-jet system such as the Tektronix 4696. This system uses a four ink reservoir system that uses special wax coated paper or mylar. It produces a 120 dpi output and requires frequent refilling of the reservoirs with frequent use or with the production of single large maps. Cost is around US $1,800. These two technologies are fast and relatively inexpensive, but are relatively low resolution and limited in the size of output products.

A newer colour printer technology is the colour thermal systems produced by Calcomp, Shinko and others. These systems use a three colour ribbon, thermal transfer technology that is relatively fast. The map is produced in three separate passes at 200 dpi resolution. These systems are also available with a printer buffer and RGB connection that can store screen dumps of whatever is on the screen. This is particularly useful in building multiple window displays for output. The systems require special thermal transfer paper and ribbons. The ribbon rolls produce about 70 maps each, at a total cost of around US $1 per map. The maps produced by these systems are less likely to fade over time. Cost is around US $10,000. This technology is somewhat more expensive, but produces much better quality output.

The finest quality hard copy devices available at the present time are the electrostatic plotters which are made by Calcomp and several other vendors. These systems use an electrostatic charge to deposit ink on the paper or mylar at up to 600 dpi resolution. These systems are available up to 36 inches and produce superb quality output. Cost is in the range of US $50,000. These systems produce very fast, large size, high quality colour output, but they are very expensive and require raster data for input.

Pen plotters

Pen plotters use ink pens to draw vector lines on paper from point to point. They hold up to eight colour ink pens. Current manufacturers include Hewlett-Packard, Houston Instruments and Calcomp, among others. Prices range from US $1,500 for smaller simple systems to $6,000 for 24 × 36 inch eight pen plotters. These systems produce fine cartographic vector output and are very low in cost, but they are relatively slow and require vector data for output.

Laser printers are relatively common, but most lack colour at the present. Colour laser printers could make a significant impact when they become widely available.

Colour digital slide camera systems

There are systems currently available that produce digital input to 35 mm slide cameras that are built into the system; Matrix camera is one such device. They hook directly into the RGB output of the colour graphics terminal and 'intercept' the signal being sent to the colour monitor. They produce very high quality slides which can also be made into prints. Such systems cost around US $6,000–8,000. Acceptable results, with experience, can also be made by directly shooting the screen with a 35 mm camera.

GIS hardware market

The GIS hardware market is currently estimated at over US $500 million per year, with a growth rate estimated at 20 to 30 per cent per year. Market estimates for GIS range from US $3.5–5.0 billion to US $11 billion by 1993. This enormous growth in hardware, software and related support, training and consulting services will have a major effect on the range and scope of options that will be available to end users, and will ensure that a wide variety of hardware systems, software and data will continue to be available. It is also certain that hardware capabilities will continue to change faster than most of us will be able to keep informed about. It will be necessary for the educational institutions to provide sufficient numbers of trained personnel to use these capabilities.

A sample GIS configuration

At the current time, a general configuration that would meet most needs for GIS use would include a 32 bit mini system with colour graphics terminal, cartridge tape, 300 Mb hard disk, colour printer and digitizing table. If the system will be using much remote sensing data, a dual density tape drive and additional memory would be recommended. Smaller configurations are acceptable for individual users who will be dealing with relatively small databases and who do not require fast turn-around, such as systems used exclusively for teaching.

Sometimes, the best system for you may be the one to which you can have access. An alternative to purchasing a GIS (for academic users) is to use a system already located in another department. GIS and remote sensing users are often located in a variety of departments, so you will have to look around, but common users include geography, geology, forestry, landscape architecture, environmental or natural resources, computer science or electrical engineering departments.

So how much will it cost to start using GIS? This is a most frequently asked and important question, but it has no easy answer. Much depends on what you want or need to do, and if you already have some of the equipment required. Software aside, if you already have a PC system and want to upgrade it to be able to use it as a minimal GIS, somewhere in the range of US $2,000–$10,000 will be required. If you want to purchase a new workstation or mini system, the cost can range from US $15,000 to $75,000 and more, depending on the number of processors, graphics terminals, amount of storage, peripherals, etc., that you need. Software can run anywhere from a few hundred dollars for public domain packages to well over US $50,000 and more.

Listed below are several hopefully typical system configurations in the micro, small mini, and large mini domains. Obviously, prices and configurations change rapidly, and I am reluctant to list even these, but in the interest of informing those who may not be at all familiar with such matters, I will. Prices are in 1989 US dollars at or around list price in the US. Obviously, significant university and other discounts are available and international prices vary considerably.

Low-end systems

A standard Apple Mac II with 5 Mb RAM, 80 Mb hard disk, 40 Mb cartridge tape, high resolution colour monitor with 8 bit video card, and keyboard costs about $8,000. It can obviously also be used for personal computing as well as GIS or image analysis work.

A typical IBM compatible DOS 386 system with 300 Mb hard disk, 1 Mb RAM, and 1.2 Mb floppy disk, with a 20 Mhz mathematics co-processor with DOS costs about $9,500. A 1/4 inch cartridge tape system is about $1,000, 640 × 480 pixel colour monitor $700, with VGA Orchid colour graphics card $575, mouse $150, and 4 Mb memory expansion another $1,500. Such a configuration of hardware alone, without printers or digitizer, runs to around $13,000–15,000.

Medium level workstations

The Sun Microsystems new SPARCstation1 offers very rapid performance with low cost. The SPARCstation1 with floating point accelerator, 8 Mb memory, 8 bit plane 16 inch colour monitor, 327 Mb hard disk and 1/4 inch cartridge tape system lists at around $19,500. Such a system runs in the UNIX environment and can also have a DOS emulator.

An INTERGRAPH Interpro 340 system with 8 Mb RAM, 19 inch colour monitor, 156 Mb internal hard drive, 355 external hard drive, 1/4 inch cartridge tape with UNIX costs $38,000.

High-end workstations

The Concurrent MASSCOMP 6400 system with 68030 processor, 68881 mathematics co-processor and floating point accelerator, 8 Mb memory, UNIX, 318 Mb hard disk and 16 inch 12 plane colour graphics device costs around $47,000. Such a system upgraded with a dual density 1/2 inch tape drive, and three CPUs and three graphics devices, costs about $86,900.

Future directions

In terms of the future of GIS hardware systems, much of the movement is from mainframes to minis, minis towards more powerful micro systems based on 386 and 68030 chips, and to standards-based (UNIX) workstations. The differentiation between micro systems and minis is rapidly blending, where micros are being upgraded in power, and minis are becoming significantly less expensive and easier to use. Networking of several systems with local area networks is also being implemented at a rapid rate. Many systems are being established with a single, large mini system with tape drives, mass storage, and input and output peripherals. These are being linked with many micros or low-end mini systems that can communicate with the large file server and use its power and storage when required, but with each personal system also having its own storage, display and computation ability. This is clearly one future direction of GIS configurations. This approach reduces the problems of configurations where many dumb ASCII terminals use the same remote mini system, with the serious delays in response that frequently happen in such a configuration.

In general, PC systems will become more like workstations, frequently working with a larger host, with networking becoming much more common. Mini systems will become much more powerful and affordable. One good trend is that many systems are now available with all components made by the same manufacturer. This has significant advantages, especially regarding maintenance issues. GIS that are used in the field, especially in the Third World, having CPU, graphics boards, monitors, etc. all made by separate vendors, cause real problems when the system goes down. Who is responsible? What gets sent where for repairs? The advent of the low-cost mini system with all parts from a single vendor will make such systems much more usable and practical in the rugged field environment and will help international and Third World management (and educational organizations) to be able to implement GIS technology realistically.

Integrated information management

The future direction of GIS hardware and software systems is towards the integration of multiple, currently separate technologies into unified systems of remarkable capability. Until recently, vector and raster GIS were separate domains, as were remote sensing and GIS, GIS and CAD, etc. These capabilities are now all being integrated into a unified GIS environment that combines these many synergistic capabilities (see Farley *et al.*, in Chapter 13, for an example of this approach).

In the remote sensing area, new remote sensing systems are scheduled to be launched by the United States, France, the European Space Agency, Canada, the Soviet Union, China and Brazil, and India within the next five years. These systems will provide frequent, diverse and high resolution data coverage of the entire world.

Five active microwave (radar) systems alone will be flown in space in the next five years. This enormous increase in remote sensing capability will certainly provide new and useful input to GIS used by archaeologists around the world. The ability to integrate fully such imagery data into a GIS context will certainly be required, and will provide archaeologists with significant new capabilities for environmental and ecological data analysis. The increased interest in global habitability and environmental issues will continue to drive new developments in integrated approaches to regional environmental and cultural studies. Potential for related applications such as primatology and human ecology is equally evident.

The direction towards newly unified, heretofore unrelated, technologies will continue at a rapid rate. New technologies, such as real-time positioning using the Global Positioning Satellites (GPS), will provide a new temporal aspect to this combination of technologies that can be combined with raster and vector GIS, remote sensing, statistical analysis and exploratory data analysis that could be termed Integrated Information Management. CAD technologies will certainly be integrated with GIS as well. The ability to map individual artifacts and scatters in the field directly into a GIS with precise coordinates will become a reasonably efficient practice. Expert systems and artificial intelligence capabilities will be integrated into existing and future software systems.

The palette of technologies that will soon be available to us will be exceedingly diverse. The use of CD-ROM storage technologies integrated with databases and video recording will open up new directions in the ways we study artifacts, documents, maps and the like in an integrated fashion. Such a system could have all artifacts from a project captured on video as individual frames, along with all field maps, field forms and other related data. These can be permanently stored on CD-ROM, with the ability to call up various combinations of the above using a relational database. Such systems go significantly beyond the current definition of GIS and will clearly improve the way we handle project information. Management and research opportunities will certainly develop that can only be dimly perceived at present. It is up to us to see that these new tools are properly used and integrated into our research and management activities, that we train future archaeologists to be able to use these technologies properly, and that we expand our theoretical and epistemological perspectives to be able to utilize the data and analytical tools which will be at our disposal.

Acknowledgements

Line drawings for this chapter were derived from materials produced by the Southwest Technical Support Center of the US Department of Agriculture, Soil Conservation Service for use in their GIS training activities. Grateful appreciation is acknowledged to Gayle TeSelle of SCS HQ for his permission to use these materials.

Reference

The GIS Source Book, 1989, (Fort Collins, CO: GIS World, Inc.)

16

The fantasies of GIS software

Ezra B. W. Zubrow

It is one o'clock in the morning of December 20. Buffalo is cold, five degrees below zero Fahrenheit with a wind chill of minus twenty five degrees. There are two feet of snow on the ground. The stars crisply twinkle in the clear night. The Niagara River is frozen solid except at the Falls. Sounds travels far on these cold nights. I hear a loud crash as if the ice is beginning to crack.

I fantasize about spring and the ice breaking up into the thousands of small icebergs that float down the river. Icebergs bring to mind GIS software. Like icebergs, GIS software is frequently attractive and even starkly beautiful. Yet GIS systems are sometimes foreign and they can be very inhospitable. As in the case of ice palaces and other frozen fantasies, GIS software may look better from a distance than close up.

GIS packages float on a veritable Arctic sea of hardware and peripherals which bump and grind as operating systems and programming languages try to communicate with each other. For the naive user, it is a treacherous polar landscape. One never knows whether your program has the driver for that obscure digitizer that the purchasing department of your university brought in from the cold. Most of the code of these GIS ice floes is hidden from view. One does not know how they work until one is deep within a project and by then it can be too late. Many of these GIS-bergs can be a tipsy platform hurling the unwary into an ocean of despair and unfinished projects.

This chapter cannot answer all your questions about software. It cannot even begin to. There are too many packages of GIS systems and they are changing too rapidly. Furthermore, there are too many types of archaeological users ranging from the computer 'illiterate' to the computer 'jock'; from the micro maven to the main-frame maniac. This chapter is a 'meta-introduction'. It has four goals. First, it provides you with some criteria by which you can begin to evaluate GIS packages intelligently. Second, it introduces you to the range of software available by quoting a very fine tabular summary by H. Dennison Parker (1989), the editor of *GIS World*. Third, it provides you with a brief summary of federal governmental activities and the names and telephone numbers of contact persons who are very familiar with their agencies' activities in GIS. The government has been and continues to be a main player in GIS software. Fourth, it provides you with the characteristics of the dream or fantasy software package.

Criteria for evaluation

Any archaeologist who is thinking of using a GIS must ask two critical questions prior to any evaluation of appropriate systems. Both questions are introspective and should be answered with as much honesty as one may muster. Each question has corollaries. First, why do you want to use a GIS? Is it because the GIS will increase efficiency? Will the 'start up' costs of a new system overwhelm the savings gained? Does GIS provide a tool which is not available in any other manner? Second, are there any alternative methods to reach one's goal? GIS tends to be a labour- and capital-intensive method by which to do research. However, once implemented and operationalized it is remarkably cost-effective and labour saving. There is one important exception to the last statement. If you have a GIS which needs continuous updating and maintenance, you will usually find that these modernizing costs are frequently more expensive than the 'start-up' costs. Since archaeological GIS databases are of this type, one should carefully consider this problem. I cannot overemphasize this point. I have seen many archaeological programmes which have begun a GIS but not completed the project due to lack of adequate continuation support.

If the answer to the first question is 'yes' and to the second is 'no' then one should try to apply the following evaluation criteria to the selection of the software.

First, check to see if anyone else uses the GIS for the same general purposes you wish to use it. If so, try to determine if they are the type who will share information or will you have to reinvent the wheel? A useful way to look into this is to seek user groups for your particular GIS.

Second, ascertain if the program or package has existed for a reasonable length of time. For most GIS one is talking about two or three years. You should check whether the GIS you are considering adopting goes through systematic upgrades. Furthermore, when the GIS upgrades are completed, make sure the company notifies the end user and makes them easily available. There is nothing worse than not being informed that your system has gone through five upgrades in the last year and has been able to do those new 'fuzzy merges' which you wanted six months ago. Often the vendor will have a newsletter or computer mail service which notifies you of such important changes. Even more importantly, be sure that your GIS has upgrades which automatically or easily upgrade your data and personal programs. Many a GIS scholar has lost months having to rewrite his application programs because the new upgrade uses a critical instruction.

Third, always check if your GIS runs on standard computers which have a long track record of continuing existence. Beware of specialized GIS which only run on specialized or a very limited set of machines unless you know what you are doing. Usually computers have a half life of two years and frequently become obsolete in three. Your software can tie your long term research into an obsolete technology if you are not careful.

Fourth, carefully consider the choice between mainframe, distributed and personal computers for your system. If you are considering mainframe and distributed systems be sure to discuss it with your system managers. GIS can be true gluttons for gobbling up system resources, in particular memory and input/output (I/O) channels. Nothing is worse for one's family or social life than to be told that the archaeological GIS may only be accessed from 2:25 a.m. to 3:15 a.m. on Friday and Saturday nights or that you can use the system for that critical research paper which will win you a MacArthur only after the central administration has run their dormitory furniture inventory.

Fifth, if all other things are equal, choose a GIS which will run on a variety of types and sizes of computer. It is best if the GIS is transparent and transportable from one machine to another. The ability to take your GIS from a small slow machine to a rapid large machine as your database grows is very valuable. It is particularly important not to have to relearn an entire program or rebuild a database when you move systems.

Sixth, there are lots of different systems for inputting and outputting data into and out of a GIS. They include keyboards, digitizers, scanners, terminals, tapes, CD-ROMs and numerous others (Madry, Chapter 15). Choose those that make sense for the size and scale of your project and choose systems with which you are comfortable. Given a choice, choose the GIS which has lots of hooks which allow you to input and output the data in the greatest variety of ways and formats.

Seventh, follow the federal government's criteria for databases. Be sure your system ensures system effectiveness (i.e. the GIS should do its intended function and provide the users with the appropriate information rapidly and easily). Be sure your system is economical and efficient (i.e. frequently one GIS may do the same thing as another five times faster and at one half of the cost in computer resources). Be sure your system protects data integrity (i.e. your GIS should have adequate controls over data input, storage, processing, output and communication so that you know where and what is happening to your data all the time). Be sure your system safeguards the information resources (i.e. your GIS should have some method to protect itself from misuse, fraud or unauthorized use, such as a password system). Be sure your system complies with the laws and regulations (i.e. your GIS should comply with the laws and regulations governing the acquisition, use and development of software). I want to emphasize this point because it is a terrible disappointment when one cannot publish the conclusions of your research because you have been enjoined from disseminating your results due to your grey copy of some GIS software.

Finally, the two most important criteria for choosing a GIS software package are 'try it out' and 'start small'. Always benchmark your GIS software. Any reputable GIS vendor will allow you to benchmark their software or will benchmark it for you prior to purchase. Only *you* will know what you want to do with the GIS so ask the vendor to demonstrate that capability with a real benchmark problem. Always try a small data set before beginning a large project. You can correct numerous problems in GIS database design and make modifications in your GIS prior to large scale investment of time, labour and money.

GIS software

GIS software comes in a wide range of sizes, functionalities and complexities. It ranges from the PC-based OSU *MAP-for-the-PC* which one can obtain relatively inexpensively from Professor Duane Marble, Department of Geography, Ohio State University, to expensive systems which can be customized by ARC/INFO and SYSTEM 9.

H. Dennison Parker of *GIS World* conducted a survey of the available software companies during 1989 which he has graciously allowed us to republish (Table 16.1). I have slightly rearranged his table to reflect what I believe are the priorities of most archaeologists (in other words, given archaeological budgets, I felt cost was a more important issue than age of system).

Table 16.1 Summary of GIS software (reproduced with permission from Photogrammetric Engineering & Remote Sensing, © 1989, The American Society for Photogrammetry and Remote Sensing, v.55, n. 11 pp. 1590–1591)

System name	Computing environment	Pricing	System type	Data structure(s)	DBMS interfaces	First installed	Number of users
AGIS	PCs/DOS	$15,000+	GIS	Vector, raster	na	1986	12
ARC/INFO	DEC, PRIME, DG, IBM, etc	na	GIS	Vector	Info, ORACLE, Ingres	1981	na
Aries	DEC VAX/VMS	$65,000	IP	Raster	na	1978	2,004
ATLAS*Graphics	PCs/DOS	$450–1,200	DM	Vector	DIF, Dbase, Lotus, etc.	1984	1,000's
Axis Mapping Info.	PCs/DOS, Sun Apollo, VAX IBM/UNIX	£7500–15,000	GIS	Vector, Raster	na	1978	25+
CRIES-GIS	PCs/DOS	$1,500	GIS	Raster	Dbase III	1978	60
Deltamap	HP9000, SUN APOLLO, SGI/UNIX	$8000–80,000	GIS	Vector, raster, TIN	Oracle, Ingres, Informix	1986	100+
Earth One	PCs/DOS	$12,000–28,000	GIS	Vector, raster	na	1986	40
EPPL7	PCs, PS-2/DOS	$500–1,000	GIS	Raster	Rbase, Dbase	1987	335
ERDAS	PCs/DOS, SUN/UNIX, VAX/VMS	$2,000	GIS, IP	Raster	Infor	1979	900+
Filevision IV	Macintosh	$495	FM, DM	Raster	na	1984	40,000
FMS/AC	PCs/DOS, SUN/UNIX, Macintosh	$2,500–7,500	GIS, FM	Vector	Dbase, etc.	1987	500
Gas, Electric, Water & Municipal FM	IBM 370.MVS, VM	na	GIS, AM, FM	Vector	IMS, DB-2	1984–89	22
Geo Sight	PCs/DOS	$4,450	GIS, AM	Vector	Dbase	1987	65+
Geo-Graphics	PCs/DOS	$2,400	FM	Vector	na	1985	na
GeoSpread-Sheet	PCs/DOS	$595–2,490	GIS	Vector	na	1989	18
GEO/SQL, MumMap 240	PCs/DOS; SUN/UNIX	$9,500	GIS	Vector	Rbase, Oracle, Ingres	1987	240

Table 16.1 Summary of GIS software (continued)

System name	Computing environment	Pricing	System type	Data structure(s)	DBMS interfaces	First installed	Number of users
GeoVision	VAX/VMS, ULTRIX, SUN, IBM-RT/AIX	na	GIS, FM	Raster, vector quadtree	Oracle	1976	47
Geovision 'GeoPro'	PCs/DOS Macintosh	$1,995–4,995	AM	Vector	SQL & DBF supported	1988	2
Geovision WOW	PCs/DOS	$595	GIS	Vector	na	1985	1,200+
GFIS	IBM/S370 architecture systems	var.	GIS	Vector	IMS/DLI, SQL/DB2	1977	180+
Gimms	Mainframes, minis (inc. UNIX), PCs/DOS, Macintosh	$1500–3000	DM, GIS	Vector, raster	Oracle, SAS, SPSS	1970	300
GISIN	PCs, PS-2S-222/DOS	na	FM	Vector	Condor	1986	5
GDS	VAX/VMS, DEC station/Utrix	$10,000+	GIS, AM	Object (vector)	Oracle, etc.	1980	800+
GRASS	Sun, MASSCOMP	$1,000	GIS	Vector, raster	na	1985	500–100
IDRISI	PCs/DOS	$50–300	GIS	Raster	Lotus, Quattro	1987	700
IGDS/DMRS	DEC VAX/VMS	$7,500–110,000	CAD-CAE FM-GIS	Vector, raster	Informix	1973	1371
IMAGE	PCs/DOS	$995+	GIS, IP	Vector	Lotus, Dbase, etc.	1989	100+
Infocam	VAX/VMS	$40,000–65,000	GIS	Raster, quadtree	Oracle	na	23
Informap	VAX/VMS	na	GIS	Vector	SQL-based	1975	na
Land Trak	PCs/DOS	$3,000–20,000	GIS	Vector	na	1983	230
Laser-Scan	DEC/VAX/VMS	£10,000–100,000	GIS	Vector	RDB	1985	150
Mac GIS (Cornell U.)	Macintosh	150	GIS, AM, FM	Vector, raster	na	1988	na

Table 16.1 Summary of GIS software (continued)

System name	Computing environment	Pricing	System type	Data structure(s)	DBMS interfaces	First installed	Number of users
MACAtlas, PCAtlas	Macintosh, PCs/DOS	$79–199	GIS	Vector, raster	na	1985	5000+
MacGIS (U. Oregon)	Macintosh	$100–300	GIS	Raster	Hypercard, etc	1987	30
Manatron GIS	Unisys/DOS, UNIX	na	GIS	Vector, raster	Oracle, Fasport, Adept, Request, etc.	1983	60+
Map Grafix	Macintosh	$8,500	GIS, AM	Vector	4th Dimension, Oracle, Double Helix, Omnix, etc.	1987	na
Map II	Macintosh	<$100	GIS	Raster	na	1989	na
MapInfo	PCs/DOS	$750	GIS	Vector	Dbase	1986	na
MatchMaker/GDT	PCs/DOS	$5,995–9,995	DM	Vector	na	1987	10
Micropips	PCs, PS-2/DOS	$745–1,490	AM	Raster		1981	250
MicroStation GIS	Intergraph/UNIX	$8,300	GIS	Vector, raster	Oracle, Ingres, Informix	1989	11
MIPS	PCs/DOS	$2,000–5,000	GIS	Vector, raster	Dbase	1987	na
MOSS	DG, Prime	(public)	GIS, IP	Vector, raster	DG/SQL, Oracle	1977	>100
Nucor GIS	PCs/DOS	$500–4,500	GIS	Vector, raster	ZIm	1988	10
Pamap GIS	Var./VMS, DOS, UNIX, ACS, CS/2	$7,500–60,000	GIS	Vector, raster	RDB, Oracle, Dbase	1983	200
Panacea	PCs/PS, 2/DOS	$500–2,000	GIS	Raster	na	1986	500
PC ARC/INFO	PCs/PS-2/DOS	na	GIS	Vector	Info	1987	na
PMAP	PCs/DOS	$895–1,600	GIS	Raster	Dbase	1987	180
SICAD	Siemens/UNIX	£20,000+	GIS	Raster, vector, quadtree	DB2	1978	250
SPANS	PCs/DOS, OS2	$8,000+	GIS	Raster, vector, quadtree	na	1985	400
StataGIS	Tektronix/UNIX, IBM, PCs/DOS	$7,000–25,000	GIS	Vector	Unify	1988	20

Table 16.1 Summary of GIS software (continued)

System name	Computing environment	Pricing	System type	Data structure(s)	DBMS interfaces	First installed	Number of users
STRINGS	PCs/DOS	$3,500–5,000	GIS/FM	Vector	Ingres, Sybase, Britton Lee	1979	150
System 600	VAX/VMS, Sun/UNIX	$10,000–50,000	GIS	Vector, raster	Ingres	1984	200
System 9	SUN/UNIX	$40,000	GIS/IP	Vector	Empress	1987	25
Territory Mgt. Sys.	PCs/DOS	$2,950–3,950	GIS	Vector, quadtree	Dbase	1988	25
Tigertools	PCs/DOS	na	GIS	Vector	na	1989	2
TIGRIS	Intergraph/UNIX Workstations	$10,000	GIS	Vector, raster	na	1988	16
Topologic	PCs/DOS, OS-2, VAX, VMS	$2,000–7,000	GIS	Raster, vector, quadtree	Dbase, RDB	1987	18
UltiMap	Apollo, AEGIS Operating System	$19,000–50,000	GIS, AM	Vector, raster	Oracle, Informix, Ingres, IMS, etc.	1974	40
USEMAP	PCs/DOS	$1,500–5,000	GIS, AM, FM	Vector, raster	Dbase III	1973	3
VANGO	VAX/VMS	$12,000–17,000	GIS	Vector	Userbase	1981	96
Zone Ranger/ GDT	PCs/DOS	$5,995–9,995	AM	Vector	na	1987	3

Table 16.2. Summary of Federal GIS activities.

Agency	GIS software	Contact names	Phone nos.
Agency for Internat. Dev.	ARC/INFO, ERDAS	Charles Paul	875-4046
Central Intelligence Agency	GRASS, ARC/INFO, others	Patricia Minard	703-482-8420
Customs Service		Douglas Smith	202-566-5325
Department of Agriculture Agric. Stab. and Cons. Service			
Forest Service	ARC/INFO, ERDAS, MOSS, others	Don Eagleston	703-235-2400
National Agric. Stat. Service	PREDITOR, GRASS	Jim Cotter	703-235-1973
Soil Conservation Service	ARC/INFO, GRASS	George Rohaley	202-447-5404
Department of Commerce Bureau of Standards	internal software	Henry Tom	301-975-3271
Census Bureau	ODYSSEY, CPS-PC	Robert Marx	301-763-5636
NOAA/NESDIS	MOSS, GRASS, others	Richard Knight	704-259-0452
NOAA/NGDC	MIXED	Dave Clark	303-497-6474
NOAA/NMFS	internal software	Hoyt Wheeland	202-673-5330
NOAA/NOS	GRASS, ERDAS	Peter Gross	301-443-8843
Dept. of Defense Corps of Engineers	GRASS, MOSS, ARC/INFO, ERDAS, others	William Klesch	202-272-1979
Engineer Topographic Labs	GRASS, MOSS, INTERGRAPH	Elizabeth Porter	202-355-2806
Naval Ocean R. and D. Agency	ERDAS, ARC/INFO, others	John Breckenridge	601-688-5224
Air Force	a	Neal Sunderland	202-767-1526
Dept. of Energy	GRASS, others	Benjamin White	202-586-3293
Booneville Power Admin.	ARC/INFO	ARC/INFO	
Pacific Northwest Labs			
Oak Ridge Labs			
Dept. of Health and Hum. Serv. Nat. Cntr. Health Statistics	SPANS	Charles Croner	301-436-7904
Dept. of Justice Drug Enforcement Admin.	FULCRUM	Ruth Torres	202-633-1153
Dept. of Interior Bureau of Indian Affairs	MOSS, AMS, COS, ARC/INFO	William Bonner	303-236-2250
Bureau of Land Management	MOSS FAMILY, IDMS, AUTO-CAD	Robert Ader	303-236-0089
Bureau of Mines	ARC/INFO	Donald Barnes	202-634-1044
Bureau of Reclamation	ARC/INFO, TEKNICAD, IDMS	James Verdin	303-236-4302

Table 16.2. Summary of Federal GIS activities.

Agency	GIS software	Contact names	Phone nos.
Fish and Wildlife Service	MOSS, SAGIS, ARC/INFO, MAP-DRAW	Claude Christensen	202-653-7526
National Park Service	SAGIS, PMAP, MOSS, GRASS, others	Phil Wondra	303-969-2590
Surface Mining, Recl. and Enf.	HCGRAM, SBSLOPE, REFLEX, others	Roger Hunt	202-343-4973
U.S. Geological Survey	ARC/INFO, GRASS, ERDAS, others	Joel Morrison	703-648-4639
Department of State			
Environ. Protection Agency	ARC/INFO	Joseph Sierra	202-382-7868
Federal Communication Comm.	internal software	Carl Brinner	202-632-3906
Federal Emerg. Mgmt. Agency	IEMIS	Robert Jaske	202-646-2865
Federal Energy Reg. Comm.		Roy Stiltner	202-357-5600
Housing and Urban Development			
Interstate Commerce Comm.			
National Aero. and Space Adm.		Richard Monson	202-453-1723
National Archives			
National Capital Planning Comm.		Frank Deter	202-724-0211
Nuclear Regulatory Comm.		Germain LaRoche	202-492-0695
Postal Service	SPATIAL II	James Bailey	202-268-3605
Tennessee Valley Auth.	ARC/INFO, INTERGRAPH	Charles Smart	615-632-1562
U.S. Information Agency			

Governmental activities and the use of GIS software

As I mentioned before the United States Government is a large player in the GIS world. They provide a large amount of the data, are a large purchaser of GIS systems, and are in many ways an indicator of the trends in GIS software. Recently, the Office of Management and Budget charged the Federal Interagency Coordinating Committee on Digital Cartography with compiling an inventory of GIS systems and their use in the federal government. The steering committee of eleven agencies suggested that Interior be the lead agency and that USGS be primarily responsible. Forty-four agencies responded to the survey (Federal Interagency Coordinating Committee on Digital Cartography, 1988).

The results can be briefly summarized. Twelve agencies use GIS on a broad scale and thirty-seven either plan to or now use GIS. Many of these agencies have classification, data collection and management standards. Most of the agencies use each others' data sets and cooperate with state agencies. There is a clear trend away from mainframe GIS to mini and micro systems. Both public domain and commercial software seem to have equal usage in the federal government. In 1989 the estimated expenditure of the federal government on GIS was approximately 65 million dollars (Federal Interagency Coordinating Committee on Digital Cartography, 1988). One can compare this expenditure to the estimated total expenditure on GIS of 135 million dollars in 1989 by Dataquest (a Dunn and Bradstreet subsidiary).

If an archaeologist is working in an area in which governmental GIS systems are also being used, it is an important criterion in considering a system to find out what work has already been done and on what system. This is particularly the case if one is doing cultural resource management. The preceding Table (16.2) presents the name of the agency, the level of work, the type of GIS, the contact person and their telephone number. It is derived from parts of several tables in the Federal Interagency Coordinating Committee on Digital Cartography's 1988 report on 'A Summary of GIS Activities in the Federal Government' (FICCDC, 1988).

Finally, let us return to my fantasy. What are the characteristics of the 'Dream GIS'? Unfortunately, there is no 'Dream GIS'. All GIS are compromises between software and hardware, I/O and processing, and efficiency and flexibility. However, if you find a GIS which is relatively transparent to many machines, usable in the field, highly relational, accessing many data formats and has a large number of analytical and visual tools, you will be nigh the GIS dream, especially if it is inexpensive. It is December 20 in Buffalo, I am looking out the window and there are palm trees swaying in the warm breezes—now that is a fantasy.

References

Federal Interagency Coordinating Committee on Digital Cartography, 1988, *A Summary of GIS Activities in the Federal Government*

Parker, H.D., 1989, GIS Software 1989: A Survey and Commentary. *Photogrammetric Engineering and Remote Sensing*, **55**, 1589–1591.

PART IV

Applications

17

Manipulating space: a commentary on GIS applications

Kathleen M.S. Allen

Introduction

With this section we turn to the practical aspects of using geographic information systems (GIS). These systems are ideally suited to archaeology because they can deal with a variety of types of data and combine them into a single large scale data and analysis system. The utility and flexibility of the GIS for archaeological data over long chronological periods and large regional areas is clear. The studies presented here portray the entire spectrum of current research on GIS in archaeological contexts. Papers present data on prehistoric and historic as well as old world and new world contexts. The chapters in this section also cover a wide range of techniques. Both raster and vector GIS are employed in innovative ways. The establishment of databases (i.e. data creation) and data interpolation are also covered.

Up to the present time, GIS have been used in a variety of contexts. At one level, the strongest development has been in the arena of cultural resource management. Studies here have concentrated on establishing predictive models for site location, building regional and site databases that are pulled into large GIS and using GIS for site and archaeological management purposes. At the other end of the spectrum are studies that apply GIS in a more theoretical manner. These include those that use GIS to model or simulate long term processes as well as those that incorporate GIS into creative studies of the cultural landscape. Chapters in this section have been ordered according to the above named themes. They move from the very practical predictive site modelling approach to landscape modelling and reconstruction.

Predictive site modelling

The earliest and most extensive use of GIS has been in the modelling of prehistoric site locations. The majority of work in this vein has been done in the context of public archaeology and has been used for planning purposes. Warren (Chapter 18) reports on a predictive model for prehistoric site location in the Ozark Hills of southern Illinois. The model is a fine-grained logistic regression generated from a vector GIS, the ARC/INFO system. This chapter illustrates some of the difficulties involved in the use of GIS since the model is only moderately successful. It predicts environmental locations with a high probability for site locations but, when tested with a cross-validation model, is only moderately better than a chance classification. Warren analyzes the model, identifies potential sources of trouble and suggests ways to improve accuracy and reliability.

Carmichael (Chapter 19) is concerned with the variability in quality and amount of data available on archaeological sites in a given region and how to make them comparable. He uses a straightforward approach to model predictively prehistoric site location for a relatively large area in north-central Montana. He creates a topographic grid using points from a Digital Elevation Model (DEM) and inter-polates a coverage using the Triangular Irregular Network (TIN) from ARC/INFO. He then identifies site and non-site locations and compares these locational points with eight environmental variables said to be important for site location. Out of this analysis comes the identification of five variables of particular importance for site location. All remaining portions of the study area are evaluated for the presence of the combination of environmental variables and locations with a high probability for site location are identified. Carmichael attains a reasonable degree of success with his model as the model correctly classifies nearly 75 per cent of the known sites. He too notes ways to improve the model, particularly by reducing the coarseness of the scale used.

The next chapter deals with this same problem of site location in a site manage-ment context but from a very different perspective. Altschul (Chapter 20) focuses on what he calls 'red flags'; sites that are costly in terms of money, time or both. Altschul suggests that rather than focus on where unknown sites are located, managers need to identify known sites that are in anomalous areas and are therefore inexplicable. He suggests that these are the ones that are likely to be most signifi-cant at least in our present state of knowledge. Altschul uses Defense Mapping Agency tapes for environmental data and a raster based GIS known as Geographic Information Management System. He identifies favourable locations for archaeological sites and compares known site locations with these favourable regions. Sites located in areas that are not favourable for sites according to the model are then identified as 'red flags'. Through further examination, Altschul is able to explain why this patterning occurs and to provide for greater understanding of the past. Altschul's contribution is an innovative approach to predictive modelling that turns it on its head.

Building site databases

The two papers in this part come out of work done for the US Department of the Army at Fort Hood, Texas. There has been over a decade of archaeological survey on this installation and an extensive data set exists. In the chapter by Williams, Limp and Bruier (Chapter 21), GIS is one component in a comprehensive system for archaeological analysis. They report on the use of a variety of techniques to classify prehistoric and historic archaeological sites at Fort Hood. They focus on site characterization using complex high dimensional groupings and the patterning of site characteristics for the purposes of management. Their research is unique since it provides extensive illustration of the utilization of several exploratory data analysis techniques in the context of a large GIS.

The next chapter by Jackson (Chapter 22) reports on the building of an historic database relating historic documents to known historic site locations. He is innovative in combining a number of kinds of historic documents including census returns, tax records and records of land ownership to create a database. He discusses procedures and problems of data entry and gives specific suggestions for novice 'GIS-ers'. Although Jackson concentrates on early stages in the establishment of the database within the context of a GIS (GRASS), the ultimate aim of the system is to evaluate historic sites for eligibility to the National Register of Historic Places.

Site and archaeological management

To some extent, all of the preceding applications have had relevance to cultural resource managers. The chapter by Hasenstab and Resnick (Chapter 23) explicitly focuses on archaeological management and the evaluation of models for site location based on field surveys. Their work evolved out of contract work to set up an archaeological database to manage Fort Drum, a 7000 acre complex near Watertown, New York. They use GIS to develop models for site location to guide field survey. They report on the development and implementation of the model and discuss stratification of the project area according to both prehistoric and historic archaeological sensitivity and disturbance factors. Part of their focus is on large historical farmsteads which are complex phenomena to analyse and have rarely been examined in such a comprehensive manner. The efficacy of the model is evaluated using systematic field survey and suggestions for its improvement are made. This is an excellent example of the use of GIS to evaluate sampling techniques as well as for archaeological management.

Model building

The model building approach in GIS as defined here focuses on the development of models for long term processes rather than for the purpose of predicting or understanding site locations. The two studies in this area utilize the same geographic area but model two very different kinds of long term processes using ARC/INFO. Zubrow's chapter (Chapter 24) is an example of the use of GIS as a modelling device. He uses aspects of network theory and data structure theory to model the spread of early contact period European populations through the cultural and hydrological landscape of New York state. He tests the modelled migration pattern against real settlement data based upon dates for the first European settlements and finds that one route explains real data better than all other alternative routes. The value of this research is that it illustrates the utility of GIS for modelling long term change and for developing and testing theory.

The chapter by Allen (Chapter 25) uses temporal data in an innovative way by actually timing the modelled growth of the trade networks. Allen models the development of early historic trade patterns in the eastern Great Lakes area using the Network system of ARC/INFO. Hydrology in the area is used for the development of the trade routes (the network) and there is a high correspondence between modelled results and archaeological data. This suggests that hydrological routes were important avenues for the distribution of trade goods and highlights the usefulness of the GIS in modelling temporal processes.

Landscape interpretations

The papers in this section take a very different approach to geographic information systems by incorporating their use within a broad based landscape approach to the archaeological record. Savage (Chapter 26) uses MAPCGI2, a raster based system, to recreate the social landscape of the Late Archaic period in Georgia and South Carolina. He subsumes his research under the broad umbrella of landscape archaeology and joins models of social organization to models of subsistence. Savage uses three models for social groupings and hypothesizes that the social landscape of the Late Archaic consisted of maximum band social territories divided into minimum band subsistence territories. He then evaluates this hypothesis by devising a series of test implications for GIS. His results suggest that within his research area there were six habitual use areas associated with hunter-gatherer minimum bands. Savage's chapter is an innovative application of GIS to the broader theoretical issues of social landscapes and spatial relationships among hunter-gatherer groups. His use of this methodology to integrate social and subsistence models holds real promise for revolutionizing the way archaeologists approach their data.

Green (Chapter 27) also uses MAPCGI2 to examine the relation between the cultural and natural landscapes of southeastern Ireland. His ultimate aim is to examine the temporal dynamics of the cultural landscape. Green's chapter focuses on using a combination of three-dimensional data entry and GIS as an approach for studying the structure of cultural landscapes. This allows him to describe the prehistoric landscape of Ireland as derived from survey data. He begins with the simple view of landscapes as composed of space (two dimensions) and the cultural and natural forms that vary over space (the third dimension). From there, the archaeological sites (as evidenced from surface scatters) are placed within their natural contexts by overlaying geographic variables (e.g. soils, topography, hydrology, etc.) and describing culture-ecology relationships. As geographic overlays are added, the co-variation of the cultural and natural landscapes can be studied. In addition, the GIS is used as a record keeping device to map the fields walked and sort sites into size and density categories. It is also used to examine the distribution of raw materials. This article clearly demonstrates the many purposes which GIS can serve in archaeological research and provides another vision of the potential of GIS for changing the way we think about the archaeological landscape.

The final chapter in this section is by Madry and Crumley (Chapter 28). This work comes out of a long term regional project conducted in east central France. The purpose of this project is to combine existing data acquired by numerous researchers into a single database for application to understanding landscapes and their evolution from the Iron Age to modern times. They apply a combined GIS and images of both airborne and satellite data with existing regional archaeological, ethnographic and environmental data. GRASS, ELAS and SPOT imaging are combined in a new way to quantify the environment of the region and model changing patterns of regional settlement.

In all, the studies in this section provide a cross-section of current applications of GIS in archaeology. They help one think about how GIS can be used in particular situations and provide a view of the actual process of data collection and manipulation. Although GIS applications are in an early stage of development in archaeology, the variety of approaches illustrated here demonstrates the great potential of this tool for the development of archaeological theory and method.

18

Predictive modelling of archaeological site location: a case study in the Midwest

Robert E. Warren

Introduction

Predictive models are tools for projecting known patterns or relationships into unknown times or places. Such models are sorely needed in archaeology. Archaeologists have documented only a fraction of the millions of sites in the New World, while thousands of sites are destroyed each year to make way for modern land development. One way to help understand and protect these sites is to create formal models capable of predicting where they are located.

Predictive modelling emerged only recently as an important component of archaeological research (Carr, 1985; Kohler and Parker, 1986). An underlying key to the success of these models is the fact that archaeological sites tend to recur in favourable environmental settings. Predictive models take advantage of these redundancies; they exploit contrasts between the environmental characteristics of places where sites do and do not occur. With appropriate data it is possible to make predictions from a relatively small sample of known locations to a much broader area.

Empirical approaches to predictive modelling may require vast amounts of environmental and archaeological data. Fortunately, recent developments in computer technology have brought within reach the capacity for handling large geographical data sets. Many institutions are developing networks of computer hardware and software designed to create, process, and display spatial data. These networks are usually referred to as geographic information systems, or GIS (Kvamme, 1989). As the papers in this volume clearly indicate, GIS are beginning to play a pivotal role in the research of many archaeologists, especially those interested in predictive modelling.

This paper has two main objectives. The first is to describe briefly a predictive model of prehistoric site location that was developed at the Illinois State Museum for an area in southern Illinois (Warren *et al.*, 1987). This model is somewhat unusual in that it uses a variety of both categorical and continuous independent variables. Unfortunately, tests indicate that the model is not as powerful as it probably should be. The second objective of the paper is to discuss the main causes of the model's deficiencies and to outline appropriate solutions. It is hoped that others will benefit from our experience with these problems.

Predictive modelling

Most archaeological predictive models rest on two fundamental assumptions. First, the settlement choices made by prehistoric peoples were strongly influenced or conditioned by characteristics of the natural environment. Second, the environmental factors that directly influenced these choices are portrayed, at least indirectly, in modern maps of environmental variation across an area of interest. Given these assumptions, it is possible to develop an empirical predictive model for any particular area, as long as the area has been adequately sampled by archaeological surveys. Several criteria can be used to judge the adequacy of surveys, the most important of which is that they consistently distinguish between places where archaeological sites do and do not occur.

The distinction between sites and nonsites is essential, as it provides the foundation upon which probabilities can be calculated (Kvamme, 1983a). In essence, this is done by computing a statistical classification model that capitalizes on the measurable environmental differences between the two groups. Such models make it possible to predict the probability that a site occurs at a given location simply by measuring an appropriate set of environmental variables. A successful predictive model is one that minimizes classification errors to such an extent that it offers a substantial gain in accuracy over null models arising from chance alone.

The practical benefits of predictive models stem from the fact that they can be applied to extensive, unsurveyed tracts of land where the actual locations of sites and nonsites are not known. Predicted distributions are useful in a variety of ways. First, they provide archaeologists not only with images of the patterns of prehistoric settlement in an area, but also with evidence of the most important environmental determinants of site location. Second, they provide land managers with expected distributions of the resources they are charged with protecting. And third, they provide planners with preliminary guides to the places where cultural resources are least likely to be affected by future construction projects.

Inductive or empirical predictive models are formal devices of pattern recognition (Warren, Chapter 9). In essence, most such models use statistical methods to extract from a sample of observations a formal decision rule, a rule that can be used to predict the composition or characteristics of future samples. One of the most powerful and widely used of the empirical methods is a set of procedures called probability models (Aldrich and Nelson, 1984). These models are well-suited for predicting the locations of archaeological sites, as they are designed to predict the responses of either-or situations (site presence versus site absence) to the interactions of independent variables (environmental measurements). The predictions themselves are expressed in terms of probabilities. Probabilities are readily interpretable and easily testable values that range between 0 (low probability) and 1 (high probability).

As noted by Carr (1985), archaeologists interested in predicting site location are abandoning the traditional methods of settlement-subsistence research in favour of procedures that are more appropriate for prediction (Scholtz, 1981; Kvamme, 1983a, 1983b, 1984, 1985; Limp and Carr, 1985; Parker, 1985). One consequence of this reorientation is a new focus on land parcels, rather than sites, as the basic unit of analysis. Other changes include a more widespread use of probability models, such as logistic regression analysis, and a healthy expansion of the environmental factors used as independent variables. Another unavoidable outgrowth of these developments is an increased reliance on computers—not just for analysis, but also for the collection of raw data and for the automated creation and measurement of variables. Computers are needed to handle the vast amounts of data required for predictive models.

One of the most powerful and flexible statistical techniques for predictive modelling is logistic regression analysis (Stopher and Meyburg, 1979; Neter *et al.*, 1983; Aldrich and Nelson, 1984). In archaeological applications, logistic regression creates a prediction formula that uses independent environmental variables of virtually any scale to predict the probability that a site occurs on any given parcel of land. The formula defines an S shaped probability curve of group membership that is oriented along an axis of intergroup discrimination (see Figure 9.4). The axis is comprised of an interaction of environmental variables that best discriminates site from nonsite locations.

A logistic model can be tested for accuracy by predicting the group member-ships of the locations used to develop the model (training sample). However, this produces overly optimistic results, as training-sample locations are not independent of the model (Kvamme, 1983b; Gong, 1986). It is more realistic to run tests using locations that are truly independent of the training sample, locations that were either unknown at the time of model development or were randomly withheld from the modelling process (test sample). In either case, accuracy is readily measured by calculating the percentages of correct and incorrect predictions along the probability scale of group membership (Warren, Chapter 9).

A case study

The case study discussed here is a predictive model that was developed by the Illinois State Museum for the western Shawnee National Forest in southern Illinois (Figure 18.1). Support for the study was provided by the USDA Forest Service, which sought a predictive tool that would be useful for cultural resource planning in the forest (Warren *et al.*, 1987).

Environmental setting

The Shawnee study area is a 91 km² tract of land. It parallels the bluff line of the Mississippi River and centres on a physiographic region called the Ozark Hills (Schwegman, 1973). This is an area of rugged, forested terrain that contains a diverse array of plant, animal and mineral resources (Spielbauer, 1976; May, 1984). It is noteworthy that these resources and landforms tend to covary from one place to another. This implies the natural environment may have had a strong influence on the settlement decisions of the area's prehistoric residents.

Archaeological background

Archaeological investigations in the Shawnee study area include a series of oppor-tunistic site surveys. Together these surveys have covered almost 12 km² of land, or about 13.2 per cent of the study area (Figure 18.2). Sixty-eight prehistoric sites have been documented. The sites average about one hectare in size and cover 5.8 per cent of the ground surface in surveyed areas.

Temporally diagnostic artifacts have been recovered from only a handful of sites, and there has been no systematic study of intersite functional variation. Hence, the Shawnee predictive model is of necessity a general formulation that may obscure underlying cultural patterns.

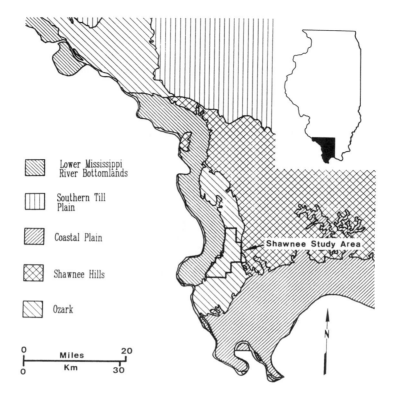

Figure 18.1. Map of the natural divisions of southern Illinois (after Schwegman, 1973), showing the location of the Shawnee study area.

Methods

Data for the Shawnee study were obtained from standard maps using a computer-based GIS. The basic sources of environmental information include detailed maps of elevation contours, stream courses, soil series and chert outcrops throughout the entire study area (Warren *et al.*, 1987). The locations of archaeological sites and surveyed areas were obtained from maps provided by the USDA Forest Service.

The first step in data collection was to digitize lines and polygons directly from the maps. Then the basic images were converted into useful data using a series of programs in the ARC/INFO package of GIS software developed by the Environmental Systems Research Institute (1986). The images were edited, a standard grid was imposed and elevation data were interpolated. The grid subdivided the study area into a regular lattice of more than 145,000 cells, each measuring 25 m on a side. Thirteen per cent of these cells have been surveyed, including 18,071 nonsite cells and 1112 site cells.

Additional GIS programs were used to transform the five primary coverages into a series of 27 secondary coverages. For instance, the elevation file was converted into six relief variables, a measure of surface slope, and the measure of surface aspect. Stream courses were buffered to create two measures of stream distance (Figure 18.3). Soil series were classified and recombined across such properties as permeability, flood frequency and landform (Figure 18.4). The 26 secondary environmental variables are listed in Table 18.1.

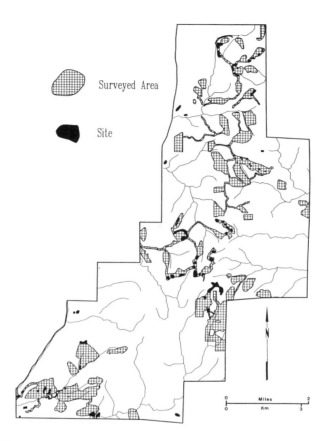

Figure 18.2. Map of surveyed areas and archaeological site locations in the Shawnee study area.

Univariate statistical tests were used to compare site and nonsite locations. The predictive model itself was created using a multivariate logistic regression program in the BMDP software package (BMD-PLR; Engelman, 1985). This is a stepwise procedure that measures the predictive power of each independent variable and creates a regression using only the strongest predictors.

Model accuracy was tested by comparing the actual with the predicted group membership of surveyed locations. Internal consistency was evaluated using training-sample locations, which include the site and nonsite cells used to generate the model. Independent cross-validation tests were run on test sample locations, or cells that were withheld at random from the process of model development (Warren, Chapter 9).

Results

Univariate statistical tests indicate there are significant environmental differences between site and nonsite locations across most of the 26 independent variables (Warren *et al.*, 1987). This suggests that the Shawnee data set is suitable for predictive analysis.

Results of the logistic regression analysis are based on a training sample of 1238 grid cell locations. These cells include a stratified cluster sample of 569 site cells and

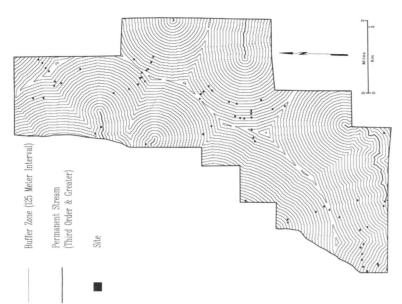

Buffer Zone (125 Meter Interval)

Permanent Stream
(Third Order & Greater)

■ Site

Figure 18.4. Map of landforms in the Shawnee study area (SOILLAND), based on the distribution of soil series. The limited extent of ridge crest soils in the northern half of the area is an artifact of different mapping techniques in Union and Alexander counties (Parks and Fehrenbacher, 1968; Miles et al., 1979).

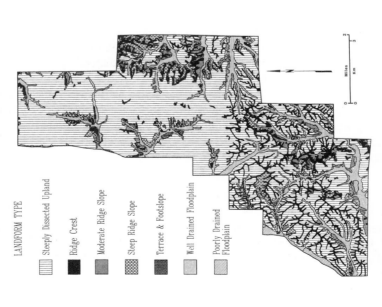

LANDFORM TYPE

Steeply Dissected Upland

Ridge Crest

Moderate Ridge Slope

Steep Ridge Slope

Terrace & Footslope

Well Drained Floodplain

Poorly Drained Floodplain

Figure 18.3. Map of distance to nearest permanent stream (PERMDIST) in the Shawnee study area. The distances were generated by creating a series of 125 m interval buffers around all third-order or larger streams in the study area.

Table 18.1 Environmental variables used to measure the locations of sites and nonsites in the Shawnee study area.

Variable	Code	Scale	Measurement interval	Number of distinct values
Elevation AMSL (dm)	ELEVATON	Ratio	1 dm	990
Total Relief (dm) in 100 m Radius Catchment	RELIEF10	Ratio	1 dm	563
Total Relief (dm) in 500 m Radius Catchment	RELIEF50	Ratio	1 dm	950
Above-Site Relief (dm) in 100 m Radius Catchment	RELFABO1	Ratio	1 dm	412
Above-Site Relief (dm) in 500 m Radius Catchment	RELFABO5	Ratio	1 dm	833
Below-Site Relief (dm) in 500 m Radius Catchment	RELFBEL1	Ratio	1 dm	416
Below-Site Relief (dm) in 500 m Radius Catchment	RELFBEL5	Ratio	1 dm	757
Surface Slope (per cent grade)	PCTSLOPE	Ratio	1 per cent	82
Surface Aspect (deviation from northerly aspect i.e. degrees east and west of north)	ASPECTSW	Ratio	1 degree	181
Distance to Nearest Stream (m)	STRMDIST	Ratio	125 m	7
Distance to Nearest Permanent Stream (m; stream rank ≥ 3)	PERMDIST	Ratio	125 m	26
Distance to Nearest Major Chert Outcrop (km)	CHRTDIST	Ratio	1 km	7
Soil Series	SOILSERI	Nominal	—	29
Soil Association	SOILASSO	Nominal	—	5
Soil Subgroup Classification	SOILGRUP	Nominal	—	9
Biome of Soil Formation	SOLBIOME	Nominal	—	2
Soil Landform	SOILLAND	Nominal	—	7
Soil Parent Material	PARENTMA	Nominal	—	4
Soil Moisture Regime	SOILMOIS	Nominal	—	2
Soil Drainage	DRAINAGE	Ordinal	—	5
Soil Permeability	PERMEABL	Ordinal	—	3
Soil Surface Runoff	SURUNOFF	Ordinal	—	7
Soil Flood Frequency	FLOODFRQ	Ordinal	—	3
Soil Erodibility (K factor)	ERODIBIL	Interval	100 K	5
Soil Productivity (Basic management; adjusted for slope and erosion)	SOILPROD	Interval	1 unit	24
Minimum Depth to Seasonal High Water Table (cm)	WATERTAB	Ratio	1 cm	3

a stratified random sample of 669 nonsite cells. Eight independent variables were selected by the program as the most powerful combination of predictors (Figure 18.5). These include three topographic variables, two hydrologic variables, two soil variables, and one lithic resource variable. The regression formula is presented in Table 18.2.

Measures of internal consistency show that the Shawnee model attains an optimum overall performance at a cutpoint probability of 0.49. It is at this cutpoint that the model correctly classifies 60 per cent of sites, 75 per cent of nonsites, and 68 per cent of all training sample locations (Figure 18.6).

However, an independent cross-validation test suggests these results are overly

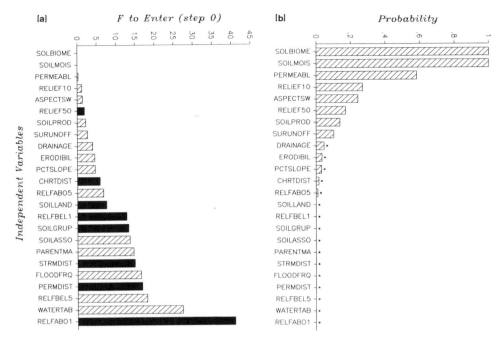

Figure 18.5. Comparison of the predictive power of all variables considered for inclusion in the Shawnee logistic regression analysis. (a) F-to-enter scores at step zero in the analysis, where scores increase with predictive power. Shaded bars denote variables that were entered and retained by the model. (b) Statistical significance of the F-to-enter scores, where asterisks denote scores significant at the 0.1 level of probability.

Table 18.2. Logistic regression formula for the predictive model of archaeological site location in the Shawnee study area.

SCORE = −0.4763	+ 0.0007(RELIEF50)	− 0.0028(RELFABO1)
	+ 0.0009(RELFBEL1)	+ 0.1819(STRMDIST)
	+ 0.0658(PERMDIST)	+ 0.1412(CHRTDIST)
	− 12.3670(SOILGRUPd1)	+ 2.4042(SOILGRUPd2)
	− 0.1301(SOILGRUPd3)	− 0.1554(SOILGRUPd4)
	+ 1.9475(SOILGRUPd5)	− 0.2807(SOILGRUPd6)
	− 1.7497(SOILLANDd1)	− 1.9227(SOILLANDd2)
	− 0.8156(SOILLANDd3)	+ 8.3751(SOILLANDd4)
	− 2.1261(SOILLANDd5)	

$$p(Site) = \frac{1}{1 + \exp(-(SCORE))}$$

Note regression coefficients for the categorical soil variables (SOILGRUP, SOILLAND) operate on design-variable codes. There is a unique coding sequence for each category (see Warren et al., 1987, p. 158).

optimistic. Cross-validation was run on a test sample of 543 site cells and 17,402 nonsite cells, all of which had been withheld from the process of model development. Optimal results were obtained at a cutpoint probability of 0.50, where the model correctly predicts 67 per cent of site locations but only 39 per cent of nonsite locations. Contingency analysis indicates these results are statistically significant. Overall, though, the Shawnee model offers only a moderate advantage over a chance classification.

(a)

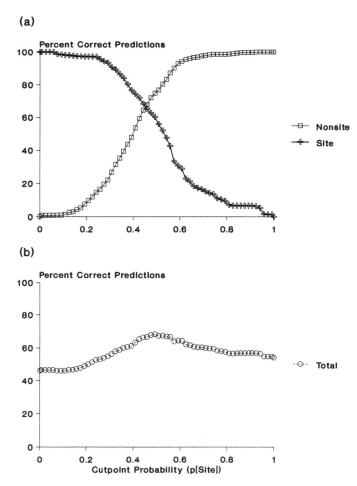

(b)

Figure 18.6. Internal accuracy of the Shawnee logistic regression, based on an evaluation of 569 sites and 669 nonsites that were included in the process of model development. The curves are percentages of correct predictions along a gradient of site-presence probability (p[Site]) for (a) sites versus nonsites and (b) the two groups combined.

The geographical implications of the Shawnee model are shown in a map of predicted site-presence probabilities across the entire study area (Figure 18.7). It is evident that sites are most probable in two distinct settings: high upland ridge crests and elevated terraces near creeks. In contrast, probabilities are low on the flood plains of creeks and on dissected valley slopes.

Problems and prospects

Many of the general tendencies implied by the Shawnee predictive model are probably valid. For instance, it is undoubtedly true that site probability is relatively high on upland ridge crests. Likewise, the potential for sites probably does increase as one moves more than about two km away from perennial streams. On the other hand, it is also true that the Shawnee predictive model currently suffers from three important limitations.

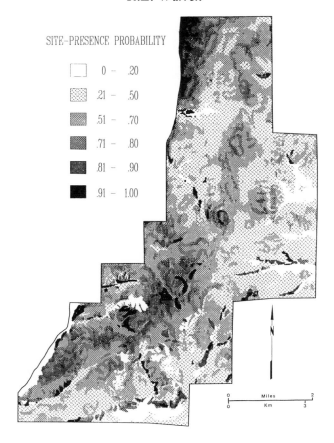

SITE-PRESENCE PROBABILITY

	0 – .20
	.21 – .50
	.51 – .70
	.71 – .80
	.81 – .90
	.91 – 1.00

Figure 18.7. Shaded contour map of predicted site-presence probability in the Shawnee study area.

Interpolation algorithms

The first limitation afflicts the nine topographic variables that were derived from the primary elevation coverage. Problems with data entry and data processing created a faulty elevation file. Data in this file are a weak reflection of the actual landscape of the study area. This problem is obvious in a 3-D plot of the Shawnee landscape (not presented here). The image portrays valley slopes not as the smooth declivities they actually are, but rather as the crude step-like tiers of a wedding cake. Further, most of the narrow to rounded ridge crests in the area are falsely portrayed as wide, flat-topped mesas. Since all topographic variables were derived from this coverage, all are to some extent erroneous.

This problem apparently has two interrelated causes. One is the algorithm used to interpolate elevations between digitized contour lines. The procedure we used scans a gridded matrix in search of cells that lack elevation data. Upon finding an empty cell it searches in eight cardinal directions for reference elevations. After finding the eight reference values (j), it applies the following algorithm to interpolate an elevation for each source cell (i):

$$Z_i = \frac{\sum_{j=1}^{8}(Z_j \times W_j)}{\sum_{j=1}^{8}(W_j)}$$

where Z_i is the interpolated elevation for the source cell; Σ_j is the sum of eight sets of reference values; Z_j is a reference elevation value; and W_j is a weight equal to $1/(D_{ij})^4$, where D_{ij} is the distance from the source cell (i) to the reference cell (j). In retrospect, it appears that the weighting factor (W_j) should have been defined as a different power function of distance (D_{ij}). The fourth power function apparently placed too great an emphasis on the elevations of nearby points. Consequently, the interpolated slope profiles between digitized contour lines are J-shaped rather than gently curved.

The second cause of the elevation problem compounded the first. Namely, elevation contours were digitized at a rather crude interval of 100 feet. This interval produces a low density of information in most midwestern environments, even in a relatively rugged area like the Ozark Hills. The large gaps between contours created an open niche for the faulty weighting factor. Together, these factors robbed much from the terrain variables used in the analysis.

As severe as the interpolation problem is, it should not be difficult to resolve. A new algorithm with a low-power weighting factor would create more realistic slopes. Supplemental elevation data would improve not just the slopes, but also the shapes of ridge crests. New data can be created by digitizing additional contour lines. It would also be possible to integrate contour data with point-elevation data produced by the U.S. Defense Mapping Agency (USDMA), although Kvamme (Chapter 10) has found that the accuracy of some USDMA coverages is poor.

Patterned residuals

The second limitation of the Shawnee model is a problem with patterned residuals. In nature, phenomena occasionally respond to stimuli in complex ways, such as when a species reacts favourably to several discrete intervals along an environmental gradient. A response like this might show up in census data as a bimodal density distribution along the gradient. It might also show up as patterned residuals if we tracked the abundance of two competing species across the same gradient.

Both of these patterns are evident in the responses of site and nonsite locations to distance variables in the Shawnee area. For example, the frequency distribution of site cells is bimodal along the variable measuring distance to nearest permanent stream (Figure 18.8). Moreover, the residuals of site versus nonsite frequencies are also bimodal (Figure 18.9a). Sites tend to be overrepresented both near streams and far away from streams, but are underrepresented at intermediate distances.

Patterned residuals are a sign that the predictions of a linear logistic regression will contain systematic error (Lewis-Beck, 1980). In the case of bimodal residuals, a linear model will tend to ignore one or another of the modes. This is quite evident in the linear bivariate regression shown in Figure 18.9b, where the regression line is insensitive to the fact that sites are overrepresented near streams. A linear logistic regression would behave in much the same way, systematically underestimating site-presence probabilities across certain intervals of the predictor variable.

This problem can be solved in several ways. The best approach may be to retain the ratio scale of problematic variables, but define them as nonlinear rather than linear functions. This can be done by entering select variables into a model as polynomial interaction terms (Blalock, 1979; Daniel and Wood, 1980). Second-power polynomials define parabolic curves with one bend, as shown in Figure 18.9c. Third-power polynomials define S-shaped curves with two bends, as shown in Figure 18.9d. With polynomials it is possible to improve the fit of a model to data and, at the same time, maintain a degree of generality in the model.

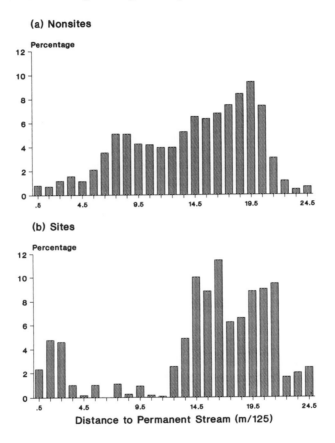

Figure 18.8. Histograms of distance to nearest permanent stream (PERMDIST) for (a) nonsites and (b) sites in the Shawnee sample area.

Categorical variables

The third limitation is a difficulty with rare or poorly represented categories of nominal-scale variables. This problem may become more common as archaeologists begin to incorporate a wider variety of variables in their predictive models. Unlike some of the more traditional statistical techniques that have been used for predictive modelling, logistic regression analysis is fully compatible with nominal, ordinal, interval and ratio scale variables. In fact, all four scales are represented in the Shawnee predictive model (Table 18.1).

The problem I am referring to here will be nothing new to anyone who has experience with contingency analysis. Rare or poorly represented categories are potentially unstable and unreliable, in the same way that low expected frequencies are unwanted in a Chi-square test. The solution is straightforward. Nominal

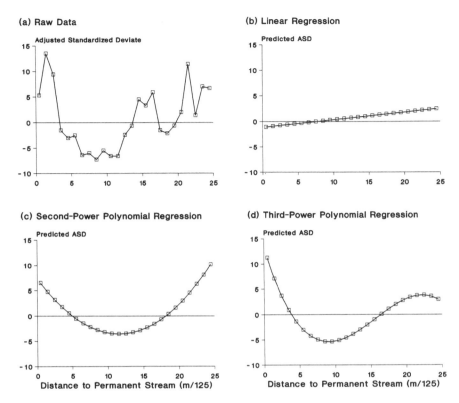

Figure 18.9. Observed and predicted site residuals across intervals of distance to permanent stream in the Shawnee study area. (a) Adjusted standardized deviates (see Reynolds, 1977) of raw frequencies show that sites are overrepresented both near and far away from permanent streams. (b) Linear regression ignores the bimodality of residuals and has no significant correlation with the raw data ($R = 0.18$; $p > 0.10$). (c-d) Polynomial regressions reflect the bimodality and correlate better with the raw data (second-power, $R = 0.67$, $p = 0.001$; third-power, $R = 0.78$, $p < 0.001$).

variables should be carefully screened before they are submitted to a logistic regression analysis. The objective of screening is to search for low and potentially unstable frequencies, either among site or nonsite locations. Weak categories can then be combined to create stronger categories, as is often done to meet the rule of thumb for expected frequencies in Chi-square analysis.

Conclusions

In conclusion, there are three general implications of the methodological problems encountered in the Shawnee predictive modelling study. First, geographic information systems play a key role in predictive modelling, and they do so to such an extent that the quality of data created by a GIS will have a strong impact upon the accuracy and value of the predictive model. Hence, it is wise to evaluate critically secondary coverages well in advance of model development.

Second, one should be aware that site-presence probability may be multimodal along some interval-scale variables. When this happens it is possible to improve the sensitivity of the variable—and hence the accuracy of the predictive model—by polynomial expansion to the second or third power.

Third, categorical variables create special problems that may affect the reliability of predictive models. One should carefully screen category frequencies and consolidate potentially unstable categories. Moreover, it would be wise to apply this procedure to the actual training sample of cases used to create the predictive model.

Due consideration of these factors can be expected to improve the accuracy and value of predictive models, especially in environmentally complex study areas like the Ozark Hills of southern Illinois.

Acknowledgements

I would like to thank Kathleen M. Allen and Timothy Kohler for commenting on the manuscript. Sheryl Oliver, Ray Druhot, and Jacqueline Ferguson were instrumental in developing the Shawnee model. Jacqueline Ferguson also assisted with the references and illustrations. This work was supported by a contract with the USDA-Forest Service.

References

Aldrich, J. H. and Nelson, F. D., 1984, *Linear Probability, Logit, and Probit Models*, Sage University Papers on Quantitative Applications in the Social Sciences, No. 07–045, (Beverly Hills, CA: Sage)

Blalock, H. M., 1979, *Social Statistics*, 2nd edition, (New York: McGraw-Hill)

Carr, C., 1985, Introductory remarks on regional analysis. In *For Concordance in Archaeological Analysis: Bridging Data Structure, Quantitative Technique, and Theory*, edited by C. Carr, (Kansas City, MO: Westport Publishers), pp. 114–127

Daniel, C. and Wood, F. S., 1980, *Fitting Equations to Data Computer Analysis of Multifactor Data*. (New York: Wiley)

Engelman, L., 1985, Stepwise logistic regression. In *BMDP Statistical Software Manual*, edited by W. J. Dixon, (Berkeley, CA: University of California Press), pp. 330–344

Environmental Systems Research Institute, 1986, *ARC/INFO Users Manual Version 3.2*, (Redlands, CA: Environmental Systems Research Institute)

Gong, G., 1986, Cross-validation, the jackknife, and the bootstrap: Excess error estimation in forward logistic regression. *Journal of the American Statistical Association*, **81**, pp. 108–113

Kohler, T. A. and Parker, S. C., 1986, Predictive models for archaeological resource location. In *Advances in Archaeological Method and Theory*, edited by M. B. Schiffer, (New York: Academic Press), Volume 9, pp. 397–452

Kvamme, K. L., 1983a, Computer processing techniques for regional modelling of archaeological site locations. *Advances in Computer Archaeology*, **1**, pp. 26–52

Kvamme, K. L., 1983b, *A Manual for Predictive Site Location Models: Examples from the Grand Junction District, Colorado*. Submitted to Bureau of Land Management, Grand Junction District, Colorado.

Kvamme, K. L., 1984, Models of prehistoric site location near Pinyon Canyon,

Northeastern New Mexico, edited by C. J. Condie, (New Mexico Archaeological Council Proceedings 6), pp. 349–370

Kvamme, K. L., 1985, Determining empirical relationships between the natural environment and prehistoric site locations: A hunter gatherer example. In *For Concordance in Archaeological Analysis: Bridging Data Structure, Quantitative Technique, and Theory,* edited by C. Carr, (Kansas City, MO: Westport Publishers), pp. 208–238

Kvamme, K. L., 1989, Geographic information systems in regional archaeological research and data management. In *Archaeological Method and Theory,* edited by M. B. Schiffer, (Tucson, AZ: University of Arizona Press), Volume 1, pp. 139–203

Lewis-Beck, M. S., 1980, *Applied Regression: An Introduction.* Sage University Papers on Quantitative Applications in the Social Sciences, No. 07–022, (Beverly Hills, CA: Sage)

Limp, W. F. and Carr, C., 1985, The analysis of decision making: alternative applications in archaeology. In *For Concordance in Archaeological Analysis: Bridging Data Structure, Quantitative Technique, and Theory,* edited by C. Carr, (Kansas City, MO: Westport Publishers), pp. 128–172

May, E. E., 1984, Prehistoric chert exploitation in the Shawnee Hills. In *Cultural Frontiers in the Upper Cache Valley, Illinois,* edited by V. Canouts, E. E. May, N. H. Lopinot, and J. Muller. Research Paper No. 16, (Carbondale: Center for Archaeological Investigations, Southern Illinois University), pp. 68–90

Miles, C. C., Scott, J. W., Currie, B. E. and Dungan, L. A., 1979, *Soil Survey of Union County, Illinois,* (Soil Conservation Service, US Department of Agriculture)

Neter, J., Wasserman W. and Kutner, M.H., 1983, *Applied Linear Regression Models.* (Homewood, IL: R. D.Irwin)

Parker, S., 1985, Predictive modelling of site settlement systems using multivariate logistics. In *For Concordance in Archaeological Analysis: Bridging Data Structure, Quantitative Technique, and Theory,* edited by C. Carr, (Kansas City, MO: Westport Publishers), pp. 173–207

Parks, W. D. and Fehrenbacher, J. B., 1968, *Soil Survey of Pulaski and Alexander Counties, Illinois,* (Soil Conservation Service, US Department of Agriculture)

Reynolds, H. T., 1977, *The Analysis of Cross-Classifications,* (NY: Free Press)

Scholtz, S. C., 1981, Location choice models in Sparta. In *Settlement Predictions in Sparta,* edited by R. Lafferty, J. Otinger, S. Scholtz, W. F. Limp, B. Watkins and R. Jones. Research Series No. 14, (Fayetteville: Arkansas Archeological Survey, University of Arkansas)

Schwegman, J. E., 1973, *The Natural Division of Illinois: Comprehensive Plan for the Illinois Nature Preserves System, Part 2,* (Rockford: Illinois Nature Preserves Commission)

Spielbauer, R. H., 1976, *Chert Resources and Aboriginal Chert Utilization in Western Union County, Illinois.* Ph.D. dissertation, Southern Illinois University, Carbondale, (Ann Arbor, MI: University Microfilms)

Stopher, P. R. and Meyburg, A. H., 1979, *Survey Sampling and Multivariate Analysis for Social Scientists and Engineers.* (Lexington, MA: Lexington Books)

Warren, R. E., Oliver, S. G., Ferguson, J. A. and Druhot, R. E., 1987, *A Predictive Model of Archaeological Site Location in the Western Shawnee National Forest.* Technical Report No. 86–262–17, (Springfield: Quaternary Studies Program, Illinois State Museum)

19

GIS predictive modelling of prehistoric site distributions in central Montana

David L. Carmichael

Introduction

The use of predictive modelling has become increasingly common in the study of archaeological settlement patterns. As part of this trend, the applications of modelling have expanded from a largely theoretical endeavour to include a variety of land/resource management concerns. In recent years, predictive models have been used as part of impact analyses to assist in the preparation of environmental assessments and impact statements (e.g., US Bureau of Land Management, 1984; Los Angeles Harbor Department and US Bureau of Land Management, 1985; US Department of Defense, 1985; US Army Corps of Engineers, 1986; US Department of Energy, 1986). In these cases, potential impacts of projects, or alternative project designs, are estimated by the number and types of sites likely to be affected.

Such analyses typically involve the definition of sensitivity zones based on vegetation or landform strata believed to reflect differential site distribution. Aside from the potential difficulty of combining the environmental parameters from a variety of studies, the resulting zones are often intuitive or subjective. Furthermore, considerable locational patterning may be obscured by using a single variable, such as vegetation, to characterize site distribution. Geographic information systems (GIS) technology provides an alternative method for modelling archaeological site location that avoids these problems.

This paper discusses the development of a predictive model of prehistoric site distribution in an 8500 square mile area of north-central Montana (Figure 19.1, inset). The size of the model area, and the necessary reliance on existing archaeological data, posed several interesting analytical problems. The effects of these conditions on the modelling process and the results are discussed.

Background

The modelling effort was part of an impact assessment conducted for a proposed United States Air Force programme in Montana. Several aspects of the assessment suggested GIS modelling as a useful technique. First, this part of the northwestern Great Plains is relatively poorly studied archaeologically. Few large scale or intensive studies have been conducted in the study area (Figure 19.1); most investigations have been of the reconnaissance type and relatively few sites have been recorded. The US-Canada border region just north of the project area is better studied, but

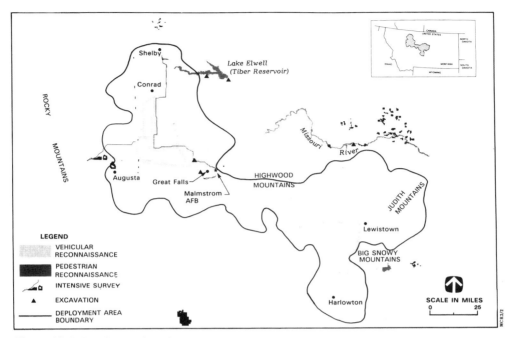

Figure 19.1. Previous cultural resources investigations in north-central Montana.

settlement patterns in that area are conditioned by glaciated terrain and would be applicable to only a fraction of the present study area.

Second, impact assessment for large projects should occur early in the planning process. However, as was the case for this project, design decisions are often not finalized until the later planning stages. Potential impacts to cultural resources needed to be evaluated before surveys could be conducted at actual construction areas. Predictive modelling provided a basis for assessing impacts, as well as a means for identifying sensitive areas to be avoided.

For the purposes of this discussion, two main aspects of the model development are of interest; the size of the study area, and the use of existing data. Although GIS technology is increasingly available to researchers, there is an associated cost. It seems reasonable that as the size of a study area (or land management unit) increases, so does the difficulty of actually ever surveying it; the larger the area, the more appropriate a modelling approach would seem to be. However, a very large study area imposes certain constraints on model precision. The present study, covering an area larger than the state of Massachusetts, necessarily involved the use of a coarse-grained model.

The geographic extent of the study area also affects the quality of archaeological data that may be used to build the model. Most GIS models reported in the literature use data collected during a single project or a series of closely related studies. In such cases, consistency in the recording of environmental conditions and in the classification of archaeological data is maintained. An additional set of challenges exists when one is faced with using existing records which include data compiled by various researchers working at different levels of intensity and with different research goals. The larger the study area, the more likely the database is to present such challenges. Large scale databases are maintained by many states and agencies; increasing interest in large scale planning, along with limitations on funding for large surveys, implies that researchers will increasingly need to use such databases. Unfortunately, many of the archival databases appear to have been designed to serve mainly as catalogues, and they may have only limited analytical utility. In many cases, procedures involving more detailed site recording and better consistency among projects will be needed if the analytical potential of large-scale regional mapping is to be fully realized.

Model development

Logistic regression techniques were used to generate a correlative model of site occurrence within a series of quadrants covering the 8500 square mile study area. Logistic regression predicts not the actual location of unknown sites but, rather, the probability that a given geographic area, with certain environmental traits, will contain a site (Wrigley, 1977; Kvamme, 1983a, 1985, 1986; Parker, 1985). Relative probabilities can be calculated for any point on the landscape where the model variables can be measured. Logistic regression has the advantage of not assuming a normal distribution in the data (Rose and Altschul, 1988, pp.214–218), and past applications of the technique have produced slightly better results than other multivariate classification methods such as discriminant analysis (Kvamme, 1988).

Although the study of prehistoric site distribution has been a primary focus of archaeological work in Montana, use of the existing database was the most challenging aspect of the present study. Existing records from north-central Montana vary greatly in the precision of their data on site type, size, age and location. Ideally, separate models would be constructed for each site type by time period, but the ages of most sites in the region are unknown. Similarly, state archivists do not consider the site type classifications in the database to be reliable. It was necessary, therefore, to limit the analysis to determining the probability of the simple presence or absence of any sites.

A total of 1120 prehistoric sites have been recorded in the study area in surveys covering approximately 1 per cent of the region. However, not all topographic settings are equally well represented in the sample, and many early investigations emphasized the highly visible site types, such as bison jumps and large tipi ring sites. For modelling purposes, a sample of 354 sites was selected from areas which have received reasonably intensive survey coverage. Survey areas were chosen which reflect, to the extent possible, the range of environmental settings in the study area. All sites in the sample occur within the known site areas shown in Figure 19.2. Site locations were mapped and digitized at 1:100,000 scale.

An important step in analyzing site distributions is the need to demonstrate patterning among sites (Kellogg, 1987). That is, it is necessary to show that the characteristics of site locations differ significantly from points scattered generally across the landscape. Known site areas were enclosed in control areas from which the environmental database was generated. Control points ($n = 4022$) were selected systematically, at 1500 m intervals, from within each control area. Control areas are intended to include settings which do and do not contain sites; for the purposes of analysis, the control points are assumed to be nonsites (Kvamme, 1983, 1985). The locational characteristics of sites and controls were compared to identify those variables which best distinguished the two groups.

Topographic variables were calculated from automated files containing a

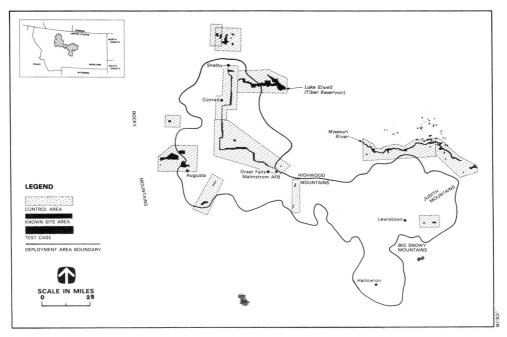

Figure 19.2. Prehistoric sites model-generation areas, north-central Montana.

combination of digitized hydrologic features and existing 1:250,000 scale Digital Elevation Model (DEM) data files. The array of points in the DEM file was converted to a coverage suitable for interpolation by using the Triangulated Irregular Network (TIN) structure of ARC/INFO (Chen and Guevera, 1987; Marozas and Zack, 1987).

Eight variables were measured or calculated (for sites and controls) which had been shown in previous studies (Kvamme, 1982, 1983, 1984, 1985) to be important predictors of hunter-gatherer site locations:

— Horizontal distance to permanent water, as measured to streams and springs rescaled from 1:62,500 USGS topographic maps;
— Horizontal distance to any water, including intermittent streams and ephemeral closed depressions rescaled at 1:62,500;
— Vertical distance to permanent water, calculated as the difference in elevation between the site and the water source;
— Vertical distance to any water, measured as above;
— Aspect, measured in degrees from UTM grid north;
— Slope, measured in per cent grade;
— Relief, as calculated by the difference between the maximum and minimum elevation within 500 m of the data point;
— Elevation as interpolated from the TIN coverage.

Univariate analyses

The known site and control groups were compared for each variable using a difference of means test (SAS Test procedure). The results of these tests (Table 19.1) indicate that site locations and control points differ significantly in their horizontal distance to permanent water, horizontal distance to any water, relief, elevation, and vertical distance to permanent water. Sites are located, on the average, closer to water sources than are control points. Sites also occur in areas of less relief than controls, and, as would be expected, there is less variability in location among sites than among controls. The five variables identified in the univariate analyses were used in a logistic regression of site and control point location.

The remaining variables were dropped from the analysis, but not simply because of their performance in the difference of means test. Indeed, the effects of a single variable in a univariate analysis could not be assumed to reflect its potential input to a multivariate model. Rather, the three were eventually omitted because of evidence that their use might produce spurious results.

Table 19.1 *Difference of means comparison of locational characteristics of sites and control points*

Variable	Group	n	mean	s.dev.	t	df	Result
Slope	Sites	353	4.18	6.59	1.36	570	No significant difference
	Control	2047	4.72	8.41			
Aspect	Sites	353	148.9	264.1	1.36	509	No significant difference
	Control	2047	168.0	240.8			
Relief	Sites	353	110.8	140.3	14.88	2247	$p \leq 0.0001$
	Control	2047	584.4	1399.8			
Elevation	Sites	353	3538	649.0	2.69	478	$p \leq 0.007$
	Control	2047	3638	641.5			
Horizontal distance to permanent water	Sites	353	2615	3614	5.73	502	$p \leq 0.0001$
	Control	2047	3821	3875			
Vertical distance to permanent water	Sites	352	142	207	3.06	474	$p \leq 0.002$
	Control	2047	179	202			
Horizontal distance to any water	Sites	353	383	494	2.72	499	$p \leq = 0.006$
	Control	2047	461	525			
Vertical distance to any water	Sites	353	38	91	-1.31	428	No significant difference
	Control	2047	32	72			

Previous studies have shown that aspect and vertical distance to water can be extremely important to hunter-gatherer site location (Kvamme, 1984). It was surprising then when these variables did not identify significant differences between sites and nonsites. Further tests were run on ratios and other second order variables incorporating these and several other landform characteristics, to investigate the effects of combinations of variables and to reduce variance. When these efforts failed to produce significant results, the DEM database for slope, aspect, and vertical distance was scrutinized. It was eventually determined (Marozas and Zack, 1987; Chapter 14) that the 1:250,000 DEM coverage produced a level of precision which, for some variables, was inadequate for the present analysis. Therefore, those variables were omitted from the logistic regression.

Multivariate analysis

Logistic regression permits estimates of relative site probabilities to be made for any point where the model variables can be measured. The technique has the advantage of not assuming a normal distribution in the data (Rose and Altschul, 1988, pp. 214–218). Past applications of logistic regression have produced slightly better predictive results than other multivariate classification techniques such as discriminant analysis (Kvamme, 1988).

Repeated analyses were conducted using the SAS CATMOD procedure. Elevation did not add significantly to the regression solution and the resulting function related site occurrence to the remaining four variables (Table 19.2).

Table 19.2 Logistic regression of prehistoric site locations, north-central Montana

Effect	Estimate	Standard error	df	Chi-square	Probability
Intercept	−0.494243	0.039123	1	159.59	0.0001
Horizontal distance to permanent water	0.0001201	0.000008	1	201.82	0.0001
Relief	0.00112131	0.000100	1	125.66	0.0001
Horizontal distance to any water	0.00012322	0.000050	1	6.07	0.0138
Vertical distance to permanent water	−0.000376	0.000150	1	6.68	0.0098

$L = 0.494243 - 0.0001201 \times$ (horizontal distance to permanent water) $- 0.00112131 \times$ (relief) $- 0.0012322$ $\times$ (horizontal distance to any water) $+ 0.00376 \times$ (vertical distance to permanent water)

The model was tested by classifying the input data as if site and control identity were unknown. Values of the topographic variables were used to solve for L at each location and the results were used to calculate site probabilities according to the logistic transformation (Kvamme, 1988, p. 371):

$$p = \frac{e^{(L)}}{1 + e^{(L)}} = \frac{1}{1 + e^{(-L)}}$$

where p is the probability, L is the logistically derived discriminant function, and e is the natural logarithm. As a result of the transformation, the predicted probability of site class membership is expressed as a value between 0 and 1. Assuming equal *a priori* probabilities for group membership, values greater than 0.5 would be interpreted as predicted site locations (Kvamme, 1983b, pp. 91–92).

Using the 0.5 cutpoint, 72 per cent of the sites and 55 per cent of the controls were correctly classified. Although these proportions are not as high as had been hoped, they are significantly better than could be expected for classification without the model (Chi-square $= 9.549$, 1 *df*, $p < 0.001$). Most of the error in the model comes from the incorrect classification of controls as sites; in other words, the errors are on the conservative side. This is the preferred error if one is attempting to avoid sites.

The few intensive surveys in the study area suggest that 0.01 is a reasonable estimate for the prior probability of site occurrence. That is, in study blocks containing sites, only about 1 per cent of the ground surface is actually site area. By applying Bayes' theorem to the model results it can be suggested that about 72 per cent of the prehistoric sites in the region should occur in 45 per cent of the available land area. It should be noted that *a posteriori* probabilities derived from the input data will provide an optimistic view of the model's performance. Nevertheless, the ability to predict nearly three-quarters of the site locations was considered sufficient for large scale planning purposes.

Model application

Topographic variables were measured at grid points distributed at 1500 m intervals across the TIN coverage for the study area. Site probabilities were calculated for each of the 11,918 sample points using the formulae shown above. The results are shown graphically in Figure 19.3 by coding the 1500 m grid units according to the values of their centrepoints. The probabilities are categorized at three cutpoints to yield four intervals of site likelihood shown as sensitivity zones. The 0.50 cutpoint represents the site likely/not likely threshold as indicated by the boundary between sensitivity zones 1 and 2. Additional intuitive cutpoints at 0.55 and 0.57 represent the boundaries between zones 2 and 3, and 3 and 4, respectively. The latter two zones serve to delineate the areas of relatively higher site potential.

High site probabilities are indicated when areas of moderate relief occur in the vicinity of water sources, including streams, springs, ephemeral drainages, and glacial kettle depressions. Areas of steep slope and high relief, as well as areas of very low relief, are identified as low probability zones. Intermediate probabilities are predicted for transitional areas such as foothills and along the Rocky Mountain Front.

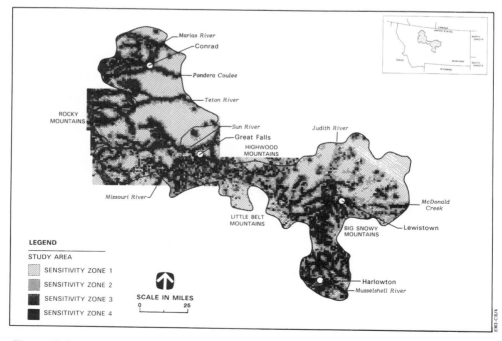

Figure 19.3. Projected prehistoric site sensitivity in the study area, north-central Montana.

Discussion

Although the predictive success of the central Montana model was not as great as had been hoped, the results are similar to some previous applications of the technique on much smaller areas. Kvamme's (1984) model of settlement patterns in Pinyon Canyon, Colorado correctly classified 76 per cent of the sites and 60 per cent of nonsite control points. However, that study area encompassed only 812 square miles, or slightly less than one-tenth the area of the Montana study.

Because a smaller area was involved, it was possible to construct a finer grained model; environmental data were measured from 1:62,500 scale maps and the prediction interval was 76 m instead of 1500 m as in the present study. In light of these differences, it is noteworthy that the Montana results are quite similar to those of the Pinyon Canyon study. There would appear to be considerable potential for the development of very large scale models. However, a number of aspects of the modelling offer areas for the improvement of model precision.

Better results could almost certainly be obtained from more detailed archival data. It has been shown repeatedly that different organizational patterns characterize different site types. The conditions affecting the locations of short term, limited activity sites could be quite different from those affecting habitation sites. The availability of even such broad classifications would be useful in modelling, provided they were applied consistently throughout the database. Reclassification by archivists might be necessary to achieve this situation. The same comments apply for time period classifications.

In addition to the archaeological database, the resolution of the DEM coverage has an important effect on the model. The only data available for very large study areas are digitized at 1:250,000 scale with a vertical resolution of ±50 m. This has a smoothing effect on the terrain as it is represented in the TIN coverage. Small scale landscape features, such as knolls and buttes, may not be identifiable at that scale. As 1:62,500 DEM files become available for more of the country, the situation should improve.

The use of more detailed map data will also permit the use of more variables in the model equation, thereby potentially increasing predictive success. For example, vantage points (locations with wide angles of view) and shelter quality (degree of protection afforded by landform or vegetation) have been shown previously (Kvamme, 1983) to be useful predictors of hunter-gatherer sites. However, the smoothing effect makes their measurement at 1:250,000 of little use (Marozas and Zack, 1987; Chapter 14). The use of second order variables could also improve the model by reducing variance.

The coarseness of the model could be improved by reducing the size of the prediction interval on the probability surface map. There would be a cost associated with increasing the number of points and, for very large study areas, there might be some limitations to the number of data points that could be manipulated, but there is potential for improvement.

One final area of interest concerns the testing of the model and its application in a management context. For a model of such large geographical extent, testing must be viewed as a long term proposition. The study area is so large, and is receiving so little attention from researchers, that tests of the model will have to come from the accumulation of data from many studies over a period of years.

As a long range management tool, the model requires much further testing and refinement. As an aid to preliminary project planning, however, the model appears to have considerable utility. The model correctly classified nearly three-quarters of the known sites in the study area; most of the error in the model came from the classification of control points as sites. In this sense, the model can be viewed as conservative, as approximating a reasonable worst case. Although use of a model does not obviate the need to do field studies prior to construction, the avoidance of areas likely to contain sites could prevent considerable disturbance of archaeological resources.

Acknowledgements

The research discussed herein was conducted under contract F04704-85-C-0062 for the US Air Force Regional Civil Engineer, Ballistic Missile Support (AFRCE-BMS/DEPV), Norton Air Force Base, California.

References

Brumley, J.H. and Dau, B.J., 1985, Proposed Historical Resource Development Zones on the Suffield Military Reserve, Alberta. Ethos Consultants, Havre, Montana. Prepared for the Alberta Energy Company, Canada

Chen, Z. and Guevera, J.A., 1987, Systematic Selection of Very Important Points (VIP) from Digital Terrain Models for Constructing Triangular Irregular Networks. Manuscript on file, Environmental Systems Research Institute, Redlands, California

Kellogg, D.C., 1987, Statistical relevance and site locational data. *American Antiquity* **52**(1), 143–150

Kvamme, K.L., 1982, Methods for Analyzing and Understanding Hunter-Gatherer Site Location as a Function of Environmental Variation. Paper presented at the 47th Annual Meeting of the Society for American Archaeology, Minneapolis, MN

Kvamme, K.L., 1983a, Computer processing techniques for regional modelling of archaeological site locations. *Advances in Computer Archaeology* **1**, 26–52

Kvamme, K.L., 1983b, *New Methods for Investigating the Environmental Basis of Prehistoric Site Locations*, (Ann Arbor, MI: University Microfilms International)

Kvamme, K.L., 1984, Models of prehistoric site location near Pinyon Canyon, Colorado. *Papers of the Philmont Conference on the Archaeology of Northeastern New Mexico*, New Mexico Archaeological Council Proceedings, Volume 6, No. 1, pp. 347–370

Kvamme, K.L., 1985, Determining empirical relationships between the natural environment and prehistoric site locations: a hunter-gatherer example. In *For Concordance in Archaeological Analysis*, edited by C. Carr, (Kansas City: Westport Publishers, Inc.), pp. 208–238

Kvamme, K.L., 1986, Geographic informations systems in archaeology. *Proceedings of the Conference on Remote Sensing and Geographical Information Systems in Management*, (Tucson: Arizona Remote Sensing Center, University of Arizona), pp. 37–42

Kvamme, K.L., 1988, Development and testing of quantitative models. In *Quantifying the Present and Predicting the Past: Theory, Method, and Application of Archaeological Predictive Modeling*, edited by W. J. Judge and L. Sebastian, (Denver: US Bureau of Land Management), pp. 325–428

Los Angeles Harbor Department and US Bureau of Land Management, 1985, Draft Environmental Impact Report/Environmental Impact Statement, Proposed Pacific Texas Pipeline Project. San Pedro, CA

Marozas, B. and Zack, J., 1987, Geographic Information Systems Applications to Archaeological Site Modelling. Report prepared for Tetra Tech, Inc. (Redlands, CA: Environmental Systems Research Institute, Inc.)

Parker, S., 1985, Predictive modelling of site settlement systems using multivariate logistics. In *For Concordance in Archaeological Analysis*, edited by C. Carr, (Kansas City: Westport Publishers, Inc.), pp. 173–207

Rose, M.R. and Altschul, J.A., 1988, An overview of statistical method and theory for quantitative model building. In *Quantifying the Present and Predicting the Past: Theory, Method and Application of Archaeological Predictive Modeling*, edited by W.J. Judge and L. Sebastian, (Denver: US Bureau of Land Management), pp. 173–256

SAS Institute, Inc., 1985, *SAS User's Guide: Statistics*, Version 5 edition, (Cary, NC)

US Army Corps of Engineers, 1986, Draft Environmental Impact Statement of the Proposed Ground Based Free Electron Laser Technology Integration Experiment, White Sands Missile Range, New Mexico, Huntsville, AL and Fort Worth, TX

US Bureau of Land Management, 1981, Final Grazing Environmental Impact Statement, Southern Rio Grande Planning Unit, Las Cruces, NM

US Bureau of Land Management, 1984, Draft Environmental Impact Report/ Environmental Impact Statement, Proposed Celeron/All American and Getty Pipeline Projects, Riverside, CA

US Department of Defense, 1985, Environmental Impact Assessment for Border Star 1985, Joint Readiness Exercise, Fort Bliss, Texas, and New Mexico, White Sands Missile Range, New Mexico, and Specified Adjoining Lands, Washington DC: United States Readiness Command

US Department of Energy, 1986, Draft Environmental Impact Statement, Conrad-Shelby Transmission Line Project, Pondera and Toole Counties, Montana, Billings, MT: Western Area Power Administration

Wrigley, N., 1976, *An Introduction to the Use of Logit Models in Geography*, Concepts and Techniques in Modern Geography Volume 10, (Norwich: Geo Abstracts, University of East Anglia)

Wrigley, N., 1977, *Probability Surface Mapping*. Concepts and Techniques in Modern Geography Volume 16, (Norwich: Geo Abstracts, University of East Anglia)

20

Red flag models: the use of modelling in management contexts

Jeffrey H. Altschul

Statistical research

Over the past decade, archaeologists involved in cultural resource management (CRM) have struggled with two interrelated problems: how to manage large numbers of often poorly documented resources and how to manage resources whose locations are not even known. These challenges have lead archaeologists and land managers down a path with many twists and turns and more than a few dead ends. During the 1970s archaeologists began to examine whether traditional studies of settlement patterns in which whole regions were surveyed could be more efficiently approached with data derived through rigorous sampling designs (Judge et al., 1975; Plog, 1976; Santley, 1979). Probabilistically based surveys aimed at elucidating patterns of settlement were undertaken (e.g., Matson and Lipe, 1975). A few studies even went beyond addressing issues of culture history or settlement dynamics, and ventured into a new arena; the prediction of site locations (Green, 1973; Hurlbett, 1977; Kvamme, 1983). Federal archaeologists and managers were quick to seize the potential of these studies. Predictive modelling flourished, becoming both a legitimate avenue of archaeological inquiry and a management tool (Cordell and Green, 1983; Brown, 1981; Darsie and Keyser, 1985; Judge and Sebastian, 1988; Kohler and Parker, 1986).

But by and large, predictive models did not live up to the expectations of either archaeologists or managers. Numerous problems arose: theoretical issues were not well thought out; applications of statistical techniques caused debate and controversy; and the proper types of data were often not available (Ambler, 1984; Berry, 1984; Tainter, 1984). While these and other problems certainly plagued the early models, in my opinion the primary reason for the waning interest in predictive modelling had less to do with any of the above mentioned shortcomings than with the changing face of CRM.

In the late 1970s the database of site locations for many regions was very meagre. Large surveys to rectify this situations were in order (e.g., Hauck, 1979a,b; Alexander et al., 1982). By the 1980s, however, the situation had changed dramatically for many of these same regions. Instead of a few sites, many Federally administered areas now had databases holding hundreds, if not thousands, of sites, which continued to grow annually. For example, the Kaibab National Forest administers approximately 1.5 million acres in northern Arizona. Each year about one per cent, or 15,000 acres, is surveyed in response to compliance needs, such as timber sales or road building, and around 100 newly recorded sites are added to the database. As of the end of 1987, there were 3,300 sites recorded in the CRM database for the forest. Ten years earlier only 567 sites were known, and these were managed by a manual site card system (John Hanson, personal communication, 1989).

The Kaibab case is not unique; it is illustrative. Archaeologists managing the resources of this forest and other Federal holdings like it do not need to know where sites will or will not be found. Generally speaking, they have done enough survey to know the broad patterns of site location. But if Federal archaeologists do not need models to tell them where sites are, what do they need?

In 1983, I asked Dr. A.E. Rogge, then Projects Archaeologist for the Bureau of Reclamation's Arizona Projects Office, what types of information Federal agencies could use from predictive models. Rogge responded that what Federal land managers would find useful was not site densities or even site locations; what managers need to know is where the 'red flags' are.

From a management perspective, a red flag is any site which is costly in terms of either time or money or both. Virtually any site eligible or potentially eligible for listing on the National Register of Historic Places could be considered a red flag. In practice, however, we want to restrict the definition to those sites which will cause undue project delay and for which data recovery will be very expensive.

The definition of a red flag also has an archaeological component. This was perhaps best expressed by the late Jennifer Jack, who until her death was a Bureau of Land Management (BLM) archaeologist on the Arizona Strip District. In response to my question about the use of modelling in management, Jack stated that she did not need a model that told her where sites were because for the most part all land potentially affected by projects is surveyed anyway. Instead, a more difficult and recurring situation is finding some forty seemingly identical lithic scatters, and then working extremely hard for a number of years to convince one's supervisors that money should be provided to do some work at one or two. The question she wanted answered was which one or two.

Both Rogge and Jack echo the sentiment of many Federal archaeologists. What is needed are not models predicting the unknown, but rather models that bring some order and direction to the huge databases that have been, and are continuing to be, amassed.

In the remainder of this paper I outline one approach that addresses these issues. This approach, termed red flag modelling, differs from previous modelling efforts less in technique than in orientation. To date, virtually all predictive models emphasize accuracy. A model is judged to be successful if it correctly predicts where sites will and will not be located 80 to 90 per cent of the time. While accuracy is important, it is not necessarily a very useful measure. Predictive models in CRM have generally been based on the observation that in most regions sites tend to cluster around certain environmental features. This observation is not new; indeed, it has been part of most field archaeologists' baggage for over a century. Sophisticated models that capitalize on this fact may be accurate, but by and large do not tell us anything we did not already know.

By extending current predictive models one step, however, we can move beyond accuracy. Instead of looking at how many sites we can predict, we can look at sites we cannot. By highlighting sites located in areas where sites are generally absent we can begin to explore portions of the archaeological record that are presently unclear. Sites in anomalous settings by definition must be the result of behaviours that do not fit current models of why prehistoric inhabitants settled where they did. Under any definition, these sites must be significant, for they more than any others have the potential of telling us something about prehistory that was heretofore unknown.

Another difference between previous predictive models and red flag models is the shift from a static to a dynamic modelling approach. Once anomalies, or red flags, are identified they become the subject of additional research. As patterns are found, many anomalies become predictable. Those sites whose locations remain anomalous grow in importance. Archaeologists want to know about these sites to further our insight into the past. Managers want to know the locations of these sites so that they can be included early in project plans.

To be a dynamic tool, modelling must be relatively simple and quick. Federal archaeologists do not have the time to conduct major modelling efforts once, to say nothing of on a continual basis. Most do not have the statistical background to conduct and interpret sophisticated multivariate techniques. The red flag modelling approach was designed to meet these constraints. We wanted a technique that required a minimal mathematical and computer background and could be completed in a few hours.

The Mount Trumbull red flag model

The example described below comes from the Mount Trumbull area of the Arizona Strip (Figure 20.1). The model study window was chosen by BLM archaeologists because the area has a rich and poorly understood archaeological record and is currently a hotbed of uranium development. The archaeological database consists of 228 prehistoric components from the AZSITE computer database. AZSITE is maintained by the Arizona State Museum and is a complete listing of sites that have received site numbers from that institution. In addition to site numbers, AZSITE contains a variety of descriptive information pertaining to the environment, location, cultural affiliation, site function and temporal components as well as management concerns, such as site integrity, ownership, and National or State Register status. Although some agencies and institutions, such as the Forest Service, have elected to maintain their own cultural resources database, and not subscribe to AZSITE, the latter is still the single largest archaeological database in the State of Arizona. As regards Mount Trumbull, the area is managed by the BLM which does subscribe to AZSITE. Thus, for this particular exercise the computerized data faithfully represents our current knowledge of site location.

Figure 20.1 presents the locations of all recorded sites and surveys undertaken in the Mount Trumbull area. The large irregular shape in the southwest portion of the study window represents the boundaries of a single survey conducted by the Museum of Northern Arizona in the mid 1970s (Moffitt and Chang, 1978). This survey is of particular interest because it was a sample survey with the explicit purpose of characterizing site location. Nearly 25 per cent of the 9000 acre area was surveyed by transects of equal width, but differing lengths that stretched east–west from one end of the survey area to the other. Site locations were then assessed against particular environmental variables. The net result was the demonstration that sites were present in greater than expected numbers in certain elevational bands, in particular vegetative communities, and so on. There was no attempt to associate site location with more than one variable at a time, nor was there any discussion of how these patterns related to cultural behaviour. In many regards, the Mount Trumbull survey is typical of sample surveys conducted on Federal lands during the late 1970s and early 1980s. Emphasis was placed squarely on pattern recognition, with no thought given to what the patterns mean.

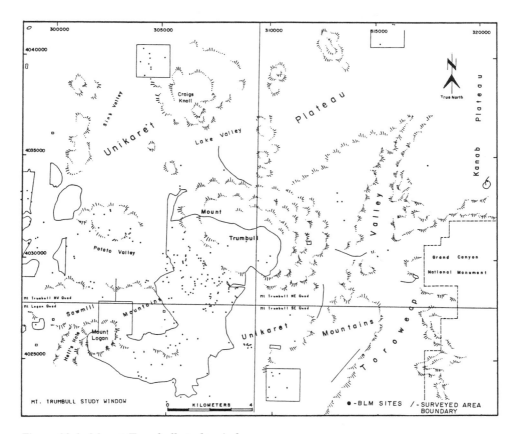

Figure 20.1. Mount Trumbull study window.

In addition to the Mount Trumbull survey, Figure 20.1 shows a variety of small to medium size areas that have been surveyed. These represent *ad hoc* surveys conducted by the BLM in response to proposed development projects. For example, the square boxes represent land exchanges between the BLM and private or public entities, while the lines reflect road or fence line surveys. Finally, it is worth noting that sites located outside survey boundaries represent fortuitous, isolated recordings by amateurs or professionals.

The environmental database for this exercise was derived from a US Defense Mapping Agency (DMA) digital terrain tape purchased from the United States Geological Survey (USGS). This tape contained elevational data at a scale of 1:250,000. Although DMA tapes are generally being replaced by finer scale (1:24,000) USGS Digital Elevation Model (DEM) tapes, the latter were not available for the Mount Trumbull area.

Three environmental variables were created. The first was elevation, which simply consisted of the data on the DMA tape. The other two, slope and aspect, were calculated from the elevational data. These are termed secondary variables or themes because they are simply mathematical transformations of another variable.

Slope was measured in per cent grade, while aspect was created by dividing the compass into eight equal quadrants for areas with exposures and a ninth value for those lying on level surfaces.

It is important to point out that different GIS are likely to use different algorithms to compute secondary variables. Thus, even though they may start with the same elevational data, two GIS are likely to produce very different slope or aspect themes (Kvamme, Chapter 10). Knowing how a particular GIS computes various secondary variables is critical in evaluation of the results.

In the case of Mount Trumbull, the GIS used was the Geographic Information Management System, or GIMS, which is owned and operated by Dames and Moore (1985). GIMS is a raster based GIS. The program first places a grid over the study window so that the area is divided into cells of equal size called pixels. The size of the pixel is selected by the user and usually represents a compromise between spatial resolution and computing power. In most archaeological cases, we want pixels to be as small as possible. In particular we want to avoid situations where sites only cover a small part of the pixel, or in which pixel size is so large that secondary environmental variables do not properly characterize the theme in question. The smaller the pixel, however, the greater the number of calculations must be performed. For Mount Trumbull, we settled on a pixel size of approximately 60×60 m.

To calculate slope and aspect using GIMS, the computer searches the eight pixels surrounding the one in question and computes an average slope or aspect score for that pixel. These scores, then, are computed over a distance of 180 m north–south and east–west and nearly 255 m in the diagonal directions. Slope and aspect, therefore, are best viewed as 'regional' variables, and not site specific calculations. It is not uncommon to find a site on the slopes of Mount Trumbull with a slope score of 30, which is actually lying on a flat bench.

Step 1: data exploration

There are four steps in developing a red flag model. The first is data exploration. Descriptive data are compiled on spatial location, temporal affiliation, and site function. The covariation of these variables with each environmental variable is explored through the use of simple associational statistics. Thus, for Mount Trumbull, by computing Chi-square tests with an alpha level of 0.05, we find that site locations are found in greater than expected frequencies in elevations between 6400 and 6600 ft, on surfaces with slopes of between 10 and 20 per cent grades, and in areas with north and northwest exposures. Sites are absent in greater than expected numbers from areas with elevations over 7200 ft, with slopes greater than 50 per cent grade, and in locations with easterly exposures. In all other conditions, sites are represented in numbers that do not differ significantly from a random distribution.

Step 2: confidence and independence

The second step involves assessing the environmental variables on the questions of confidence and generalization. First, we want to determine whether survey coverage of each environmental zone is adequate to have any faith in the resulting distributions. For example, only 1.2 per cent of the area below 6000 ft has been surveyed, although this area encompasses nearly 65 per cent of the entire study window. In contrast, nearly 29 per cent of the area between 6600 and 7200 ft has been surveyed, even though this region constitutes less than 8 per cent of the study window. These observations bear directly on our confidence in generalizing the survey results. While no sites have been found below 5000 ft, we cannot argue that sites were not located below 5000 ft because only 0.1 per cent of this area has been surveyed. On the other hand, we can be quite confident that the survey results in the 6600 to 7200 ft zone can be generalized to the entire zone.

It is important to point out that confidence in this context is not of a statistical nature. The total surveyed area represents a conglomerate of different surveys. Thus, we cannot assess the bias in the sample and cannot realistically compute variances or confidence intervals. Confidence as used here is a subjective term. As a general guideline for Mount Trumbull, we had confidence in the survey results of a zone when over 10 per cent of it had been surveyed *and* the amount of the zone that had been surveyed was proportional to the zone's overall representation in the study window.

The second aspect to the environmental assessment is determining the degree of statistical independence between the three environmental zones. In many areas environmental variables strongly covary. Well drained soils are found with particular vegetative communities, within a narrow range of slopes, and on specific topographic features. Using interrelated variables in the same model is likely to inflate the model's apparent power.

There are various statistical methods, such as partial correlation or multiple regression, that will insure that two or more variables are statistically independent. These methods tend to be rather complex and their use would seriously jeopardize our key objective, which is to create a modelling procedure that can be used by non-statistically oriented archaeologists. A much simpler, although not as rigorous, approach is to simply evaluate the maps visually. Figures 20.2 to 20.4 are computer drawn maps of the area viewed in Figure 20.1 and present the three environmental themes, elevation, slope and aspect, used in this exercise. By comparing them, one can gain an impression of whether shaded areas on one map are related to shaded areas on the other maps. In a few seconds one can see that while in certain areas, such as the mountain crests (shown as black in Figure 20.2), the variables may be related, overall there seems to be a fair degree of independence.

After visual examination, the nature of the relationships can be pursued statistically. For elevation and slope a Pearson's *r* correlation coefficient was computed on the basis of the pixel's scores. The *r* score was 0.23 which indicates that there is no significant relationship between the two variables. For analyzing the relationship between aspect and the other two variables, contingency tables could have been created. Chi-square and Phi-square values could be computed on these tables to determine whether a significant relationship existed, and if so how strong the relationship was (see Altschul and Jones, 1990, for an example). Even without them, however, it appears safe to conclude that aspect is not strongly related to either of the other two variables, and thus all three of them can be used in the model.

Step 3: the favourability map

The third step is to create a map showing the relative probability of each pixel to contain a site. This map, termed a favourability map, is developed by creating a favourability scale for each variable from +1 (zones which have more sites than expected by chance), 0 (site distributions show no difference from random), and −1

J.H. Altschul

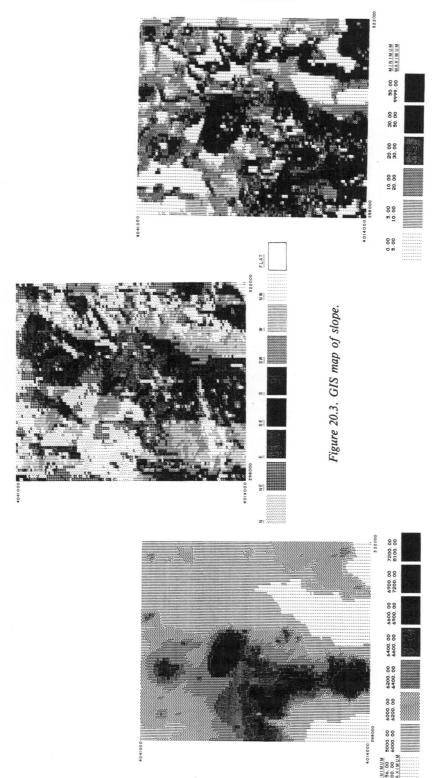

Figure 20.4. GIS map of aspect.

Figure 20.3. GIS map of slope.

Figure 20.2. GIS map of elevation.

(sites are less frequent than expected by chance). Favourability scores are summed and favourability zones are defined. For Mount Trumbull, zones with scores of +1 to +3 were coded as 'favourable', those summing to 0 were coded 'neutral', areas scoring −1 were defined as 'unfavourable', and those with scores of −2 or −3 were coded as 'very unfavourable'. A favourability map showing these zones was then created (Figure 20.5).

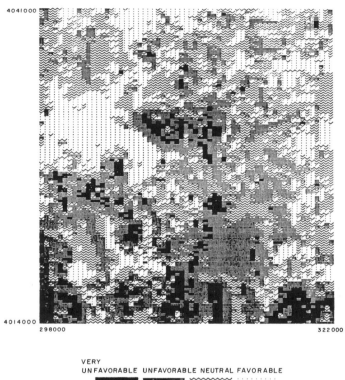

Figure 20.5. Favourability map.

A favourability map is a useful tool. Figure 20.5 clearly illustrates that the western slope of Mount Trumbull was a particularly favoured location for site placement. Other 'favoured' locations include Craigs Knoll overlooking Lake Valley and the Unikaret Plateau (top of Figure 20.5), a corridor stretching the entire length of the Toroweap Valley (right side of Figure 20.5), the western slope of Mount Logan (bottom left corner of Figure 20.5), and the saddle between Mount Logan and Mount Trumbull (bottom left corner of Figure 20.5). This information is useful both for archaeologists and managers. At a glance, managers can determine the likelihood of encountering sites on a particular development project. For archaeologists, the map represents a compilation of the relationship between environmental variables and site location. Previous investigators had found the same relationships, but had no means of presenting the results graphically (Moffitt and Chang, 1978). The favourability map developed for Mount Trumbull provides the same information in a much more concise fashion. The importance of a western exposure is made much more dramatic in Figure 20.5 than in a series of Chi-square results. One can hypothesize that site placement on these slopes is related to agricultural potential. Indeed most of the sites in these areas are small one to five room pueblos that are generally interpreted as fieldhouses or storage facilities. The favourability map, then, can be used to point future research in promising directions.

The red flags

This is the point at which most predictive models stop. Red flag models, however, have one more step. Once the favourability map is created, site locations are cross tabulated with favourability zones. In our example 93 habitation components, or 71 per cent, were found in favourable locations, 20, or 15 per cent, were found in neutral areas, 17, or 13 per cent, were located in unfavourable settings, and only 1, or 0.8 per cent, was found in a very unfavourable location. The distribution of temporary camps and unknown site type components follow the same trend.

While it is reassuring that over 70 per cent of all component locations can be predicted with just three variables, our interest at this point focuses on those components that fall in the unfavourable or very unfavourable settings (Figure 20.6). In examining the results we first divided the sites by time period. Anasazi and undated components have approximately the same proportion of red flags, about 13 or 14 per cent. In contrast, protohistoric components, which are all affiliated with Paiute occupation of the region, had nearly triple the proportion of red flags. In studying the distribution of Paiute red flags, and of Paiute sites in general, we found that the occupation of the Mount Trumbull study window during the Protohistoric Period was focused on the east flank of the Unikaret Mountains and the Toroweap Valley. The Anasazi occupation followed a very different pattern, concentrating on the western slopes of Mount Trumbull and the Unikaret Mountains. The placement of Paiute sites as red flags is understandable in terms of the construction of the model. The model is heavily weighed with Anasazi data and therefore is a better predictor of this culture than any other. While understandable the finding is still intriguing. Why would the Paiutes concentrate on one side of the mountain and the Anasazi the other?

One answer was suggested in the discussion of the favourability map. The Anasazi may have been much more dependent on agriculture than the Paiutes. This hypothesis is certainly consistent with current regional interpretations (Altschul and Fairley, 1989). It is also consistent with the data from Mount Trumbull. Paiute settlement is more or less uniformly distributed throughout the area; a pattern often characteristic of a hunter-gatherer culture. In contrast, Anasazi settlement follows a clustered pattern in which a set of circumscribed areas contain large numbers of sites and the rest of the region is largely avoided. This pattern is likely to emerge through a subsistence strategy heavily focused on one set of resources; in this case, agriculture.

A second interesting aspect to the red flags concerns the Anasazi habitation components (Figure 20.7). Of the 23 Anasazi red flags, 18, or nearly 80 per cent, were habitation components. This percentage is substantially higher than the overall percentage of habitation components (60 per cent). Of the 18 habitation red flags, nine consist of components with five rooms or less. Most of these small components were probably used as fieldhouses. Nine of the components, however, are relatively substantial habitations, with four being C or L shaped pueblos and two others containing over 20 rooms in one block.

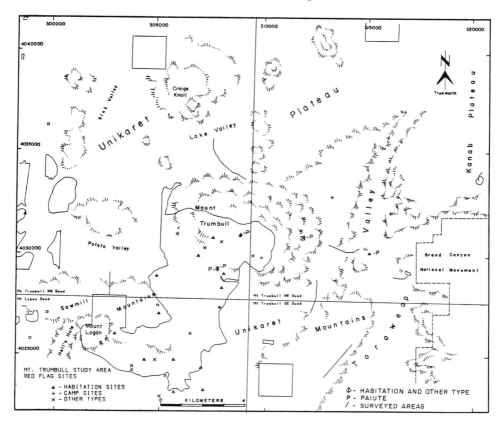

Figure 20.6. Mount Trumbull red flag sites.

The distribution of the larger habitation components is interesting. Three components form a tight spatial cluster along the southwestern base of Mount Trumbull. Two of these sites also contain Paiute components suggesting that the desirability of these locales transcended time and culture. In studying the USGS quadrangle, however, there is nothing at these locations that is especially noteworthy. There are no springs in the immediate area nor even a reliable water source. At present we have no good explanation for these sites; they are truly enigmatic. It is reasonable to assume, however, that a cluster of large multicomponent sites is indicative of strong culturally patterned behaviour. Further modelling efforts might focus on these sites by using a different set of environmental variables. Instead of using the regional variables employed in this analysis, site specific variables might be developed based on microtopographic and hydraulic factors. Regardless of what archaeologists do, the research value of this area is high and to managers it should be shown as a 'dark' red flag.

A second set of habitation components also poses an intriguing problem. These are distributed along the southern edge of the modelled area. What distinguishes these components from other habitation components is not so much the site specific setting, but the fact that they are among the largest components in the region and each one is placed in a relatively isolated setting. Unlike the vast majority of sites located to the north which are found on the western flank of Mount Trumbull, these larger sites are situated on the southern slopes of Mount Logan. Thus, we have two very different settlement patterns represented; a large number of generally small habitation sites on the west slopes of Mount Trumbull and the small number of spatially isolated large habitation sites on Mount Logan. Numerous hypotheses could be formulated to account for the situation. One possibility is that most of the inhabitants are living in the Mount Logan sites and are using the Mount Trumbull sites as temporary fieldhouses. Another possibility is that the patterns represent two separate groups. These groups could either have been contemporaries or separated in time.

Although plausible and worthy of future consideration, these hypotheses are not central to this discussion. The point here is that these patterns were not intuitively obvious. Previous researchers, including those that recorded both sets of sites, did not notice this situation. As an analytical tool, the red flag model and the study of outliers proved to be a very useful approach.

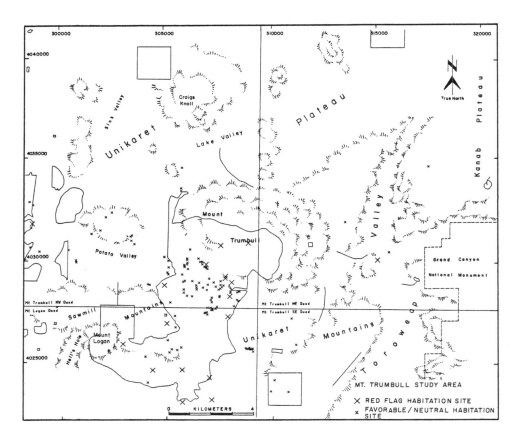

Figure 20.7. Mount Trumbull habitation sites.

These sites also point to another part of the model that needs refinement. In this study, habitation sites were defined as any site exhibiting architecture. Habitation sites ranged from one room features to 20 room pueblos. The fact that the locations of the largest habitation sites were anomalous suggests that these sites were functionally or culturally distinct. Habitation site classes, therefore, need to be refined. Although in hindsight it may seem obvious that 1–5 and 10–20 room pueblos are not functional equivalents, the size boundary was not at all clear to previous researchers or myself until the red flag model was completed.

Conclusions

The modelling approach advocated here is an outgrowth of previous work in site locational studies. No new statistics have been added and no new computer techniques are required. The basic changes are ones of philosophy: instead of looking at what we can predict, we look at what we cannot; instead of viewing models as end products, we view them as dynamic analytical tools.

Red flag models are viewed as analytical tools that can be used in everyday management. The process is not something that is done once, with the product being used as is for a number of years. My goal was to develop a process so that when an archaeologist is informed of a timber sale or mineral lease, for example, within a few hours he or she could have a set of results which would allow them to know where they should expect sites and where they should be particularly careful for sites in unusual or unexpected settings. The red flag model meets this goal.

Acknowledgements

I want to thank John Hanson, A.E. Rogge, Gary Stumpf, Rick Malcolmson, Jennifer Jack and Kathleen Allen for comments on earlier versions of this paper. I also want to acknowledge Dan Marino for performing the computer operations and Kathe Kubish for illustrating the results. This work was performed under Forest Service Contract 53-8371-6-0054 funded jointly by the Kaibab National Forest and the Arizona office of the US Bureau of Land Management.

References

Alexander, R.K., Hartley, J. D. and Babcock, T. F., 1982, *A Settlement Survey of the Fort Carson Military Reservation*. Report submitted to Interagency Archaeological Services, Contract No. C3572(78), Denver

Altschul, J. H. and Fairley, H., 1989, *Man, Models, and Management: An Overview of the Archaeology of the Arizona Strip and the Management of its Cultural Resources*. (Albuquerque, NM: USDA Forest Service and USDI Bureau of Land Management)

Altschul, J. H. and Jones, B. A., 1990, *Settlement Trends in the Middle San Pedro Valley: A Cultural Resources Sample Survey of the Fort Huachuca Military Reservation*. (Tucson, AZ: Statistical Research) Technical Series 19

Ambler, J. R., 1984, The use and abuse of predictive modelling in cultural resource management. *American Archeology*, **4**(2), 140–145

Berry, M. S., 1984, Sampling and predictive modelling on federal lands in the west. *American Antiquity*, **49**, 842–853

Brown, M. K., 1981, *Predictive Models in Illinois Archaeology: Report Summaries*, (Springfield, IL: Illinois Department of Conservation, Division of Historic Sites)

Cordell, L. S. and Green, D. F., (editors), 1983, *Theory and Model Building:*

Refining Survey Strategies for Locating Prehistoric Heritage Resources (Albuquerque, NM: Cultural Resources Document No. 3, USDA Forest Service, Southwest Regional Office)

Dames and Moore, 1985, GIMS System Description. Ms. on file, Dames and Moore, Phoenix, AZ

Darsie, R. F. and Keyser, J. K., (editors), 1985, *Archaeological Inventory and Predictive Modelling in the Pacific Northwest* (Portland, OR: Studies in Cultural Resource Management No. 6. USDA Forest Service, Pacific Northwest Region)

Green, E. L., 1973, Location analysis of prehistoric Maya sites in Northern British Honduras. *American Antiquity*, **38**, 279–293

Hauck, F.R., 1979a, *Cultural Resource Evaluation in Central Utah 1977* (Salt Lake City, UT: Utah Bureau of Land Management Cultural Resource Series No. 3)

Hauck, F.R., 1979b, *Cultural Resources Evaluation in South Central Utah 1978–79* (Salt Lake City, UT: Utah Bureau of Land Management Cultural Resource Series No. 4)

Hurlbett, R. E., 1977, *Environmental Constraint and Settlement Predictability, Northwestern Colorado*, (Denver, CO: Colorado Bureau of Land Management Cultural Resource Series No. 3)

Judge, W. J., Ebert, J. I. and Hitchcock, R., 1975, Sampling in regional archaeological survey. In *Sampling in Archaeology*, edited by J. W. Mueller, (Tucson, AZ: University of Arizona Press), pp. 82–123

Judge, W. J. and Sebastian, L., (editors), 1988, *Quantifying the Present and Predicting the Past: Theory, Method, and Application of Archaeological Predictive Modeling* (Denver, CO: Bureau of Land Management)

Kohler, T. A. and Parker, S. C., 1986, Predictive models for archaeological resource location. In *Advances in Archaeological Method and Theory*, Volume 9, edited by M. B. Schiffer, (New York: Academic Press), pp. 397–452

Kvamme, K. L., 1983, A Manual for Predictive Site Location Models: Examples for the Grand Junction District, Colorado. Draft submitted to the Bureau of Land Management, Grand Junction District, CO

Matson, R. G. and Lipe, W. D., 1975, Regional sampling: a case study of Cedar Mesa, Utah. In *Sampling in Archaeology*, edited by J. W. Mueller, (Tucson, AZ: University of Arizona Press), pp. 124–143

Moffitt, K. and Chang, C., 1978, The Mount Trumbull archaeological survey. *Western Anasazi Reports*, **1**, 185–250

Plog, S., 1976, Relative efficiencies of sampling techniques for archaeological survey. In *The Early Mesoamerican Village*, edited by K. V. Flannery, (New York: Academic Press), pp. 136–158

Santley, R. S., 1979, Sampling strategies and surface survey in the Basin of Mexico. In *The Basin of Mexico: Ecological Processes in the Evolution of a Civilization*, edited by W. T. Sanders, J. R. Parsons, and R. S. Santley, (New York: Academic Press), pp. 491–532

Tainter, J. A., (editor), 1984, Predictive modelling and the McKinley Mine dilemma. *American Archeology*, **4**(2), 82–146

21

Using geographic information systems and exploratory data analysis for archaeological site classification and analysis

Ishmael Williams, W. Fredrick Limp and Frederick L. Briuer

Introduction

Geographic information systems (GIS) are increasingly being used for the analysis and management of archaeological site data (Farley, 1987; Hasenstab, 1983; Kvamme, 1983, 1984, 1986, 1989; Kvamme and Kohler, 1988; Lafferty et al., 1981; Limp, 1987; Limp et al., 1987; Parker, 1985). The great majority of such studies have focused on the question of site distribution evaluation, in particular, on the question 'Where are archaeological sites and why are they there?' In the following we will consider another type of issue—that of site characterization or, in somewhat more loaded terms, site classification. By integrating the geographic information system with relational databases and powerful exploratory data analysis tools, complex 'high dimensional' groupings and patterning of site characteristics can be studied (Limp et al., 1987). In the following we will discuss how such an approach was implemented utilizing an extensive data set from the Fort Hood military installation. These data included the results of more than a decade of ongoing archaeological survey, as well as a comprehensive GIS environmental database gathered over eight years of intensive effort.

The archaeological and environmental database

The site characteristics database

Over the past ten years, Ft. Hood has conducted an inventory of archaeological sites over approximately 330 square miles or 90 per cent of the 339 square mile US Army Installation. Using in-house staff and outside contracted expertise, this intensive cultural resource survey effort has documented 2300 prehistoric and historic sites in a long-range multi-phased programme to identify and preserve the archaeological resources at Ft. Hood (Carlson et al., 1986, 1987, 1988; Dibble et al., 1984, 1985; Guderjan et al., 1980; Koch and Mueller-Wille, 1987; Roemer et al., 1985; Skinner et al., 1981, 1984; Thomas, 1978). This ongoing effort has focused on a number of management and research goals including systematic site documentation, archival and oral history studies of historic resources (Carlson, 1984, 1987; Jackson, 1982; Jackson et al., 1982a, 1982b), site impact studies (Briuer, 1981b; Briuer and Shaw,

1984; Carlson and Briuer, 1986), site settlement (Briuer, 1981a), site classification (Carlson, 1984) and other problem-oriented archaeological research (Briuer, 1981a, 1983). The Ft. Hood Archaeological Resource Management programme includes close coordination with and monitoring of military construction and training activities to avoid impacts to and preserve cultural resources.

The collection of site characteristics data has been guided by the *Standard Operating Procedure For Field Surveys* (Briuer and Thomas, 1986), a formal set of specifications to be followed by archaeologists surveying on the installation which sets forth minimum systematic site recording standards to assure adequate site documentation and comparability between survey efforts. The Standard Operating Procedure (SOP) provides guidelines for the conduct of site discovery surveys, transect site sampling procedures, site mapping, and general description and documentation. A dictionary of landform terms, a list of standardized mapping symbols, a number of forms for recording site observations, and examples are included in the SOP. The research and management continuity attained by the use of standard field surveying procedures has resulted in the critical evolution of recorded site information involving both qualitative and quantitative features. The site characteristics data have been refined and upgraded over the years with each cumulative survey, monitoring and data recovery project in a way that is not generally characteristic of inventories for Federal projects. Analysis of site data in the course of conducting the systematic surveys over the years has provided opportunities for critical evaluation of the operating procedures with respect to management and research goals which have been addressed in subsequent surveys. This has sometimes involved revisits to earlier recorded sites to raise documentation up to the level of other sites, to fill in the gaps between sites, or to reconcile inconsistencies between different site recorders. The acquisition of archaeological data has been a process of investing in higher and higher quality information rather than a set of discrete compliance events.

Most of the site attribute data recorded during the field surveys is now stored and maintained in a Dbase III computer file at Ft. Hood. Copies of these Dbase files have been sent to the Arkansas Archeological Survey where they were transferred to the INFORMIX relational database management system. Transferring the data from the multiple Dbase files into the larger and more capable INFORMIX system allows the easy integration of all data about each site into an accessible format.

The prehistoric site characteristics database files include observations on over 60 variables for each site (Table 21.1) and include basic site information (site number, area, project number), morphological information such as the presence of lithic scatter, burned rock mound, shelter, midden, etc., and the sample transect information on counts of specific tool types. The counts of diagnostic point types (i.e., Angostura, Bell, Bulverde, etc.) may include another ten or twelve variables for each site depending on the artifact density and the number of components at each site. For the sake of brevity we have not shown each point type that may be found in the Ft. Hood area other than the first four common ones (i.e., Angostura, Bell, Bulverde, Castroville), but most of the sites are multicomponent and contain a fairly wide range of diagnostic points.

The historic site characteristics database contains observations on 37 categories for each site (Table 21.2) including area, density of artifacts, date ranges for diagnostic artifacts, and presence of morphological features such as bricks, rootcellars, rubble, and other structural remains. The beginning, ending, and mean dates were computed from temporally diagnostic artifacts and structural remains documented on the site.

Table 21.1 Prehistoric site characteristics database

Site number	Total biface scrapers
Area of site	Total modified materials
Cultural affiliation	Total dart points
Project number	Total arrow points
Presence of lithics	Total blanks
Presence of burned rock scatter	Total retouched flakes
Presence of burned rock mound	Total retouched blades
Presence of rockshelter	Total sidescrapers
Presence of quarry	Total end scrapers
Presence of midden	Total gravers
Presence of cave	Total burins
Presence of mounds	Total other unifaces
Presence of rock art	Total cores
Presence of sinkhole	Total hammerstones
Presence of burial	Total choppers
Estimate of burned rock	Total one-sided manos
Total sample transects	Total two-sided manos
No. transects without burned rock	Total other manos
No. transects with low burned rock	Total one-sided metates
No. transects with moderate burned rock	Total two-sided metates
No. transects with high burned rock	Total other metates
Size of debitage sample transect	Total groundstone other (1)
Debitage count	Total groundstone other (2)
Total tools	Total groundstone other
Total ecofacts	Total miscellaneous tools
Total tools on transects	Total Angostura points
Total biface type 1	Total Bell points
Total biface type 2	Total Bulverde points
Total biface type 3	Total Castroville points
Total borer-perforator	Total (other diagnostic) points

Table 21.2 Historic site characteristics database

Site number	Concrete walk	Roofing material
Area	Corral	Rubble
Total artifacts	Concrete tank	Concrete slab
Beginning date	Dam	Paving blocks
Ending date	Depression	Stone wall
Mean date	Dip tank	Structure
Bricks	Dump	Stock tank
Bridge	Fence	Tiles
Rootcellar	Foundation	Trees
Cemetery	Other building	Trough
Chimney	Other feature	Well
Cistern	Piers	Water tank
Windmill		

The environmental database

The environmental data for Ft. Hood are accessed through the Geographical Resources Analysis Support System (GRASS), a comprehensive geographical information system (GIS) developed for the Army installations by the US Army Construction Engineering Research Laboratory (Westervelt, 1989). GRASS-GIS is

an integrated set of tools to manage land resources by providing means of inputting, storing, and manipulating data which are stored as categorical environmental
map layers consisting of spatially discrete cells like a checkerboard across the region
of interest. The GIS tools provided by GRASS permit digitizing, image processing,
data analysis such as neighbourhood, proximity, and logical analysis, map overlay,
map algebra, expert system analysis, and data presentation. These tools allow information derived from such diverse sources as satellite imagery, paper maps and data
generated by other computers to be brought together as input to land use questions
such as the locational patterns of archaeological sites. It is the capacity of GRASS
to process coordinate point information such as UTM site locations within each of
the map layers that makes GIS such an important tool in cultural resource
management.

Environmental data layers used in the site study consist of 31 map layers which
include most of the basic map layers in the Ft. Hood GRASS data set such as elevation, slope, aspect, vegetation, soils, geology, roads, and hydrology plus a number
of map layers derived from these basic map layers (Table 21.3). These include an
erosion status map, a tripartite land zone map, agricultural, grazing, and wildlife
capability maps, soil ph levels, pre-settlement (Range) vegetation, historic community
zones, and others. The GRASS proximity analysis tools have also been used to
generate maps expressing the distance of areas on the installation from drainages,
communities and roads. Other map layers constructed at Ft. Hood contain important management information such as military training areas, training functions, and
training frequency. Figure 21.1 illustrates the pre-settlement vegetation, a reclass of
soil types, as produced by GRASS. Figure 21.2 is the prehistoric site distribution
plotted against streams. The final step in setting up the data for site analysis involves
combining the site characteristics data with the GIS environmental data and
transporting both to the 'S' statistical software system.

Table 21.3 GRASS environmental data layers for archaeological sites

Site number	Modern vegetation
UTM east	Openland wildlife potential
UTM north	Rangeland productivity
Distance to roads	Range vegetation (pre-settlement)
Distance to communities	Agricultural limitations
Aspect	Grazing potential
Low order drainage basins	Cotton potential
Composite soils	Oats potential
Soil types	Sorghum potential
Elevation	Wheat potential
Erosion status	Low pH range of soil
Geology	Training intensity
Landzones	Large order drainage basins
Slope	Archaeological predictions
Project survey areas	Training functions
Training areas	Training frequency

Exploratory data analysis in S

The 'S' is a software system that was developed at Bell Laboratories (Becker and
Chambers, 1984). It is designed to provide facilities for exploratory data analysis
(EDA), an approach to data analysis pioneered by Tukey (1977) and others (Hoaglin

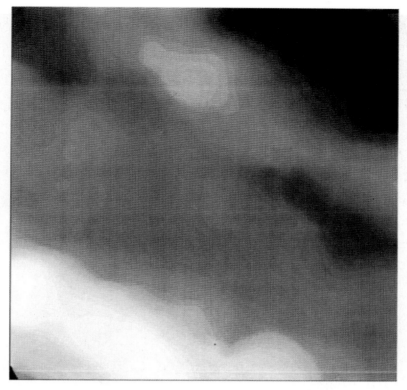

Plate II. Contoured surface of project area (ERDAS, Stine and Decker, Chapter 12)

Plate I. 512×512 subset of KEYLARGO.LAN. (ERDAS, Stine and Decker, Chapter 12)

Plate III. Three dimensional view northwest towards Key Largo island. (ERDAS, Stine and Decker, Chapter 12)

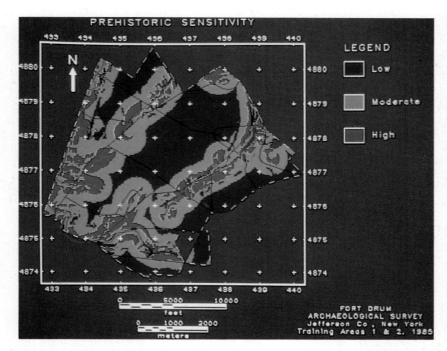

Plate IV. Prehistoric sensitivity of Fort Drum, New York project area. (Hasenstab and Resnick, Chapter 23)

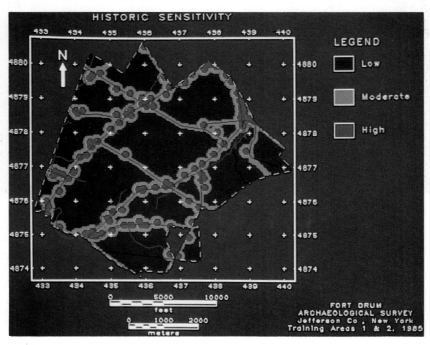

Plate V. Historic sensitivity of Fort Drum, New York project area. (Hasenstab and Resnick, Chapter 23)

Plates

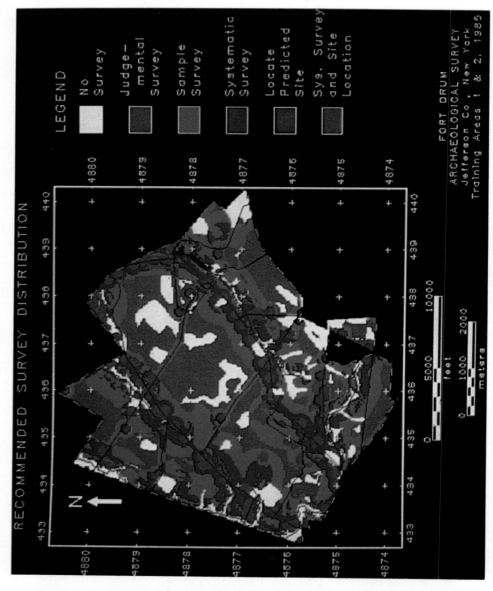

*Plate VI. Recommended archaeo-
logical survey distribution for the Fort
Drum New York project area.
(Hasenstab and Resnick, Chapter 23)*

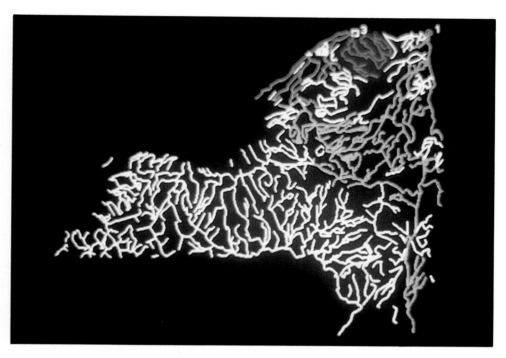

Plate VII. Hydrology of New York State. (ARC/INFO, Zubrow, Chapter 24)

Plate VIII. Hudson Valley Model: 17th century migration and transportation routes, New York State. Coloured lines are predictions of the migration model, white lines are waterways not taken. (ARC/INFO, Zubrow, Chapter 24)

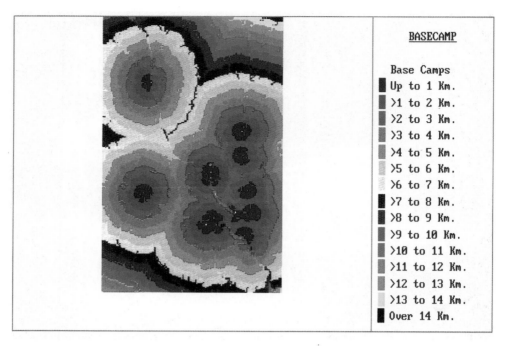

Plate IX. Distance from base camps, over rough terrain and through hydrology, Savannah River Valley, South Carolina. (MAPCGI2, Savage, Chapter 26)

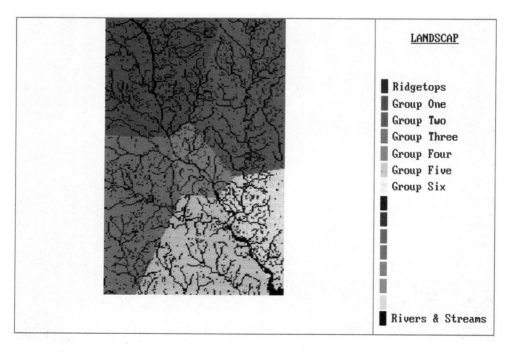

Plate X. Hypothesized hunter-gatherer habitual use areas, Savannah River Valley, South Carolina. (MAPCGI2, Savage, Chapter 26)

Plates

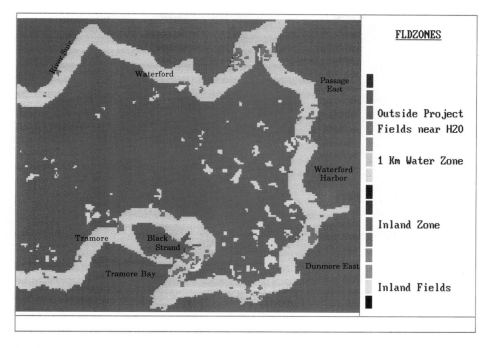

Plate XI. Distribution of prehistoric sites marking those within 1 km of the coast or river shore, Waterford, Ireland. (MAPCGI2, Green, Chapter 27)

Plate XII. GRASS display of enlarged SPOT image of the area around Autun, France, with ancient Roman roads radiating from the city overlaid in yellow. Forests are green, pasture fields pink, bare agricultural fields light blue, the city light blue. White slash at lower left is a jet contrail. (GRASS, Madry and Crumley, Chapter 28)

Plates

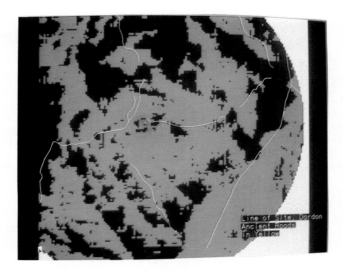

Plate XIII. Line-of-sight calculation from top of Mt. Dardon hillfort. Areas within sight are red, out of sight are black, ancient road system in yellow. Note that roads tend to follow paths within the line-of-sight. (GRASS, Madry and Crumley, Chapter 28)

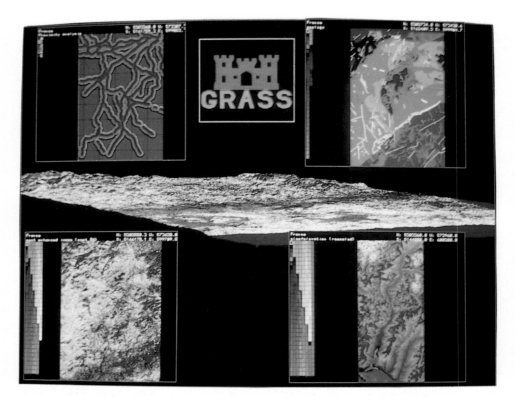

Plate XIV. 3-D perspective of the Arroux valley area, with SPOT satellite layer at lower left, distance to ancient roads at upper left, geology at upper right, and elevation at lower right (GRASS, Madry and Crumley, Chapter 28).

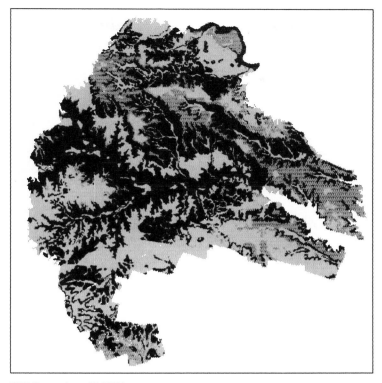

```
SCALE:     1 : 280957
                       3474750.00
WINDOW:   600300.00              643000.00
                       3430250.00
```

1 20% Oak-Savannah
2 Tall Grass Prairie
3 25 % Hardwood
4 Scattered Trees

Figure 21.1. Pre-settlement vegetation at Ft. Hood, a reclass of soil data showing mid-nineteenth century range vegetation communities.

et al., 1983) that utilizes iterative graphical modes of data screening and analysis. As defined by Tukey (1977) and others (Hartwig and Dearing, 1979), EDA is an approach that involves searching for patterning in a data set with the goal of gaining insights into the nature and causes of the data's total structure, particularly the unanticipated relationships that may occur. In contrast to the deductive method in which hypothesis testing is employed to confirm statistically a test implication or expected structure, EDA emphasizes flexible, open minded exploration to facilitate the discovery of unexpected patterns that suggest new solutions and problem areas. It is particularly effective in data sets with 'high dimensionality'; that is, data sets in which each observation has many data values associated with it. Archaeological site

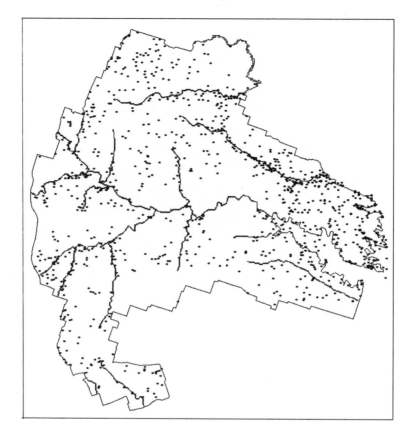

Figure 21.2. Prehistoric site distribution at Ft. Hood compared to stream locations.

data such as the Ft. Hood prehistoric sites with some 90 total data categories for each site are an example of such high dimensional data.

EDA involves iterative, stepwise examination and visual inspection as well as evaluation of both orthodox and alternative representations of data structure. In recognition of the multidimensional complexity of data analysis, EDA places emphasis on graphic representations of data, including display features such as colour-coded overlays of data and three-dimensional plot rotations of multivariate relationships, to bring the human brain's full visual processing capabilities into the gestalt of pattern recognition. Rather than using data to evaluate models as in the deductive confirmatory method, EDA employs multiple mathematical models and the re-expression of the data on various scales of measurement to explore and evaluate the nature of data structure from multiple perspectives. EDA does not preclude hypothesis testing since inductive exploration can serve as an entry-level step in a multistage investigation where the relevant relationships within the data structure are assessed prior to hypothesis formulation and deductive analysis. However, by not limiting pattern discovery to prior, sometimes poorly understood, notions of variable relationships formulated outside the data set, EDA offers great promise in providing an alternative to the constraints and rigidity of confirmatory hypothetico-deductive methods where complex processes must be viewed as simple cause and effect relationships.

Modules for exploratory data analysis

As part of the project, modules have been written in S in a menu form to allow statistical analysis of the combined site characteristics and GRASS environmental data sets (Parker, 1989). These S modules provide access to most standard exploratory data analysis tools organized by level of analysis in three primary menus: univariate, bivariate and multivariate (Table 21.4), along with other support menus. The univariate menu currently presents the user with six commonly used options. Notable among these options is the one comparing histograms of site locations and the histograms of all cells in the study area and a QQplot (quantile–quantile) comparing the structure of data from all site locations to all cells in the study area. The bivariate menu presents six options including Matplot (a type of bivariate plot), and Matplot-with-overlays which overlays observations from two or more data sets in different colours. Both Matplots provide options to identify and label observations by site number or other label which is very useful for identifying outliers and other interesting points.

The multivariate menu includes nine analysis procedures for treating three or more variables simultaneously (Table 21.4). The brush option which combines, on one screen, three dimensional rotation of data swarms, histograms, pairwise plots and site tables, and permits the identification and highlighting of individual sites simultaneously on all of these graphical displays, is one of the most useful S tools for the exploration of site variability. The cluster analysis module includes four options for obtaining the distance metric (euclidean, maximum, manhattan and binary) and three options for the clustering method (average, single and complete linkage). The principal components analysis includes pairwise plots of all components generated, the factor matrix and other standard output, plus an option of choosing two components from principal components analysis for interactive identification and labelling of observations. The modules in all three menus above automatically provide titles that identify the data set and data layers utilized and, where appropriate, allow for the interactive placement of additional captions and legend.

Table 21.4 S-GRASS statistical modules

Univariate	Bivariate	Multivariate
QQnorm plot	Matplot	Star plot
General stats	Matplot w/overlays	Star plot label
Site data histogram	Bivariate scatterplot	Chernoff's faces
Cell data histogram	Regression overview	Brush 3-D spin
Histogram comparison	Jittered regression	Pairwise scatterplots
QQplot of sites to cells		Cluster analysis
		Distance options:
		Max, manhattan, binary and euclidean
		Cluster options:
		Average, single, complete
		Principal components
		PCA plots
		Discriminant function analysis

The menu presentation of the S modules serves to organize the steps involved in choosing a site data set, specific GRASS and site characteristics variables for each data set, an analysis method, specific analysis options, and interaction with the graphical output in terms of a series of questions that the user answers at the keyboard. We have included below an example from one session with the multivariate menu to demonstrate how the menu version of S works. The session begins when the user types the command '?M' which initiates the macro for the multivariate modules (Figure 21.3).

As this demonstration session illustrates, complicated analyses can be conducted by the user without detailed knowledge of S programming language at all. However, for those who are interested, a number of other data management and analytical tools in S are available for use outside the menu. These include functions that sort, subset, append, remove, round off and transpose data values in arrays and matrices, perform arithmetic (adding, subtracting, multiplying, etc.) and Boolean operations, and those that perform mathematical and trigonometric functions on data sets. Other functions and S macros compute Chi-square and other probability distributions, variance, covariance, correlation values, contingency tables, produce pie charts, two-way plots, stem and leaf displays, contour plots, three-dimensional perspective plots, etc. A particularly helpful set of macros used for documenting analysis during the Ft. Hood study are diary, sink and stamp. Diary stores S commands used in a session in a file while sink routes the output from the session to another file which is stamped with the current date and time and any other optional string comments written by the user about the output (Becker and Chambers, 1984).

Pilot study in historic site analysis using GRASS and S

In the following sections, we will illustrate how such tools can be used in the study and management of the Ft. Hood sites. For the purposes of this paper, we have focused on the historic archaeological resources and one aspect of the analysis dealing predominantly with site characteristics data and agricultural land capability estimates. The following serves to illustrate, by example, the premises of the approach. This report of analysis results is not a comprehensive treatment—this can be found in the complete report (Williams *et al.*, 1989).

The historic archaeological resources at Ft. Hood consist of 1,165 sites ranging in date of occupation from the mid-nineteenth century to the middle of the twentieth century. These sites consist predominantly of architectural and archaeological remains of communities of farmers and ranchers who settled the region in the late nineteenth and early twentieth century. Of these, 790 sites, documented thoroughly enough to permit assessment of age, function, and integrity, are the subject of the following analysis with GRASS and S. As noted above, data recorded in the field on these sites include information on the extent, age and composition of archaeological deposits, and the presence of structural remains of houses, farm buildings, and other features such as enclosures, troughs, dip tanks, windmills, etc.

Datable archaeological material recovered during the field investigation of the sites has provided approximate occupation ranges for the historic sites. The occupation span of Ft. Hood ranges from initial settlement in the mid-nineteenth century to 1954 when the present configuration of the installation was established. This 100 year span has been divided into five general periods from initial settlement to the establishment of the base: Mid-nineteenth century (prior to 1880), Late nineteenth century (1880–1900), Early twentieth century (1900–1920), Depression Era (1920–1940), and Mid-twentieth century (1940–1954).

```
> ?M

Data layers available are:

       "Distance to Drainages"      "Distance to Communities"
[ 3]   "Distance to Roads"          "Aspect"             "Low Order Basins"
[ 6]   "Composit Soils"             "Elevation"          "Erosion Status"      "Geology"
[10]   "Land Zones"                 "Slope"              "Soils"               "Survey Areas"
[14]   "Training Areas"             "Training Frequency"                       "Vegetation"
[17]   "Openland Wildlife"          "Range Productivity"
[19]   "Range Wildlife"             "Range Vegetation"
[21]   "Agricultural Productivity"  "Grazing Potential"
[23]   "Cotton Potential"           "Oats Potential"     "Sorghum Potential"
[26]   "Wheat Potential"            "Low Ph Range"
[28]   "Training Intensity"         "Large Order Basins"
[30]   "Archeological Predictions"                       "Training Functions"

Multivariate Analysis Options Are:
1 - Stars?
2 - Star Labels?
3 - Faces?
4 - Pairs?
5 - Correlations?
6 - Brush?
7 - Cluster?
8 - Principal Components?
9 - Principal Component Plots?
10 - Discriminant Function Analysis?
11 - Exit from menu
Which one?  7
Lowest number to cluster ( >= 1):  1
Highest number to cluster ( <= maximum):  50
Enter data layers to be used to cluster sites
Use control-d when all data layers are entered
1:  1
2:  2
3:  3
4:  4
5:  5
6:  6
7:  7
8:  8
9:  9
10:  10
11:  11
12:  12
You have chosen the following data layers
       "Distance to Drainages"      "Distance to Communities"
[ 3]   "Distance to Roads"          "Aspect"             "Low Order Basins"
[ 6]   "Composit Soils"             "Elevation"          "Erosion Status"      "Geology"
[10]   "Land Zones"                 "Slope"              "Soils"
Choose the distance metric to be used for clustering

1 - Euclidean
2 - Maximum
3 - Manhattan
4 - Binary
Which one?  1
Choose the clustering method desired

1 - Average linkage
2 - Single linkage
3 - Complete linkage
Which one?  3
```

Figure 21.3. Demonstration session of the multivariate menu part of S-GRASS.

In addition to the information recorded in the field at each of these historic sites, oral histories and limited cartographic and archival data have been collected on the community levels of social and economic organization during the historic period. This has permitted the tentative identification of the approximate centres and date ranges of 53 dispersed rural communities that existed in the late nineteenth and twentieth centuries in and around Ft. Hood. This community information has been encoded into the INFORMIX database and has been included as a GRASS data layer for analysis with the historic archaeological sites. Figure 21.4 is a GRASS display illustrating the settlement pattern during the early twentieth century, the period of peak historic occupation, showing the distribution of all historic sites overlaid on the map of historic communities. The vacant zones in the centre and southern portion of the map are the artillery range and cantonment, respectively, where no survey has been conducted. In the remaining analyses, this area where no survey has been conducted has been 'masked out'. Using tools in the GRASS system, environmental data from this area has been excluded from analysis so as not to bias the results. In general, the distribution of sites and communities appears fairly even over the installation except perhaps for a recognizable linear alignment reflecting the close orientation of sites to roads. The distribution of communities appears to reflect a degree of equal spacing with a few exceptions where close community centres may be a product of collapsing communities from different time periods. Since community centres are spaced approximately 3 to 6 km apart (Figure 21.4), it would seem logical that sites might be associated in community residential groups of about that diameter assuming that sites are affiliated with the nearest community centre. While this may be so, the actual association of the sites to these community centres is not a clear cut division of the sites into clusters. Instead, the community residential groups of sites would appear to be closely packed with sites on the periphery of community centres closely overlapping with neighbouring community centres.

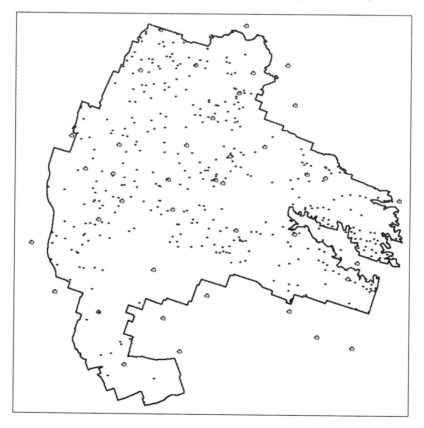

Figure 21.4. Historic settlement during the period 1900–1920 with historic sites marked as black boxes and community centres marked with circular icons.

Analysis of site characteristics data for communities

The preliminary oral history at Ft. Hood and the general pattern observed throughout the southeast is that families consider themselves residents of a community region surrounding a cluster of local commercial and public services which serve the local populace (Briuer, personal communication). As a general rule, households consider themselves members of the closest community centre that can be reached by road. Some exceptions to this pattern undoubtedly existed as a result of transportation networks, kin and other social influences, church and school affiliations, and the uneven distribution among communities of commercial services such as hardware stores, mills, blacksmith shops, post offices and farm produce wholesalers. It is probable that households may have counted themselves as residents of a local community yet may at times have frequented more distant, higher order communities out of necessity to gain access to services not available locally. For instance, strategic access to the railroad transportation network made communities like Killen important hubs for the collection of commercial crops like cotton.

Thus, it is useful to distinguish between the residential area of a community, the community centre where services (the church, post office, school and store) are clustered, and the various orders of communities or towns which households may patronize. The development of hierarchical orders of hamlets, communities and towns (Newton, 1967) is a complex social and economic process that is not yet understood well enough to include as a component of the historic site analysis at Ft. Hood. For purposes of the Ft. Hood study, analysis will be confined to exploring the residential bases of the local communities where the expected close spatial correlation of households of a community to the nearest community centres can be feasibly addressed as a GIS problem.

Figure 21.5 is a plot of the communities with the sites of each community of interest derived in GRASS shown by unique plotting symbols (icons). For this report we will focus on three of the communities shown on the map of training areas: Antelope, Sparta, and Eliga. The site density and size of the communities range between 20 to 30 sites distributed over an area 2 to 3 km in diameter. A series of communities for the different historic time periods has been derived and some comparative analyses run for the different geographic areas covered by Ft. Hood. A detailed report of these studies is included in Williams *et al.* (1989). At this time, we will focus analysis in S on all historic sites and the three communities dating to the early historic period (1900–1920) shown in Figure 21.5.

Site analysis in S

In S the site characteristics data collected during fieldwork and the environmental information derived from GRASS are combined for exploratory analysis and the generation of site clusters. S provides a number of analytical tools for exploring the variability of archaeological sites. Some of the most useful for classifying sites are the multivariate display (faces and stars) and clustering tools (cluster analysis and principal components analysis). These two tools can be used together to explore the nature and composition of historic sites with respect to the presence/absence of both categorical and continuous variables.

The star plot function is particularly useful for displaying the presence/absence of variables relating to structural features on sites such as chimneys, foundations, piers, roofing material, wells, windmills, water tanks, troughs, and the like. The star display works by plotting variables as rays from a central point with the present state designated by a ray and the absent state shown with no ray. For variables in which the data are expressed as continuous values rather than presence or absence, the length of the ray approximates the value for each site. In this example plot of the

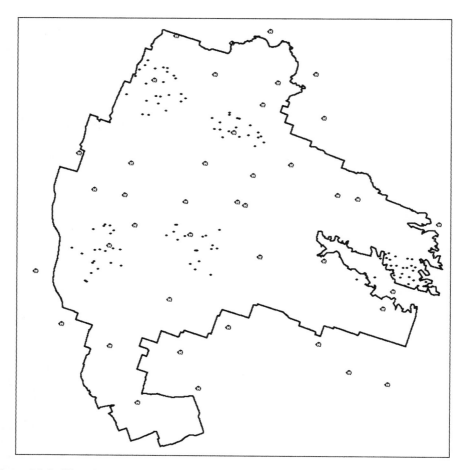

Figure 21.5. Historic sites associated with selected community centres.

architectural features from site 1511 from the Antelope community (Figure 21.6), seven out of eight of the major architectural features (bricks, roofing material, piers, foundation remains, a cistern, and a chimney) are present. The site does not include evidence of a rootcellar so the star plot does not show a ray in this position. The different configurations of presence and absence of categorical variables results in patterns of stars for each site that can be compared visually. By running stars on a group of historic components (Figure 21.7), the set of components can be examined to reveal the degree of similarity between sites, a task made easier by visually comparing the various star patterns generated. This particular group of sites is from the Antelope community. The procedure used for the derivation of the community is discussed later in this report.

   Star plots of the sites from the Antelope community (Figure 21.7) show a wide range of star patterns that illustrate the relative complexity of these sites with respect to the presence of these eight architectural features. Close inspection of the 21 star plots reveals a number of apparent trends in the complexity of sites. A basic core of architectural features shared by all sites consists of the presence of a chimney, bricks, roofing material and piers. Sites 1194, 1198, 1506, 1508, 1543, 1555, 1586 and 1955 possess only this basic core of architectural traits. The next most complex pattern includes the addition of an architectural feature to the basic core including foundation material (site 1497) and rubble (sites 1205, 1509, 1520, 1547, 1548, 1561, 1596 and 1601). The other star plots represent increasingly elaborate variations of the star configuration as a result of additions to the basic pattern such as the presence at site 1209 of the basic core plus a rootcellar and rubble.

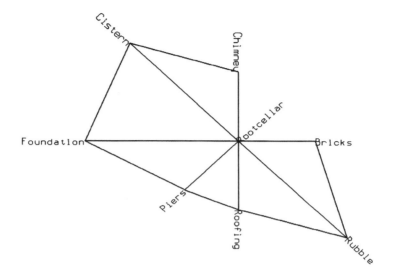

Figure 21.6. Star plot of Antelope community site 1511 with presence of variables (clockwise from upper vertical) rootcellar, bricks, rubble, roofing, piers, foundation, cistern and chimney.

While the star plots suggest some general tendencies in the combinations of structural features on historic sites, a cluster analysis using the same categorical presence/absence variables produces groups of sites that reflect the star configurations. Figure 21.8 is a dendrogram produced by the cluster analysis of the structural feature data from the Antelope community sites. The cluster option used was complete linkage using euclidean distances. A comparison of the dendrogram with the star plots (Figures 21.7 and 21.8) shows that most of the sites (16 out of 21) fall into two basic clusters which fall at the lower right of the dendrogram. One cluster made up of sites 1194, 1198, 1506, 1508, 1543, 1555, 1586 and 1955 represents the simplest set of present structural features (chimney, bricks, roofing and piers). The second cluster made up of sites 1205, 1509, 1520, 1547, 1548, 1561, 1596 and 1601 includes all of the structural features that the first cluster had with the addition of rubble and foundation material. A third small cluster consisting of only sites 1197 and 1202 is identical to the second cluster with the addition of foundation material while site 1209, an outlier, is also identical to the second cluster but with the addition of a rootcellar.

Thus, by moving back and forth between the star plots and the cluster analysis, it is possible both to define groups of sites using hierarchical clustering techniques while at the same time using the star plots to monitor the rules of group membership. The iterative capabilities of S also allow this star plot/cluster analysis screening of categorical variables to be run over and over again with different combinations of variables that change the criteria for the definition of clusters.

In addition to the site characteristics data analyzed above, information concern-

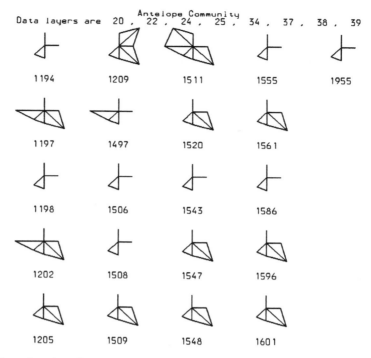

Figure 21.7. Star plots for all Antelope community sites with variables (clockwise from upper vertical) rootcellar, bricks, rubble, roofing, piers, foundation, cistern and chimney.

ing the environmental context of site settlement such as aspect, slope, distance to roads, distance to streams and land capability are important variables that can be factored into the process of site classification. As discussed earlier, one important aspect of land use is the dimension of land productivity that serves as the economic base of the inhabitants of Ft. Hood; that is, the capability of the land to produce subsistence crops, commercial crops or forage for grazing animals.

A plot of cotton potential and oats potential shows that the Ft. Hood sites are unevenly distributed with respect to the potential yields of these two crops (Figure 21.9). The sites fall into several clusters arranged in a positive linear pattern suggesting that both cotton and oats potentials are highly correlated across Ft. Hood and exhibit a wide range of potential yields on different soils. The plots for other crops such as wheat and sorghum are also positively correlated with each other, and with cotton, to suggest a general pattern of land capability consisting of parcels of land with varying degrees of farming quality from very poor potential to relatively high. The fact that the areas that rank the highest in terms of commercial cash crops like cotton and wheat also rank fairly low with respect to limitations on their use is another important measure of varying farmland quality. From the results of the environmental assessment of land use, we can identify sites in the low productivity class that are situated on range land, sites in the middle range of land productivity situated in the pasture/range zone, and sites situated in the most productive land where cropland/pasture land use may have occurred.

The utility of these rough measures of agricultural productivity for gaining insights into land use during the historic period of occupation bears some further discussion. The use of land capability estimates derived from Soil Conservation

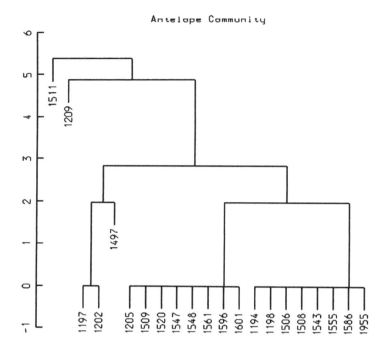

Figure 21.8. Dendrogram of cluster analysis of all Antelope community sites with the variables rootcellar, bricks, rubble, roofing, piers, foundation, cistern and chimney.

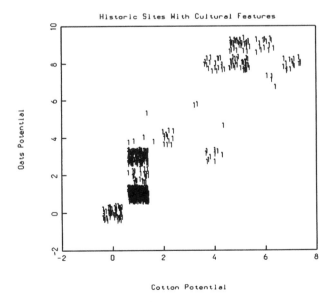

Figure 21.9. Matplot of cotton and oats potential for all historic site locations.

Service data is useful in land use studies because it permits some level of quantification in comparisons of one land parcel to another or one farmstead to another. However, there are certainly limitations in their use in predicting the mix of agricultural crops that a region may pursue since the influence of such factors as farming traditions, levels of technology and the marketplace in determining the development of agriculture are important factors in the makeup of a region. Farmers were, in most instances, well aware of the agricultural capability of their land particularly in comparison to their neighbours' parcels and did not make decisions in a vacuum. While they might be obliged to follow a course dictated by the high demand and price of cotton, for instance, the size of the yield per acre and relative success of one farmer compared to his neighbour is going to be vastly different across Ft. Hood. In an economic sense, land is capital to be invested and if the distribution of capital is unevenly distributed in area and quality, then we can expect corresponding differences in the makeup of the sites that were occupied. The intent of the present study then is to use the land capability measures to get at the basic underlying dimension of land use potential in terms of the value of the resource base at the local and regional level and to establish a basic reference point for understanding the range of options available to the inhabitants of the area.

One point that should be mentioned is the potential bias involved in estimating land potential based only on figures for the land occupied by the site rather than considering the larger parcel held and farmed by the resident of the site. This dilemma endemic to historic site studies is analogous to the tricky problem of defining the catchment area utilized by Native American groups residing at prehistoric sites. We are on somewhat more solid ground in defining use areas on historic sites since landholdings can be derived from deed records, from oral history, from evidence of property lines in the field, or by consulting records of average farm size in government sources. Such a database on historic land holdings is not available for the Ft. Hood area, but this question will be addressed using GRASS tools that allow for neighbourhood analysis of a defined area surrounding site locations.

For instance, a site occurrence report on cotton potential can be run using the site locations (Table 21.5), the site locations plus the 8 cells surrounding the site, or the site location plus the 24 cells surrounding the site, and so on. The site occurrence report indicates the occurrence of each class of the environmental variable of interest (here cotton potential) over all cells in the study area, the percentage occurrence of each class, the distribution of the sites (here the historic sites), if they were distributed proportionally in each environmental class, the actual site distribution, and a Chi-square and its associated degrees of freedom. Note that the repeated use of a Chi-square measure should be adjusted for the repeated trials nature of the test but the measure as presented can be viewed as a useful relative indicator of the significance of the deviation between expected and actual site frequencies. Running a site occurrence report on the site location, a cell 100 metres square, amounts to an area of 2.5 acres. If the 8 cells surrounding the site are included, the site area expands to a block of 9 cells or 22.24 acres, and including the 16 cells surrounding that block defines a site area of 61 acres. Comparing the Chi-square results from the three site occurrence reports shows that the area defined around the site does make a significant difference in the outcome of the tests with the Chi-squares lessening in size, indicating less unexpected site occurrence in the various categories of cotton yields. By increasing the area of historic sites from 2.5 acres to 22.24 acres the one significant Chi-square test of 48.163 (Table 21.5), that for category 2 (200 lb/acre), is nearly halved to 26.629. If historic sites are assumed to encompass an area of approximately 61 acres, then the land parcels surrounding the sites are found to be distributed fairly randomly across the landscape with respect to cotton potential since there are no high Chi-squares. Obviously, the definition of the unit of analysis makes considerable difference in the outcome of our tests of site association to dimensions of the environment. While we cannot hope to replicate completely the complex nature of historic landholdings in the historic sites analysis, we can use the GRASS tools, as demonstrated above, to provide a check on the reliability of our site groupings with respect to environmental variables such as land capability.

Table 21.5. Site occurrence report of cotton potential for all historic sites using the site location cell option.

Site characteristics	Cells cover	% cover	Expected sites	Actual sites	Chi-square	Degrees of freedom
(0)	121985			96		
(1) 0 lb/acre	55672	81.8	794.6	759	1.596	1
(2) 200 lb/acre	258	0.4	3.7	17	48.163	1
(3) 250 lb/acre	518	0.8	7.4	3	2.611	1
(4) 350 lb/acre	3566	5.2	50.9	45	0.683	1
(5) 400 lb/acre	5187	7.6	74.0	93	4.858	1
(6) 450 lb/acre	1216	1.8	17.4	21	0.765	1
(7) 550 lb/acre	1613	2.4	23.0	33	4.324	1
Totals	68030	100.0	971.0	971	63.000	6

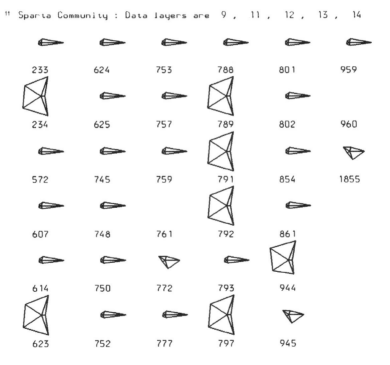

Figure 21.10. *Star plots of all Sparta community sites for variables agricultural productivity, wheat potential, sorghum potential, oats potential and cotton potential.*

A star plot (Figure 21.10) of five variables measuring land capability (agricultural productivity, wheat potential, sorghum potential, oats potential and cotton potential) for the Sparta community sites illustrates the range of site variability with respect to farmland capability. Three basic star configurations are apparent. The first group is typified by sites 233, 572, 607 and others marked by low potential yields for cotton, oats, sorghum and wheat and high restrictions on their use for farming. The second group is characterized by a star configuration much like group one, and is marked by low to moderate scores in all categories of land capability and includes sites 772, 945 and 1855. The third group includes sites 234, 623, 789 and others characterized by high potential in cotton, oats, sorghum and wheat, and low restrictions on agricultural capability. The group defined by sites 761 and 960 is located on soil types that are so low in land capability that they are not rated by the US Soil Conservation Service (SCS).

A cluster analysis of the above farmland capability variables replicates the four basic categories of land types noted above. For all practical purposes then the Sparta community sites appear to fall into four broad categories with respect to land capability: (1) locations ideal for cropland and improved pasture use, (2) areas with moderate suitability for cropland that would fall into the pasture/range land use zone, (3) locations that are poorly suited for farming and probably utilized as rangeland, and (4) areas that are prohibitively unsuitable for farming, and most certainly used for rangeland.

Further exploratory analysis of environmental data

While the community site analysis has revealed some important trends in the association of historic sites, it left unresolved some aspects of site patterning that involve a more complex process of data screening and exploratory data analysis. In order to deal adequately with the complex nature of the Ft. Hood environmental and site characteristics data, the analysis proceeded in an interactive fashion with considerable input back and forth between the univariate, bivariate and multivariate dimensions in an attempt to isolate the relevant factors influencing site location and composition. In this section, we will describe in greater detail how the GRASS and S tools are being used to accomplish these objectives for some of the proximity data, topographical data and agricultural land capability data.

Site proximity analysis

In order to look closer at the spatial structure of sites and community centres, a proximity analysis was run in GRASS which in effect partitions the data according to defined distance measures (0 to 100 m, 100 to 200 m, and so on) from some mapped reference point or feature of the landscape such as a stream. With this done, it is then possible to run a site occurrence report to assess the density of sites at various distances and test the significance of any trends. In the Ft. Hood analysis we considered site distance with respect to community centres, roads and drainages, basic features of the cultural and natural landscape with which settlement is often correlated. Using these distance maps and the site occurrence reports we were quickly able to assess the site data to determine if these sites were evenly distributed across the range of locational options available in the study area, or if there were indications of selective use of particular configurations such as slope, aspect, or proximity to drainages. This relationship between the sites and the installation as a whole can be examined visually with histograms and QQplots to compare the shape and ranges of the two distributions and statistically with Chi-square tests.

In the following discussion we will present only the results for the proximity to drainage portion of the analysis. A review of histograms comparing distance of sites

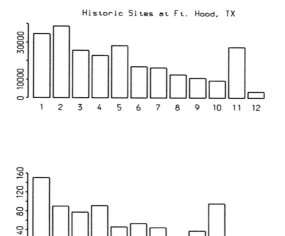

Figure 21.11. Histograms of distances to drainages for all sites and all cells.

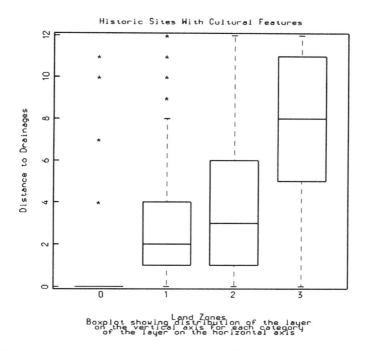

Figure 21.12. Boxplot of distance to drainages for all historic sites in the three landzones.

to drainages to those for all cell values at Ft. Hood provides some insight into whether site locations conform or depart from the range of variation present in general at Ft. Hood. The histograms for distances to drainages (Figure 21.11) displays bimodal trends for both sites and cells. However, the bimodal pattern for site distance to drainages does not indicate strong clustering of sites away from streams since the modal trends may be merely a reflection of the greater number of cells and therefore a higher chance for sites to fall in these categories. A boxplot of site distance from drainages against landzones (Figure 21.12) shows that the bimodality evident from the histogram of distance to drainages can be understood in terms of landform morphology and the different hydrological characteristics between lowlands, uplands and intermediate uplands. Note that in the boxplot there is a separate box for sites in each environmental zone; the dimensions of each box visually display the range of the distance to drainage for sites in the specified land-zone, with the top and bottom of the box defining the upper and lower quartiles, the horizontal line through the box the median, and the vertical lines that go up and down from the box define the extremes of the data. As the boxplot shows, the lowlands zone (category 1), as might be expected, exhibits the shortest site distances to drainages, followed by category 2 (the intermediate uplands), with the uplands having the greatest site distances from drainages in the installation. It is clear that the high frequency of sites between 500 to 1000 m from drainages as shown in the Figure 21.11 histogram can be accounted for almost entirely by the sites in the uplands, while the short distances to drainages are due almost entirely to sites in the lowlands and intermediate uplands. A site occurrence report from GRASS which includes a Chi-square test of site proximity to drainages (Table 21.6) shows a greater than expected frequency of sites within 50 m of drainages (category 1) with Chi-square value of 10.86 and a lower than expected frequency at locations between 250 and 300 m (category 6) where a Chi-square of 8.18 was computed.

Table 21.6. Site occurrence report of distance to drainage for all historic sites.

Site characteristics	Cells cover	% cover	Expected sites	Actual sites	Chi-square	Degrees of freedom
(0) Location from which distant	128417			202		
(1) 0.00 to 50.00 m	8787	14.3	123.4	160	10.860	1
(2) 50.00 to 100.00 m	9537	15.5	133.9	134	0.000	1
(3) 100.00 to 150.00 m	6326	10.3	88.8	86	0.090	1
(4) 150.00 to 200.00 m	5855	9.5	82.2	91	0.938	1
(5) 200.00 to 250.00 m	7030	11.4	98.7	92	0.457	1
(6) 250.00 to 300.00 m	4111	6.7	57.7	36	8.179	1
(7) 300.00 to 350.00 m	4135	6.7	58.1	56	0.074	1
(8) 350.00 to 400.00 m	3181	5.2	44.7	40	0.488	1
(9) 400.00 to 450.00 m	2665	4.3	37.4	35	0.157	1
(10) 450.00 to 500.00 m	2298	3.7	32.3	27	0.861	1
(11) 500.00 to 1000.00 m	6866	11.1	96.4	93	0.121	1
(12) Greater than 1000 m	807	1.3	11.3	15	1.187	1
Totals	61598	100.0	865.0	865	23.412	11

Looking at all the data there were at least indications of subtle clustering of historic sites to within 3 km of community centres, of sites to within 50 m of drainages, and of sites to within 200 m of roads.

While the pattern of site proximity to community centres, roads and drainages appears subtle at best, we can examine the proximity measures further to assess whether there are any combinations of distances to pairs of either roads and communities, communities and drainages, or roads and drainages. Such bivariate patterns might more closely reflect the historic settlement behaviour at Ft. Hood than did the univariate assessment of the three variables individually. To get a quick visual sense of the interaction of these proximity measures, bivariate plots were created of site distance from communities against site distance from drainages (Figure 21.13), site distance from communities against site distance from roads, and site distance from roads against site distance from drainages. In general, these three plots showed a fairly even distribution with respect to these three variables with no clear discrete grouping of sites at any particular combination of distances, at least not for all historic sites considered together. The cluster of sites in the lower left of the three plots at coordinate points 0,0, are simply sites for which no proximity data are available and do not represent real patterning with respect to distance.

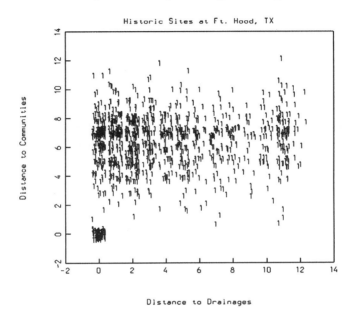

Figure 21.13. Matplot of distance to drainages and distance to communities for all historic sites.

Analysis of topographical data on site location

A number of other tests of the association of sites and communities were run on the GRASS environmental variables, but only a couple of examples will be reported here. To determine whether sites and communities exhibit a preference for any particular landform type, a site occurrence report was run. This report showed that slightly less than half of the sites fall in the intermediate uplands, about 30 per cent occur in the lowlands, and about 20 per cent are located in the uplands. A Chi-square test of site location and the three land zones shows a very high incidence of sites in the lowlands, 246 would be expected by chance while 331 were reported.

A boxplot of the beginning date of all sites for each of the three landzone categories shows that the period of initial site occupation is comparable across the lowlands, intermediate uplands and uplands (Figure 21.14a). A plot of the ending dates of sites (Figure 21.14b) shows a similar pattern of occupation termination between the three landzones. It is also interesting that the beginning dates of site

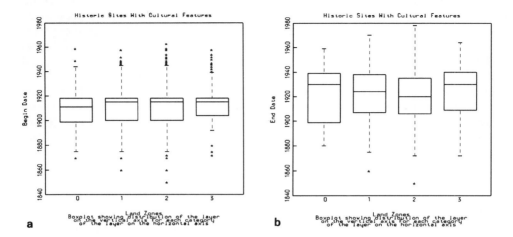

Figure 21.14.(a) Boxplot of beginning dates for all historic sites in the three landzones; (b) Boxplot of ending dates for all historic sites in the three landzones.

occupation are much more narrowly constricted to a twenty year interval (1900–1920) than are the termination dates which span 30 to 35 years between 1900 and 1940.

Agricultural land use

One of the most important aspects of the land to the historic settlers in the Ft. Hood area was the potential of the soil to sustain agricultural production including both farming and grazing. Since the quality of the land varies considerably across Ft. Hood, this dimension of the landscape probably played a major role in the development of the land use and site settlement patterns that emerged. Certainly, the early settlers employed some criteria based perhaps on written and oral accounts, past experience and trial and error and were able, after a time, to adapt to the constraints of the area.

Land quality can be assessed using modern measures developed by the SCS using potential crop yields, estimates of grazing production and agricultural limitations to generate some relative ranking of the agricultural value of the land. The variable Range is the natural annual production of plant growth (leaves, twigs and fruits) for the range type of pre-settlement vegetation in annual pounds per acre under normal conditions. The variables Cotton, Wheat, Oats and Sorghum are measures of potential yields in bushels or pounds per acre. Land cap. is the SCS land capability class which measures the suitability of soils for most kinds of field crops according to limitations, risk of damage such as erosion and response to management. The variable Graze is a measure of the capacity of an acre of the soil to sustain animal foraging expressed in animal unit months (AUM) or the amount of forage or feed required to feed one animal unit (one cow, one horse, one mule, five sheep, or five goats) for 30 days (McCaleb, 1985). In Table 21.7, seven means of measuring land capability derived by the SCS have been used to rank the 24 soil types found at Ft. Hood in ascending order of suitability from high (1) to low (7, 8, 9, etc.).

A number of trends are evident. In general, very good soils like Bosque, Krum and Lewisville rank high in almost every category and poor soils like Cho, Eckrant

Table 21.7. The agricultural capability of soil types ranked for production and suitability. A value of 1 indicates highest capability.

Soil type	Range	Cotton	Wheat	Oats	Sorghum	Land cap.	Graze
Bastil	4	4	1	3	3	2	2
Bolar	3	—	2	6	7	3	5
Bosque	3	1	2	2	3	1	2
Brackett-Topsey	5	—	5	7	—	4	6
Cho	9	—	6	8	—	4	9
Cisco	5	4	4	7	7	3	4
Crawford	3	4	3	5	4	3	3
Denton	3	3	1	2	2	2	2
Doss-Real	7	—	—	—	—	6	9
Eckrant	8	—	—	—	—	7	—
Evant	6	—	—	—	—	3	8
Frio	5	2	2	3	2	2	2
Houston-Black	1	2	3	3	6	2	1
Krum	1	3	3	1	3	2	2
Lewisville	2	2	2	1	1	2	1
Lindy	3	5	3	4	7	2	4
Minwells	5	—	2	5	5	2	3
Nuff	4	—	—	—	—	6	—
Purves	8	—	5	6	8	4	7
Real Rock	8	—	—	—	—	7	—
Seawillow	5	6	—	6	7	3	5
Slidell	3	4	3	2	3	2	2
Topsey-Pidcoke	5	—	5	—	—	4	8
Wise	4	—	—	—	—	4	6

and Real Rock rank low in every measure of land capability. Upon closer examination, it is also apparent, however, that some soils rank high in some variables but low in others. For example, Bolar ranks high in wheat but low in oats and Houston-Black ranks low in sorghum but high in grazing (Table 21.7). There are also numerous cases of soils that do not rank high in anything but do rank midway for all crops such as Crawford, Lindy and Slidell. To get a clearer idea of the variation in land capability of the soil types beyond the subtleties of soil rankings, it is necessary to examine and compare the actual yields estimated by the SCS.

The yields for these soil types are shown in Table 21.8 along with the range zone (RZ), the pre-settlement vegetation type supported by the soil type, the range wildlife rating (RW) and openland wildlife rating (OW). The variables from Table 21.7 have been abbreviated in Table 21.8, and the soil types are presented with their SCS designation (e.g. BaB for Bastil fine sandy loam). The consecutive numbers in the column under 'No.' are labels assigned to each soil type to serve as an identifier in graphical analysis plots from the S statistical system.

The bivariate patterns exhibited by cotton, oats and grazing capacity seem to illustrate the basic structure of the soil data. As noted in Table 21.8, some of the soil types show high to moderate scores for a number of yield estimates, as indicated by the cluster of observations in the upper right-hand corners of the plots, and other soil types rank high to moderate on one variable and low or 0 on the other. For example, the plot of cotton and oats showed a cluster of seven cases with 0 yields for both cotton and oats, a cluster of five cases with 0 yields for cotton and low to moderate yields for oats (between 20 and 50 on oats scale), two cases rating moderate for both cotton and oats (the middle of the plot), and about 10 soil types

Table 21.8. Yield estimates, wildlife capability and range zones for the Fort Hood soil types.

No.	Soil	Rng	Ctn	Wht	Ots	Srgm	Grz	LC§	RZ§	RW§	OW§
01	BaB	4500	350	40	60	70	7.0	2	1	1	1
02	BgB	5000	0	35	40	40	5.0	3	2	2	2
03	Bo	5000	550	35	65	70	7.0	1	3	1	1
04	Btc2	4000	0	20	30	0	4.0	4	4	2	2
05	Chb	2000	0	10	20	0	2.0	4	5	3	3
06	CoB2	4000	350	25	30	40	5.5	3	1	1	1
07	CwB	5000	350	30	50	65	6.0	3	6	2	2
08	DeB	5000	400	40	65	75	7.0	2	2	2	1
09	DrC	2800	0	0	0	0	2.0	6	4	2	2
10	EcB	2500	0	0	0	0	0.0	7	7	2	2
11	EvB	3500	0	0	0	0	3.0	3	6	2	2
12	Fr	4000	450	35	60	75	7.0	2	3	2	2
13	HoA	6000	450	30	60	45	7.5	2	8	2	1
14	KrB	6000	400	30	70	70	7.0	2	2	2	1
15	LeB	5500	450	35	70	80	7.5	2	2	2	1
16	LyB	5000	250	30	55	40	5.5	2	6	2	1
17	MnB	4000	0	35	50	60	6.0	2	1	1	1
18	NuC	4500	0	0	0	0	0.0	6	2	2	2
19	PrB	2500	0	20	40	25	3.5	4	5	3	2
20	ReF	2200	0	0	0	0	0.0	7	4	3	3
21	SeC	4000	200	0	40	40	5.0	3	2	2	2
22	SlB	5000	350	30	65	70	7.0	2	8	2	1
23	TpC	4000	0	20	0	0	3.0	4	2	2	2
24	WsC2	4500	0	0	0	0	4.0	4	2	2	2

§ LC = land capability, RZ = range zone, RW = range wildlife rating, OW = openland wildlife rating

rating high estimates for both cotton and oats. Some exceptions, such as oats and wheat, showed a near linear configuration indicating high correlation, as do range production and grazing.

A star plot of the yield estimates (cotton potential, range production, grazing potential, sorghum potential, oats potential and wheat potential) reflects this pattern of high, moderate and low land capability (Figure 21.15). The large star patterns made up of six radians are the soil types with the best yield estimates. These high rating soil types include Bastil, Bosque, Crawford, Denton, Frio, Houston-Black, Krum, Lewisville and Slidell. The other star plots suggest at least two patterns consisting of small star plots made up of less than six radians such as soil types Bolar, Brackett-Topsey and Evant. A cluster analysis of these same variables produces at least three major groups of like soil types. Cluster 1 is made up of the large six radian star plots for soils rating high on all SCS measures, Cluster 2 is composed of those soils symbolized by very small plots made up of only 1 to three short radians indicating very low scores on all yield and Cluster 3 consists of soil types that fall in the middle of the range of yield estimates. Table 21.9 shows the soil type cluster assignments derived from the cluster analysis of the six land capability estimates.

The three soil type groups revealed by the cluster analysis of the six yield estimates are sensitive to overall productivity of the soils. The farming limitations, which measure the limitations of the land for most types of crop production also generally conform to the clusters derived for overall productivity. That is, the soils that rank high in overall production also pose few limitations to agricultural manage-

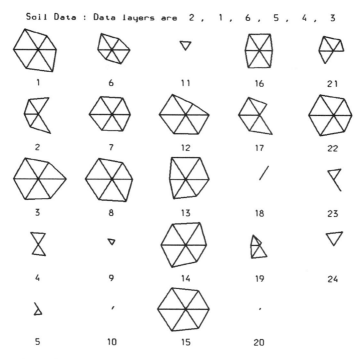

Figure 21.15. Star plots of cotton potential, range production, grazing potential, sorghum potential, oats potential and wheat potential (counterclockwise from right horizontal radian) for the 24 soil types at Ft. Hood.

Table 21.9 Soil type groups derived from cluster of cotton, wheat, oats, sorghum, grazing, and range production

Cluster	% cover	Soil types	Characteristics
1	16.1	Bastil, Bosque, Crawford, Denton, Frio, Houston-Black, Krum, Lewisville, Slidell	Rate high for yield estimates in all six categories, and low in farming limitations.
2	61.4	Cho, Doss-Real, Eckrant, Evant, Nuff, Real-Rock, Wise	Rate low for yields in one to three of the six categories and 0 for all others, and moderately low farming limitations.
3	22.5	Bolar, Brackett-Topsey, Cisco, Lindy, Minwells, Purves, Seawillow, Topsey-Pidcoke	Rate moderate for three to six yield estimate categories, and moderately low to high farming limitations.

ment. As the column for per cent cover shows, the productive soils (Cluster 1) make up only 16.1 per cent of the Ft. Hood installation while low productive soils (Cluster 2) constitute over 61 per cent (Table 21.9). Even if the moderate ranking soil types (Cluster 3) are added to the high yielding soils in Cluster 1, the percentage of good agricultural land makes up only 39 per cent of the project area. This suggests that farm or ranch operations situated on the Cluster 2 type low yielding soils are at a distinct disadvantage compared to the operations on high to moderately productive soils.

264   I. Williams, W.F. Limp and F.L. Briuer

However, one problem with the cluster analysis solution is that the importance of ranching operations might be masked by the heavy weight given to farming operations as a result of the use of cotton, wheat, oats and sorghum yield categories and only two variables (grazing and range production) that measure livestock forage potential. Grazing potential can be reassessed by looking at the yield potential of grazing and range production again and comparing it with the three soil groups derived from the cluster analysis. The bivariate plot of these two variables shows a wide range of forage potential among the 24 soil types. A logical partition of this plot would appear to be two groups of high versus low grazing and range potential. The high ranching potential ranges between 5.0 and 7.5 on the grazing axis and 4000 and 6000 on the range axis and includes 14 of the soil types. The area on the plot of low ranching potential is confined to the soil types falling between 0 to 4.0 on the grazing axis and 2000 and 3500 on the range axis and consists of 10 soil types.

By imposing this new partition onto the three clusters obtained in Table 21.9, four new soil type groups are generated that measure the potential for both farming and ranching (Table 21.10). In general, the ideal locations for farming are also the best locations for ranching based on the forage yield estimates. Not included in this evaluation are the SCS descriptions of the current general use of these soil types and their assessments of the suitability of the soils for development as cropland, pasture or rangeland. While this information may be directly relevant to the soil type groups to be derived in this section, the uncritical projection back to the nineteenth century of modern land use patterns for the Ft. Hood area would not serve us in our effort to discover historic land use trends. Instead, the soil types will be assessed below relying as closely as possible on the objective quantitative measures of soil capability.

Table 21.10 Soil type groups assessed for farming and ranching potential

Cluster	% cover	Soil types	Characteristics
1	16.1	Bastil, Bosque, Crawford, Denton, Frio, Houston-Black, Krum, Lewisville, Slidell	High farming and high ranching potential.
2	2.2	Bolar, Cisco, Lindy, Minwells, Seawillow	Moderate farming and high ranching potential.
3	20.3	Brackett-Topsey, Purves, Topsey-Pidcoke	Moderate farming and low ranching potential.
4	61.4	Cho, Doss-Real, Eckrant, Evant, Nuff, Real-Rock, Wise	Low farming and low ranching potential.

The results of comparing the cluster analysis of soil productivity with the modern agricultural uses for these soils described by the SCS suggests that the derivation of current land use patterns can be broken down into a few basic principles: (1) high soil productivity in both grazing potential and crop potential = cropland use primarily or secondarily pasture use; (2) moderate to low land productivity in crop potential and high to low grazing potential = pasture use primarily or rangeland if limitations preclude development of improved pasture; and (3) low land productivity in both crop and grazing potential and land limitations on pasture development = rangeland use. To avoid projecting too much the actual land use development of these soils prior to modern times, the four clusters can be regrouped into three categories; (1) high land capability, (2) moderate land capability, and (3) low land capability. Figure 21.16 displays the distributions of these newly defined zones at Ft. Hood.

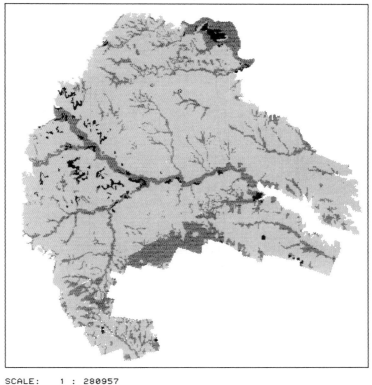

SCALE: 1 : 280957
                                3474750.00
WINDOW: 600300.00   643000.00
                                3430250.00

1 cropland/pasture
2 pasture/range
3 range

Figure 21.16. Land use zones defined at Ft. Hood.

The land capability ratings are the more reasonable projections into the nineteenth and early twentieth centuries. The actual land use assignment for these three groups may have varied depending on market fluctuations, levels of agricultural technology, the development of transportation networks and other factors. We know that, for instance, the extent of grazing use of the land was very high initially in the settlement of the Southwest as free range livestock production was one of the first adaptations to become established in this region (Jordan, 1970; McDonald and McWhiney, 1975; Owsley, 1945). Once a market for crops developed, the high land capability of Group 1 would have made it the most feasible area in which to expand into crop production. Had the high demand for crops prompted further expansion of farming into the Group 2 soils, the difference in yield potential would still justify the separation of the soil groups into these two discrete groups.

A univariate analysis of the site occurrence in the land use zones by time period shows fairly even use of all three land use zones from the mid-nineteenth century up to the 1950s. The site occurrence reports were run on sites dating to five time periods: (1) prior to 1881, (2) between 1880 and 1901, (3) between 1900 and 1921, (4) between 1920 and 1941, and (5) sites dating after 1940. The 24 earliest sites on record at Ft. Hood were distributed within the expected range, and there is little deviation from this pattern up to the late nineteenth century. By the early twentieth century, land use zone 2, the pasture/range group, shows a greater than expected density of historic sites with a Chi square test score of 24.8. The periods between 1920 and 1941 and 1940 to the 1950s show no increase in settlement in this zone, and the other two zones demonstrate no significant distributional pattern deviation from the expected either. With the exception of the major concentration of use of the pasture/range zone during the early twentieth century, site distribution appears to be relatively even for all land use zones during the five major historic periods.

The search for high dimensional site clusters

The first stages of this analysis involved exploring the structure of sites with respect to three dimensions of variability: architectural features, farming/ranching features, and agricultural land potential. Examination of these three sets of variables produced site clusters that can be related directly to factors such as (1) the architectural composition of the domestic house component such as the presence of rootcellars, piers, chimney remains, etc.; (2) the nature of agricultural support features such as dip tanks, corrals, troughs, fences, etc.; and (3) the agricultural potential of the land in terms of potential yields of oats, cotton, sorghum, and other plants. Up to this point it has been useful to break out the 21 variables into these three sets as a means of isolating and assessing the unidimensional patterns of historic site occupation in the Antelope Community. At this point, however, it is desirable to evaluate the multidimensional structure of the sites by running analyses on all the 21 architectural and agricultural potential variables together.

As a first step, star plots for the Antelope Community sites were generated on the 21 variables (Figure 21.17). The similarities between the stars shows that a number of sites match up closely with respect to the 21 variables. For instance, sites 1194, 1198, 1508, 1543, 1586 and 1955 have identical star patterns because they share the characteristics of having stock tanks, water tanks, bricks, chimneys, piers, roofing material, and rate high in agricultural limitations. Sites 1205, 1547, 1561 and 1596 share the same traits as the six sites above plus they also contain building rubble. Beyond these clear similarities, there are several sites that overlap closely in their star plots and are nearly alike such as sites 1506 and 1548 which include, in addition to the 7 variables noted above, the presence of concrete tanks. The difference between these two sites is the presence, on site 1548, of building rubble which is not documented at site 1506. Other patterns of near similarity indicate that at this level of variability, many sites exhibit high overlap with other sites, but do not fall into neat clusters that can be easily recognized visually.

In order to deal adequately with the complexity of site variability at this multidimensional level, a cluster analysis of the Antelope Community sites for all 21 variables was run (Figure 21.18). The dendrogram shows the two clusters of identical sites noted above in the lower right side of the figure (i.e. Cluster 1 = sites 1194, 1198, 1508, 1543, 1586 and 1955; Cluster 2 = sites 1205, 1547, 1561 and

Antelope Community

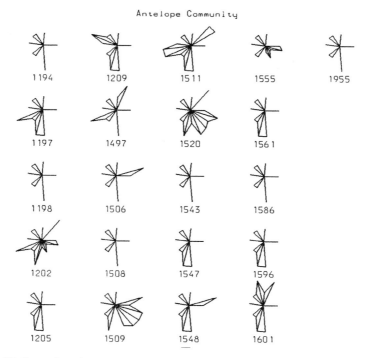

Figure 21.17. Star plots for all Antelope community sites for all 21 variables.

1596). Three other site pairs are also shown on the right side of the dendrogram: sites 1509 and 1520 (Cluster 3), sites 1202 and 1555 (Cluster 4), and sites 1506 and 1548 (Cluster 5). Examination of the star plots (Figure 21.17) reveals that these three site pairs are similar in that they closely overlap with respect to most variables, but are not identical because they also possess other qualities not shared by the other site.

The remaining five Antelope Community sites (sites 1511, 1601, 1209, 1497 and 1197) are unique and do not cluster closely with any of the other 16 sites. They can be ordered in terms of site similarity with respect to the other sites discussed above by examining where they branch out of the dendrogram from the clusters noted above. For example, site 1197 falls out of the dendrogram between the first two large clusters, designated Clusters 1 and 2, and Cluster 5 is made up of sites 1506 and 1548 (Figure 21.18). By looking back at the star plots in Figure 21.17, we can see that site 1197 indeed closely resembles the sites in Cluster 2 in that it shares all of the traits of this cluster, but in addition includes foundation material and building rubble which is not a characteristic of Cluster 2. Site 1197 also closely overlaps with site 1548 of Cluster 5, in that both sites share eight variables but they differ in the addition of other variables such as the presence on site 1548 of concrete tanks and on site 1197 of foundation material.

Assessment of the relationships between the other four nonclustering sites with site clusters occurring on adjacent branches of the dendrogram can also be made to evaluate the similarity structure of the Antelope Community sites. We might then choose to force the cluster of site 1197 in either Cluster 2 or Cluster 5 if we feel that the differences between these sites are negligible for our purposes. The same can be

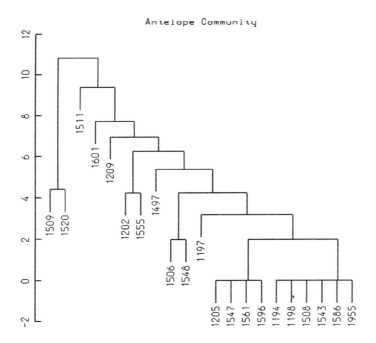

Figure 21.18. Dendrogram displaying results of cluster analysis of all 21 variables for the Antelope Community sites.

done for the other anomalies that do not fall neatly into one of the primary clusters. This is a management decision that can be made based on our firm understanding of the significance of all of the variables and the weight we want any single one of them to have on the definition of site groups. Such a decision about the importance we wish to place on single variables can be made from an informed base after we have explored the multidimensional nature of the sites through exploratory data analysis.

Summary of analysis of environmental data

This review of historic site variability with respect to some of the GRASS environmental variables has revealed a number of patterns in the data relating to higher or lower than expected site occurrences in certain zones. For instance, sites show a slightly to moderately significant trend toward close proximity to drainages, roads and community centres. In addition, site occurrence away from drainages, roads and community centres is less than would be expected by chance. These trends generally fit the expected pattern, based on our understanding of historic migration and settlement patterns, of close attraction to these basic features of the natural and cultural landscape. Two points are evident from the proximity analysis: (1) it is clear that historic sites are not strongly clustered with respect to drainages, roads and community centres, but instead are situated to make use of other aspects of the landscape probably related to agricultural land productivity, the space requirements of which are probably the main factor accounting for the even spacing of sites from each other; and (2) there is only slight indication of close clustering of sites in the core area of the community centres which may be a reflection of the low degree that clustered specialized services and other facets of 'town' development were ever attained in these community centres. Closer examination of the degree of development of a core area in the proximity of the community centres may allow us to stratify the community centres better with respect to this dimension. We expect that some of the community centres developed beyond the stage of the dispersed rural service centre to serve additional economic and social needs met only by a smaller number of higher level nascent 'towns'. Such a development might be indicated by either greater complexity or density of sites in the core area or by the expansion of the distance of these community centres from other centres as the rising towns eclipse nearby community centres. If such a pattern exists, then it would suggest that differentiation of community centres should be considered. A potentially significant pattern is shown in historic site settlement with respect to landform zone with the lowlands exhibiting a very high incidence of site occurrence while the uplands and intermediate uplands show fewer sites than would be expected by chance. It is probably not coincidental that the lowlands constitute the most productive agricultural landzone at Ft. Hood, while the uplands and intermediate uplands are dominated by the low land capability zone used as rangeland. It is also interesting that there is a significant increase in site settlement during the early twentieth century in the lowlands-pasture/range zone. Whether this represents a response to the cropland (farming) potential of this zone or simply expansion of ranching is unknown. Elsewhere at Ft. Hood, site settlement over time appears to be fairly even over the five time periods of occupation.

The three part agricultural land use division of the landscape into cropland/pasture, pasture/range and range zones suggests potentially very different agricultural strategies, that are very important dimensions of the historic sites. The assessment of agricultural land capability suggests that land use may be dominated by a basic range and pasture utilization (ranching) since this type of land makes up over 80 per cent of the total surveyed area at Ft. Hood. While there are areas of high land productivity that would have supported crop production (16 per cent of the area), the extent to which such agricultural activities actually took place in these specific areas is unknown. We do know that Ft. Hood was characterized by both ranching and farming and that this land use was discretely partitioned across the area.

GIS and EDA

While the other papers in this volume rightly emphasize the roles and benefits of geographic information systems, we emphasize here that the GIS should be seen as only one component in a comprehensive system for archaeological analysis (see also Farley *et al.*, 1988, and this volume). In addition to the GIS, powerful EDA software can be used to expand our understanding of past and present landscapes. As GIS use becomes more frequent, researchers and managers will have to deal with large, high dimension data sets. Commonly used analytical tools today are not adequate for such data. Fortunately, systems such as S and powerful relational databases are also now available and can be linked to the GIS. Hardware with the graphic display and processing capabilities required by such data sets and software are now increasingly available. While such situations are cause for optimism, we must caution those who see such an approach as a panacea. In order to use such a system effectively, massive investments must be made in the acquisition and review of high quality data. At Fort Hood this investment has been continuing over more than a decade and represents some two and a half million dollars and many person-years.

Acknowledgements

The field research on which this paper was based was supervised by Frederick Briuer. Mr. Emmett Gray, Director of the Fort Hood Environmental Unit, is responsible for the development of the GRASS environmental data sets. Dr. Sandra Parker of the University of Arkansas, Department of Industrial Engineering, developed the S software used in this analysis with support provided by the Construction Engineering Research Laboratories. Funding for the research reported here was provided through the Construction Engineering Research Laboratories by Fort Hood, US Army Forces Command, and the Department of the Army. The support of Dr. Constance Ramirez, Federal Preservation Officer, Department of the Army; Dr. James Cobb, Command Archaeologist, Forces Command; and Mr. Bill Goran, US Army Construction Engineering Research Laboratories, is much appreciated. The opinions expressed in this report are those of the authors and do not necessarily reflect the opinion of the US Army.

References

Becker, R. and Chambers, J., 1984, *An Interactive Environment for Data Analysis and Graphics*, (Belmont, CA: Wadsworth Advanced Book Program)

Briuer, F., 1981a, Deer and water as potential determinants for the location of prehistoric sites. In *Archaeological Survey at Fort Hood, Texas — 1979*, edited by S. Skinner *et al.*, (La Jolla, CA: Science Applications), appendix D

Briuer, F., 1981b, Identification and measurement of impacts upon cultural resources at Fort Hood. In *Initial Survey of Archaeological Resources at Fort Hood, Texas — 1978*, edited by S. Skinner *et al.*, (La Jolla, CA: Science Applications), pp. 26–29, 56–70

Briuer, F., 1983, Incorporating problem-oriented research into archaeological resource management strategies. *Proceedings of the American Society for Conservation Archaeology*, pp. 6–19

Briuer, F. and Shaw, I., 1984, Fiscal year impact research. In *Archaeological Survey at Fort Hood, Texas — 1979*, edited by S. Skinner *et al.*, (La Jolla, CA: Science Applications), appendix E

Briuer, F. and Thomas, G., (editors and compilers),1986, Standard operating procedure for field survey. *Research Report No. 8*, (Fort Hood, TX: Archaeological Resource Management Series)

Carlson, D. and Briuer, F., 1986, Analysis of military training impacts on protected archaeological sites at West Fort Hood, Texas. *Research Report No. 9*, (Fort Hood, TX: Archaeological Resource Management Series)

Carlson, D., Briuer, F. and Bruno, H., 1983, Selecting a statistically representative sample of archaeological sites at West Fort Hood, Texas. *Research Report No. 8*, (Fort Hood, TX: Archaeological Resource Management Series)

Carlson, D., Carlson, S., Bruier, F., Roemer, E. and Moore, W.,1986, Archaeological surveys at Fort Hood: the FY83 eastern training area, *Research Report No. 11*, (Fort Hood, TX: Archaeological Resource Management Series)

Carlson, S., 1984, Ethnoarchaeological studies at a 20th century farmstead in central Texas: the W. Jarvis Henderson site (41BL273), *Research Report No. 12*, (Fort Hood, TX: Archaeological Resource Management Series)

Carlson, S., 1987, *An Assessment of the Research Potential of Historic Sites at Fort Hood, Texas*. Society for American Archaeology, Annual Meeting, Savannah, GA

Carlson, S., Carlson, D., Ensor, B., Miller, E. and Young D., 1987, Archaeological surveys at Fort Hood: the FY85 northern sector, *Research Report No. 14*, (Fort Hood, TX: Archaeological Resource Management Series).

Carlson, S., Carlson, D., Ensor, B., Miller, E. and Young D., 1988, Archaeological survey at Fort Hood: the FY85 northwestern training area, *Research Report No. 15,* (Fort Hood, TX: Archaeological Resource Management Series)

Dibble, D., Briuer, F. and Mishuck, E., 1984, Archaeological survey, FY80, spring, *Research Report No. 3*, (Fort Hood, TX: Archaeological Resource Management Series)

Dibble, D., Briuer, F. and Mishuck, E., 1985, Archaeological survey, FY80, fall, *Research Report No. 4*, (Fort Hood, TX: Archaeological Resource Management Series)

Farley, J., 1987, The land manager's dilemma: monitoring multidimensional resource associations in a three dimensional world. *Proceedings U. S. Army Corps of Engineers Sixth Remote Sensing Symposium*, Galveston, TX, pp. 198–221

Farley, J., Limp, F. and Lockhart, J., 1988, *The theoretician's workbench, integrating GIS, relational databases, exploratory data analysis and remote sensing*, Society for American Archaeology, Annual Meeting, Phoenix, AZ

Guderjan, T., Thomas, G. and Cramer, H., 1980, *Existing Data Inventory of Cultural Resource and Paleontological Information, Fort Hood, Texas*, (Marietta, GA: Soil Systems)

Hartwig, F. and Dearing, B., 1979, *Exploratory Data Analysis*, (Beverly Hills, CA: Sage Publications)

Hasenstab, R., 1983, *The application of geographic information systems to the analysis of archaeological site distribution*, Society for American Archaeology, Annual Meetings, Pittsburgh, PA

Hoaglin, D., Mosteller, F. and Tukey, J., 1983, *Understanding Robust and Exploratory Data Analysis,* (New York, NY: John Wiley and Sons)

Huckabee, J., Thompson, D., Wyrick, J. and Pavlat, E., 1977, *Soil Survey of Bell County, Texas,* (Austin, TX: U. S. Department of Agriculture in cooperation with Texas Agricultural Experiment Station)

Jackson, J., 1982, Okay: an Archaeological Reconstruction and Settlement Pattern Analysis of a Dispersed Hamlet in Bell County, Texas, Unpublished Masters Thesis, University of Texas, Austin

Jackson, J., Dibble, D. and Briuer, F., 1982a, Archival information search and archaeological survey for the proposed aircraft maintenance facility, Robert Gray Army Airfield, Bell County, Texas, *Research Report No. 6,* (Fort Hood, TX: Archaeological Resource Management Series)

Jackson, J., Dibble, D. and Briuer, F., 1982b, Final report of the archival research on the 'Maberry Community'. *Research Report No. 7.* (Fort Hood, TX:

Archaeological Resource Management Series)

Jordan, T., 1970, The Texan Applachian. *Annals of the Association of American Geographers*, **60**, 409–427

Koch, J. and Mueller-Wille, C., 1987, Archaeological surveys at Fort Hood, the FY85 delivery order No. 4. *Research Report No. 16*, (Fort Hood, TX: Archaeological Resource Management Series)

Kvamme, K., 1983, Computer processing techniques for regional modeling of archaeological site locations. *Advances in Computer Archaeology*, **1**, 26–52

Kvamme, K., 1984, Models of prehistoric site location near Pinon Canyon, Colorado. In *Papers of the Philmont Conference on the Archaeology of Northeastern New Mexico*, edited by J. Condie, (Santa Fe, NM: New Mexico Archaeological Council Proceedings), pp. 349–370

Kvamme, K., 1986, The use of geographic information systems for modeling archaeological site distributions. In *Geographic Information Systems in Government*, edited by B. Opitz, (Hampton, VA: Deepak Publishing), pp. 345–362

Kvamme, K., 1989, Geographic information systems in regional archaeological research and data management. In *Archaeological Method and Theory, Volume 1*, edited by M. Schiffer, (Tuscon, AZ: University of Arizona Press), pp. 139–203

Kvamme, K. and Kohler, T., 1988, Geographic information systems: technical aids for data collection, analysis, and display. In *Quantifying the Present and Predicting the Past: Theory, Method, and Application of Archaeological Predictive Modeling*, edited by J. Judge and L. Sebastian, (Washington, DC: US Government Printing Office), pp. 493–547

Lafferty, R., Otinger, J., Scholtz, S., Limp, W., Watkins, B. and Jones, R., 1981, *Settlement Predictions in Sparta: A Locational Analysis and Cultural Resource Assessment in the Uplands of Calhoun County, Arkansas*, Research Series No. 14, (Fayetteville, AR: Arkansas Archaeological Survey)

Limp, W., 1987, The identification of archaeological site patterning through integration of remote sensing, geographic information systems and exploratory data analysis. *Proceedings US Army Corps of Engineers Sixth Remote Sensing Symposium*, Galveston, TX, pp. 232–262

Limp, W., Parker, S., Farley, J., Johnson, I. and Waddell, D., 1987, *An Automated Processing Approach to Environmental Resource Management for Military Installations*, (Champaign, IL: Construction Engineering Research Laboratory)

McCaleb, N., 1985, *Soil Survey of Coryell County, Texas*, (Austin, TX: US Department of Agriculture in cooperation with the Texas Agricultural Experiment Station and US Department of the Army)

McDonald, F. and McWhiney, G., 1975, The antebellum southern herdsman: a reinterpretation. *Journal of Southern History*, **41**, 147–166

Newton, M., 1967, The Peasant Farm of St. Helena Parish, Louisiana: A Cultural Geography, Unpublished Ph.D. dissertation, Louisiana State University, Baton Rouge

Owsley, F., 1945, The pattern of migration and settlement on the southern frontier. *Journal of Southern History*, **11**, 147–176

Parker, S., 1985, Predictive modeling of site settlement systems using multivariate logistics. In *For Concordance in Archaeological Analysis: Bridging Data Structure, Quantitative Technique and Theory*, edited by C. Carr, (Kansas City, MO: Westport Publishers), pp. 173–207

Parker, S., 1989, *Using the S-GRASS Software*, (Champaign, IL: Construction Engineering Research Laboratories)

Roemer, E., Carlson, S., Carlson, D. and Briuer, F., 1985, Archaeological survey at Fort Hood, Texas, FY82 range construction projects. *Research Report No. 10*, (Fort Hood, TX: Archaeological Resource Management Series)

Skinner, S., Briuer, F., Thomas, G. and Shaw, I., 1981, Initial archaeological survey at Fort Hood, Texas, FY78. *Research Report No. 1*, (Fort Hood, TX: Archaeological Resource Management Series)

Skinner, S., Briuer, F., Meiszner, W. and Shaw, I., 1984, Archaeological survey at Fort Hood, Texas, FY79. *Research Report No. 2,* (Fort Hood, TX: Archaeological Resource Management Series)

Thomas, G., 1978, A survey and assessment of the archaeological resources of Fort Hood, Texas, *Bulletin of the Texas Archaeological Society*, **49**, 195–240

Tukey, J., 1977, *Exploratory Data Analysis,* (Reading, MA: Addison-Wesley)

Westervelt, J., 1989, *An introduction to the GRASS software,* (Champaign, IL: Construction Engineering Research Laboratories)

Williams, I., Briuer, F. and Limp, F., 1989, *A quantitative assessment of the archaeological resources at Fort Hood, Texas*, (Champaign, IL: Construction Engineering Research Laboratories)

22

Building an historic settlement database in GIS

Jack M. Jackson

A number of the other papers in this volume relate various aspects of the development and use of the Geographic Resources Analysis Support System (GRASS) at Fort Hood, Texas. This paper outlines a concept for building a database relating historic documents to the historic era sites found on the Fort. Conventional surface survey of the 339 square mile military reservation has located over a thousand sites relating to the occupation of the area by Anglo-American farmers and ranchers during the century between 1842 and 1942. In contrast to the long term agricultural occupations of Europe and Eastern North America, it is this very short sequence encompassing the rapid transition from horse-drawn vehicles, through the steam locomotive to automobiles and trucks, which makes the Fort Hood data most useful. The profound effects of these energy usage changes on settlement patterns and on the economics of agriculture are of considerable interest. The very short settlement sequence, truncated by the US Army takeover in 1942, telescopes these economic and technological changes into a very manageable sequence with a minimum of outside influences.

The area of rural Texas where Fort Hood is located has been characterized by Jordan (1970) as 'Texan Appalachia', because it was settled by the earliest English speaking pioneers who generally peopled the upland regions of the southern Appalachian mountains. These folk have been studied and puzzled over by several generations of scholars from several disciplines, because of their persistent resistance to the forces of industrialization. The treatment of this cultural variant in mainstream America has ranged from the admiration of musicologists (Sharp and Karpeles, 1983) who collected largely pristine Elizabethan ballads from folk singers in the hills to the contempt of the historian Arnold Toynbee (1947, p. 149) who characterized the Appalachian folk as the people who lost civilization. The immediate impact of the construction of Camp Hood in 1942 was studied by the social anthropologist Oscar Lewis (1948). His study of the various cultural groups in Bell County, Texas is no less valuable and insightful than his better known works on the peasant villages of Mexico (Lewis, 1951, 1961). In short, the research value of the historic period archaeological sites on the Fort Hood reservation is far greater than one might expect from the examination of a few shovels full of soil from one of the house locations. Indeed, the research value lies more in the changing spatial relationships and changing economic functions among the sites as a whole than in the artifacts in and on the soil.

This paper explains how the author attempted to use many of the available historic materials in an integrated approach to the reconstruction of historic settlements located on the reservation in some very modest studies done a few years ago (Jackson, 1981, 1982; Jackson and Briuer, 1989). Most of this work was done using manual methods and was quite limited in scope because of the drafting table and light table work involved. Because most readers are already quite familiar with these tasks; they serve not only to illustrate the power of the GRASS system as a tool, but also allow us to relate each task, such as map overlaying, to a well tried pre-GIS methodology like placing two maps on a light table and tracing a third from selected features of the two.

Relating data such as manuscript census returns on the composition of families and the nature of their agricultural practice reflected in cattle and crop statistics back to individual sites points up the primary difficulty of the application of a site-centred approach to settlement studies in the historic period. Most such studies involving prehistoric sites rest on the simple, but often undeclared, assumption that all activities related to a particular residential location took place within walking distance. This applies to large villages of horticulturist-hunters as well as to the occupants of rock shelters. On the theoretical or Euclidean plane this results in a construct of circular areas surrounding each site from which various commodities were procured. The need for these sites to be close, but not too close, to surface water is one of the most serviceable and long-held assumptions of settlement pattern archaeology (Skinner *et al.*, 1981, pp. 25ff). Paynter's (1982, pp. 26ff) discussion of the literature relating to settlement pattern models may be useful to those interested in the theoretical aspects of spatial models and how they relate to historic period sites. The main point of interest to us here is that changing transportation technology changes the pattern of settlement.

When Anglo-American settlers arrived in this area of Texas in the 1840s they brought with them a culture that used oxen and horses as traction power and insisted on surveyed rectilinear blocks of privately owned land. Only those who arrived first on the scene had the luxury of selecting parcels with natural springs or streams. The later arrivals had to dig wells or collect rainwater in cisterns. The system of land grants in Texas was a great deal different from orderly one-statute-mile squares containing 640 acres laid out on the cardinal directions. This 'township-range' survey method prevails in most of the Great Plains, and even in some parts of western Texas. The Republic of Texas inherited a system of land grants that originated with the Spanish. This system was used to grant lands to the first English speaking settlers, even when Texas became a province of the Mexican Republic. Many features of this system were retained by the Republic of Texas after the successful revolt against Mexico in 1836. When Texas joined the United States by treaty in 1846, all of her public lands were retained by the state and were granted to settlers by the state land office. Consequently most land parcels in the area of Fort Hood were surveyed using the *vara* as a unit of linear measure. This somewhat variable Spanish unit was finally standardized by the state land office at 33.3 inches, or slightly less than both the metre and the yard.

The Texas State Land Office still issues its county land grant maps at a scale of 2000 *varas* to the inch. This is close to, but not quite equal to, the more commonly used one mile to the inch scale used in many other historic maps. However, transfer of data from one of these maps to a 1:24,000 USGS standard topographic map with a Universal Transverse Mercator (UTM) metric grid is a job to try the patience of an experienced cartographic draftsperson. Consequently, I have done many such maps myself, unless they were intended for publication. Most of these maps were done manually at a drafting table. Geography departments commonly have a variable scale projector that makes these tasks easier, but in practice gaining access to the machine is more time consuming than doing a quick manual transfer of a few property lines. Some experiences during the early years of the massive Fort Hood pedestrian archaeological survey convinced the writer that land ownership data were vital to the accurate recording and interpretation of historic sites. The cost of doing this sort of analytical work at the drafting table apparently helped to convince Dr. Briuer, the staff archaeologist at the Fort in 1981–82, that investment in the development of the GRASS system was worthwhile. Before describing the procedures for loading such data into the computer, it might be useful to review some of the lessons that were learned between 1980 and 1982.

Primitive methods of 1980 to 1982

During the fall of 1980 the Texas Archaeological Survey of the University of Texas at Austin conducted a series of archaeological surveys at Fort Hood. One relatively small section of the area to be surveyed was known to have been part of a small farming hamlet known as Okay. A cluster of historic sites had been recorded and were thought to represent a centre of settlement.

The writer was assigned the task of deriving from archival sources the identity and function of these historic sites as well as completing a systematic survey of the surrounding land. Three primary source maps were used to complete this task. First, the Army had compiled a land ownership map showing from whom each parcel of the reservation was acquired in 1942, whether by purchase or by legal condemnation. This map gave the names of the owners and indicated the location of some roads, cemeteries, schools and other public facilities. However, no structures were depicted. Second, the 1936 county highway map showed all public roads and all structures which were visible from those roads. Streams were also depicted on this map, but were found to be very approximately placed. Third, the Texas General Land Office map of Bell County depicting the bounds and dates of patent on each of the original grant parcels was also used. Data from all three maps had to be transferred to a conventional topographic map with a UTM grid such that sites and other geographic features could be related to the more abstract road and ownership maps.

Census data from the agricultural and population schedules of 1880 were also related to individual land parcels as well as tax, deed and probate records from the county court house. Rural Route delivery maps from the archives of the US Postal Service were obtained from the National Archives and proved useful in the identification of tenants who did not own the land where they lived. The resulting sketch map shown in Figure 22.1 was one of the products of this research. It served in both the required contract report to the Army (Jackson, 1981) and in my M.A. thesis (Jackson, 1982a). The contract report has since been abridged and published (Jackson and Briuer, 1989). It was found that without the aid of archival data detached features of a single homestead were easily misinterpreted as separate sites and that single sites later traversed by a roadway were also recorded as different sites. Spatial separation in the current landscape was a poor guide to functional unity. One of the sites from the Okay community was later investigated through systematic sampling and the excavation of two filled cisterns. The identification of the last owner of the house allowed the project archaeologist to locate and interview several members of the family who had lived at the site (Carlson, 1984).

A similar project was undertaken at a slightly later date to investigate what was presumed to be another small community. The site included a dam with a pond on a small stream, a cemetery and a large number of masonry features. On the ground, it was quite easy to assume that all the features were roughly contemporary and that a number of homes and other structures had once been present. Archival research quickly corrected that impression. The site was, in part, the remains of the head-quarters of what had once been a large sheep ranch. After the Army took over the land, first a picnic area, then a boy scout camp had been laid out along the margins of the large pond that had been originally for watering sheep. The presumed community that had been so obvious to the field recording team literally dis-integrated in the research process. This report has also been abridged and published (Jackson and Briuer, 1989). The lesson learned from both of these projects was that the accurate functional interpretation of historic era sites requires far more data than the ordinary field crew can find on the ground and record on a site form. The definitive evidence in this case was contained in an aerial photograph taken in 1938 for the Soil Conservation Service. It showed a house and a large barn shed complex, but no town.

A second conclusion was reached from the man-hour costs involved in these two projects. Simply put, detailed archival studies of every historic site on the reserva-tion were so expensive as to be prohibitive. Based on that finding two actions were taken. First, an interim minimal land ownership and property boundary sketch map was developed for each one-kilometre square quadrat and sent to the field with the field party doing the survey. This made it easier for the field crews to interpret what they found and draw site boundaries on a more informed basis. The task of generating these little 'keyhole' maps involved constructing a one-kilometre grid over the existing land acquisition maps. Drawing such a grid manually is not difficult unless one requires some degree of accuracy. The greater the degree of accuracy desired the harder the task becomes, particularly when the original real estate plat map was drafted with less than optimum precision. The final result of this exercise was a 5×8 record card with a sketch map of the real estate bounds found in a one-kilometre square quadrat. The card also indicated the original earliest patent date on each parcel of land and the names of the owners at the time the Army acquired the property. Structures depicted on the 1936 highway maps were also shown. This procedure proved to be both useful and economical enough to become a part of the standard operating procedure for the pedestrian survey.

The second action was not taken solely on the basis of the difficulty of mapping in the two projects mentioned above, but on the basis of similar experiences in a number of other projects at many different Army installations. These varied experiences all spurred the development of the GRASS system by the Construction Engineering Research Laboratory of the US Army Corps of Engineers. Because Fort Hood shared in the funding of the project and had a relatively large inventory of recorded archaeological sites, most of the GRASS procedures used in cultural resource management have been developed and tested using the Fort Hood database. The next section of this paper outlines the concept, plan and procedures for building a fully integrated historic sites database which will allow site evaluations of the type outlined above with far less manual manipulation of data.

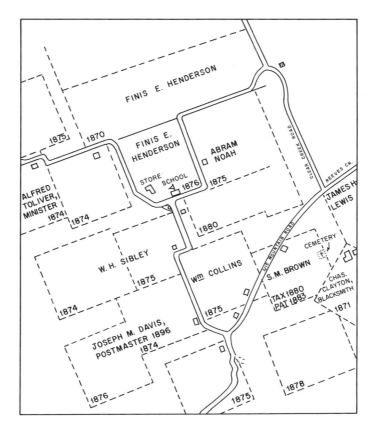

Figure 22.1. Sketch map of the dispersed hamlet of Okay, Texas (Jackson, 1981, 1982a).

Historic database concept

One of the difficulties in writing about historic settlement patterns and a major problem in structuring a database is that both spatial and temporal changes must be taken into account. There is also some difficulty with the availability of data. For instance, manuscript census schedules only become available to the public after 70 years. The census of 1910 is therefore the latest that is now available. The 1920 census should be released sometime after the summer of 1990, but it will be more than 20 years before the 1940 schedules can be made a part of the database. Thus, at the outset a 32 year gap exists between the final 1942 land owner map and the latest available census. Fortunately, the Army has complete abstracts of title prepared on each tract they acquired in 1942. These abstracts include every deed or other legal document relative to ownership from the original grant up to 1942. Those records have been located and are being microfilmed. Some idea of the data volume involved can be visualized from the 54 standard record storage boxes used to pack away the abstracts. These abstracts can be used to work backward from the 1942 ownership map and construct an ownership map for any given year.

Map digitizing

The maps of the original land grants for Coryell and Bell Counties are also available. Because these are printed at a scale of 2000 *varas* to an inch, they serve well to illustrate how the GRASS system is told about the scale of any map. However, whatever the scale of a particular map, GRASS requires that the scale be expressed as a unit free fraction, such as 1:24,000, the standard scale of the USGS 7.5 minute series. Thus 2000 *varas* multiplied by 33.33 (inches per *vara*) equals 66666 and the unit free fraction is expressed as 1:66,666. The next thing the GRASS program needs to create a map layer would be at least four register points. These are points which can be identified both on the new map to be digitized and on the base map of the installation. In practical terms these points turn out to be things like cemeteries which appear on both maps, cross roads and odd angles in the installation boundary. One identifies each point in UTM coordinates from the key board, then 'registers' each point in turn from the digitizing table. The GRASS software at this point detects any inconsistency between the scale specified and the scale implied by the identification of the register points. The software calculates the scale implied by the registered points and gives the user the calculated unit free fraction. One can then choose either of the scales or look for possible errors in the UTMs specified for the register points. In practice, hand drafted maps prepared from surveys done at different times by several crews are seldom found to be exactly to the declared scale. GRASS also calculates a rather complex statistic which can be read as 'close enough' or 'too far off' for each of the points registered. The user has the options of dropping points with high error coefficients, checking and correcting the given UTM of the point, or re-registering the point with the mouse. Sometimes all three actions may be required. It may be a good idea to pick at least five points to register, work carefully and drop the point with the highest error coefficient.

Point mode

After the scale is selected one must enter corners between straight lines to define each line of the map that is to be entered into the GRASS system. In physical terms this means putting the cross hairs of the mouse on the digitizer table at each of the four corners of a square or rectangular property and pushing a button on the mouse. The system needs to be told if the point is just a point on that line where there is some slight change of direction, or a node where two or more lines meet. The graphics monitor is run at the same time so that the operator can see each line as it appears on the screen. The operator then can visually compare the emerging image on the screen with the map on the table. If areas such as property boundaries are digitized the graphics monitor will display 'unsnapped' nodes, where slight deviations in the operation of the mouse have failed to close the bounds of an area. These may be corrected later in an editing session, or one may choose to change the tolerance of the software for slight manual errors. As a general rule a 24 metre tolerance at a scale of 1:24,000 forces a good degree of accuracy without creating too many tiny errors that must be edited out later.

Streaming mode

For curved lines such as stream courses or early roads which meander across the map the system offers a streaming mode. This mode is selected at the main work station from a menu. Once this mode is selected then a given button on the mouse signals the beginning of a streaming mode entry. One then carefully traces the course of the stream or road with the cross hairs of the mouse and pushes another button on the mouse to signal the end of each streaming entry. As with entries in the point mode all nodes must be the end of a stream entry. The software collects points in a continuous stream, then calculates how many of the points will be needed to define the line fully. Depending on the threshold set by the user the stream is edited down to the essential points. The course of the resulting irregular line appears on the graphics monitor as it was traced with the mouse. During a digitizing session one needs to operate three separate devices: the digitizing table, a graphics monitor and the system monitor with a keyboard.

Aerial photographs

Aerial photographs can be abstracted and digitized into the system in very much the same way as a map. Obviously there is little need to digitize the precise position of every tree, rock and cattle trail which appears on a photograph, but features which are of interest can be input to the system. For instance, the 1938 aerial photograph of a particular farm or ranch complex can be used to digitize the position of each building and those features can be aligned with features recorded by archaeologists on their survey forms. As discussed elsewhere in this volume the choice of scale and level of detail for a particular map layer must consider the capacity of the storage medium and computing capacity available at a particular installation. Trying to build a map layer that is too detailed may make it either impossible or tediously slow to use with simple hardware. Readers contemplating a minimal investment in hardware should not be misled into thinking that all they need is a computer—any computer. One can also use a scanning process to input photographic images. Storing and processing digital images at high resolution can be very expensive. Currently, only the index sheets for the 1938 aerial photographic series are planned for input by scanner.

Secondary map construction

The procedure of digitizing a map creates a vector map file in GRASS. This vector map has the outlines of all areas plus such lines as streams and roads which do not function as the boundaries of areas. Each line or area can be labelled either individually or in groups. The land grant map was labelled to indicate the name of the original grantee and the date of the grant.

Because GRASS is a raster based system there are a number of statistical and data checking routines which are designed to work only with cell file maps. GRASS provides an automatic conversion of a labelled vector file into a cell file. This routine is called 'vec.to.cell'. Once the map layer has been converted various quality checks can be made such as calculation of the acreage of each of the mapped parcels. Most people with some experience of putting nineteenth century cadastral maps into a GIS computer have concluded that the world needs to be resurveyed. Do not despair; chances are that the bulk of the errors detected are errors made in the field by survey parties, not errors the operator makes in digitizing the map prepared from their work.

Once the original land grant map is entered with the dates of the original patent on each parcel it is a reasonably straightforward matter to construct a base layer for a given census year by simply omitting from that layer all of the lands that had not yet been claimed at the specified date. This base map can then be further modified using the abstracts microfiche to account for such matters as heirs dividing large grants and single owners buying up several small holdings. Since all deeds and other legal instruments make reference to these parcels by the name of the original grantee or a state-assigned county abstract number, a detailed ownership map corresponding to the census year can be constructed by modifications made either through the graphics monitor, or through the digitizer. It is against these maps that files on the persons, crops and animal stocks of each parcel can then be built.

Perhaps the most obvious use of such a map layer as the 1880 land ownership map is as an overlay on the historic sites map. While it is not always safe to presume that ownership was established at the same time occupation began, sites are seldom older than the original patent date. Early land grants in Texas were made under several laws that often resulted in lands being surveyed and privately owned many years before they were actually settled. Other types of grants were made on the basis of an affidavit certifying that the claimant had built a home and cultivated the land. These considerations add a considerable degree of complexity to the construction of such a database in Texas. Elsewhere on the Great Plains where the strict cardinal direction township-range grid was used and only the federal homestead laws applied, this is a far simpler matter. However, in the eastern tidewater areas of the United States and in Europe, land ownership and settlement patterns can be much more complex than the sequence of 100 years with which we are dealing at Fort Hood.

The fact that this short sequence involves some 339 square miles, but only about 400 land-owning families in 1942, and includes both the transition to railroad transportation in the 1880s and the equally significant transition to automobile and truck transportation begun in the 1920s, makes the data both brief and quite handy for testing a number of theories. Once the body of data is fully developed it can be transferred via telephone line from computer to computer. Even in the interim, the microfiche and microfilm can be copied and the maps that have been put into computer format can be copied and transferred with far greater ease than would have been possible with 54 file boxes of abstracts and several hand drawn maps. Such a database becomes portable in the sense that it is no longer necessary for the researcher to travel to where the files are kept and overheat a photocopier or a note pad. Because of the massive size of the database involved in such a project, any project that used it and published the results could only afford to publish very brief summary tables of the primary data. Historians, geographers, agronomists and archaeologists all have different uses for various subsets of the primary data. Automation of those data in GRASS will facilitate not only access to the primary data by a variety of scholars, but more importantly allow the author and other cultural resource managers to make much more informed and sophisticated decisions about the relative significance of archaeological sites of the historic period.

Naturally much of the site related data is stored as an auxiliary text file, not in map form as a data layer. As currently configured only the centre point of each site is stored as a data layer. We are currently engaged with Construction Engineering Research Laboratory (CERL) in upgrading the sites data layer to include the boundaries of each site. A decision has not yet been taken to attempt storing directly as a data layer a map of the site itself generated from the survey notes and the 1938 aerial photographs. As currently envisioned the site related data file would contain only a reference to the photo sheet where those images are located in hard copy. However, if the speed with which storage media and computing power has increased over the last five years continues for a few more years, we may cease to think of paper records as 'permanent' and begin to use magnetic or compact disc storage for archival storage of data. Although I am not yet prepared to contemplate discarding two file cabinets full of site records or the multi-drawer map case and hundreds of yellowing paper maps that dominate the archaeological laboratory at Fort Hood, the idea of being able to lend the whole database to some researcher at a university a thousand miles away via a computer to computer telephone hook up essentially justifies and validates the massive effort involved in gathering and evaluating the data.

So what?

The essential trick in putting it all together as an integrated and automated database is to create a system that truly makes such a mass of data useful. Far too many of the early database systems that have been designed to store and retrieve archaeological data work very well only if you use them like file cabinets, looking at one or two records at a time. In any other mode these systems often overwhelm the user at retrieval, like trying to get a sip of water from a fire hose. Although we seldom think of it this way, an ordinary scaled map is a highly efficient information storage medium which is very easy to use. The essential power of GRASS and other GIS applications is in the fact that they combine the power of the computer with the essential information efficiency of the map. In the historic settlement database we are building at Fort Hood we are trying to make maximum use of this powerful combination. If the effort is moderately successful, it can be used as a model to be improved upon. If not, some valuable lessons will have been learned.

The initial and primary use of the system will be in the evaluation of historic era sites for US National Historic Register eligibility. One could, for instance, ask the system to produce a map of all of the sites which are located on land claimed before the year 1850. The map produced would not by any means prove that the mapped sites all represented early frontier occupations, but would provide a short list of sites to be examined for other evidence of early occupation. A cluster of grants appearing on such a map with a cluster of sites would certainly be a strong indication of the early settlement pattern of the region. One could also use other GIS layers such as the soils maps and streams maps to examine the question of early settler preference. Did the early settlers really prefer alluvial soils near streams? With a GRASS one can begin to formulate quantified answers to such questions.

A whole range of spatial questions which Paynter (1982) has begun to explore as well as questions related to the effects of increased energy usage on cultural systems (Jackson, 1988) can be formulated and answered. It is virtually certain that a primary result of these initial explorations will create as many new questions as new answers to old questions. Such a powerful tool for the manipulation of information always has that effect. The publication of these first faltering steps toward the use of GRASS in cultural resource management as a tool for assessing the significance of historic period sites is intended as an aid to that creative systemic process, not as a rote formula to be repeated again and again.

References

Carlson, S. B., 1984, Ethnoarchaeological Studies at a 20th Century Farmstead in Central Texas: The W. Jarvis Henderson Site (41BL273). Fort Hood Archaeological Resource Management Series, *Research Report Number 4*. United States Department of the Army

Jackson, J. M., 1981, Archival Information Search and Archaeological Survey for the Proposed Aircraft Maintenance Facility, Robert Gray Army Airfield, Bell County, Texas. Fort Hood Archaeological Resource Management Series, *Research Report Number 6*, United States Department of the Army

Jackson, J. M., 1982a, Okay, the Archeological Reconstruction and Settlement Pattern Analysis of a Dispersed Hamlet in Bell County, Texas. Unpublished MA Thesis, The University of Texas at Austin

Jackson, J. M., 1982b, Archival Research on the Mayberry Community. Fort Hood Archaeological Resource Management Series, *Research Report Number 7*, United States Department of the Army

Jackson, J. M., 1988, The Self Organization of the American Frontier: A Theory of Social Evolution for Historical Archeology. Unpublished Ph.D. dissertation, The University of Texas at Austin

Jackson, J. M. and Briuer, F. L., 1989, Historical Research and Remote Sensing: Applications in Cultural Resource Management at Fort Hood, Texas, Fiscal Year 1981. Fort Hood Archaeological Resource Management Series, *Research Reports 5, 6, and 7*, United States Department of the Army

Jordan, T. G., 1970, The Texan Appalachia. *Annals of the Association of American Geographers,* **60**, 409–427

Lewis, O., 1948, *On the Edge of the Black Waxy, a Cultural Survey of Bell County, Texas.* (St. Louis, MO: Washington University)

Lewis, O., 1951, *Life in a Mexican Village: Tepozlan restudied.* (Urbana, IL: University of Illinois Press)

Lewis, O., 1961, *The Children of Sanchez, Autobiography of a Mexican Family.* (New York: Random House)

Paynter, R., 1982, *Models of Spatial Inequality: Settlement Patterns in Historical Archeology.* (New York: Academic Press)

Sharp, C. and Karpeles, M., 1983, *Eighty Appalachian Folksongs, collected by Cecil Sharp and Maud Karpeles.* (New York: Faber and Faber)

Skinner, S.A., Briuer, F., Thomas, G. and Shaw, I., 1981, Initial Archaeological Survey at Fort Hood, Texas: Fiscal Year 1978. Fort Hood Archaeological Resource Management Series, *Research Report Number 1*, United States Department of the Army

Toynbee, A. J., 1947, *A Study of History.* (abridged edition) (New York: Oxford University Press)

23

GIS in historical predictive modelling: the Fort Drum Project

Robert J. Hasenstab and Benjamin Resnick

Introduction

Cultural resource management survey is aimed at maximizing recovery of archaeological information from areas slated for development, while at the same time gathering a representative sample of the variability of cultural resources present. Given the variable intensity with which a project area could be tested, in conjunction with restrictions in archaeological funding, the distribution of archaeological survey effort throughout any project area necessarily becomes a decision making process. The decision to investigate certain areas or resources more intensively than others must be based on an informed analysis of each project area's composition, resource content and regional setting, within a framework of regional anthropological research questions. In addition to cultural resource managers, land development planners also need to make decisions based on archaeological survey information. They must be informed of acreages and locations of archaeologically sensitive areas, and of the relative significance of archaeological sites, based on population estimates for sites within the project area and within the region. Because planning must often be done promptly, and always accountably, it is necessary to employ methods for assessing archaeological sensitivity which are rigorous, quantifiable and replicable.

Geographic information systems (GIS) offer the ability to employ such methods. GIS make possible the analysis of large amounts of archaeological and environmental information over broad expanses of space at a high level of resolution. GIS techniques permit the implementation of sophisticated models for predicting the distributions of cultural resources. Finally, GIS permit the testing of hypotheses concerning past human settlement, through prediction and subsequent field survey.

This paper reviews the application of GIS methods to cultural resource survey through a case study in the northeastern United States. In particular, it examines a model developed for the prediction of historic archaeological resources in a large project area, the Fort Drum military reservation. The paper will evaluate the effectiveness of the model in focusing survey efforts and in identifying historic archaeological sites.

Background

Project definition

Nearly seven thousand acres of rural land were to be disturbed in association with the installation of the 10th Mountain Division, Light Infantry, at Fort Drum, Jefferson and Lewis Counties, New York (Figure 23.1). In compliance with federal cultural resource legislation, the National Park Service Mid-Atlantic Regional Office and the U.S. Army contracted with Louis Berger and Associates Inc. (LBA) to provide cultural resource management services. To facilitate the planning of survey, LBA subcontracted with the senior author in 1985 to develop a GIS-based predictive model of cultural resource sensitivity for the 6,700-acre project area.

Figure 23.1. The project area at Fort Drum indicated on a USGS quadrangle. Reproduced from USGS (1982).

Project goal

The chief goal of the model was to focus archaeological survey efforts on those areas most likely to contain surviving archaeological resources. A model was developed based on the expected distribution of both prehistoric and historic archaeological sites and on distribution of previous ground disturbance throughout the project area. The three distributions were overlaid using GIS techniques to define resource preservation potential areas. Field survey methods were then prescribed so as to maximize coverage of areas with a high potential for containing intact archaeological resources, while sampling representatively those areas with little or no potential for containing these resources (Foss, 1987).

Anticipated cultural resources

The project area is located along the Black River within the St. Lawrence Valley Lowlands physiographic province. The foothill belt of the Adirondack Mountains—containing steep bedrock outcrops interspersed with small drainages—lies to the east. Very little is known of prehistoric cultural resources within the project area; however, a model of prehistoric archaeological sensitivity was developed for Fort Drum based on the locations of known prehistoric sites throughout the Jefferson County region (Klein *et al.*, 1985, pp. 6–18). The latter study formed the basis of the predictive model developed for prehistoric cultural resources here; sites were expected to occur most likely within 250 m of a water source, on slopes less than five per cent and on soils of moderate drainage or better.The prehistoric model will not be discussed here, except to say that it provided additional historic survey coverage in areas not particularly sensitive to historic sites.

Anticipated historic resources included settlements dating from the period following the American Revolution through the beginning of the twentieth century when the Fort Drum military base was established (Powell, 1976; Friedlander *et al.*, 1986). The northwestern portion of New York known at the time as North County was first purchased in 1791 by an Alexander Macomb. Macomb's partner, William Constable, eventually sold off lands to friends in France, and by the 1790s the village of Long Falls (presently Carthage) was settled on the Black River roughly ten miles upstream from the project area. In 1798 the state of New York rescinded the right of French to own land, and the occupants of Long Falls returned home—leaving behind several cabins, a saw mill, and several roads (Powell, 1976, pp. 117–119). In 1802, James LeRay purchased Tract No. 4 of Macomb's land, which includes the present project area. Between 1806 and 1808, the LeRay mansion was built, which still stands in the project area near the present St. James Lake (Figure 23.1). The mansion served as a nucleus for additional settlement, leading to the emergence of the village of LeRaysville (Figure 23.2). Between 1797 and 1800, pioneers from New England arrived in Jefferson County in search of agricultural lands. They settled on small family farms, and practised a diversified agriculture involving a variety of crops and livestock. The earliest towns in the area focused on water power; by 1820 over one hundred saw mills and fifty grist mills were operating in Jefferson County, and several mill villages, including LeRay, Watertown, Philadelphia, Carthage, Champion and Castorland, were established. By the mid-nineteenth century, the introduction of railroads and factories facilitated a shift in the region's economy; dairying became the lead industry as numerous cheese and butter factories and creameries were established (Schlebecker, 1975). Numerous small rural villages soon emerged in and around the project area, focusing on mills, and supporting a variety of government and service facilities.In 1909, the United States Government purchased the initial lands constituting the project area, and established Pine Camp, which later became Fort Drum. Dairy farming continued within the project area until 1941, when the US Government purchased all lands contained within Fort Drum.

Historic maps of the project area date to the latter part of the nineteenth century (eg., Beers and Beers, 1865; Robinson, 1888). Unlike historic sites associated with the eighteenth century occupation of the area, sites dating to the dairy farming era should be documented on the historic maps.

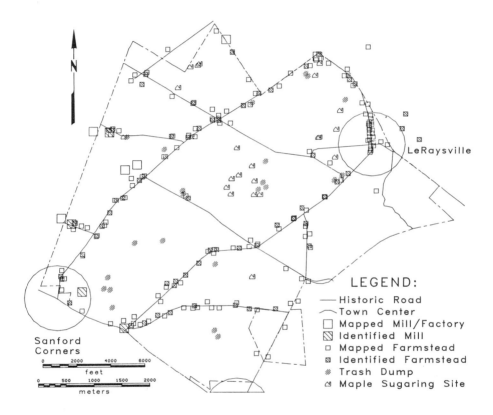

Figure 23.2. The distribution of historic sites within the project area. Shown are the locations of historically mapped and field identified sites.

Methodology

Overview

The study began with the development of the predictive model for historic archaeological sites. This involved the compilation of environmental data, historically mapped site locations and modern ground disturbance areas (Foss, 1987). These data were encoded into a GIS as a series of maps of discrete sensitivity zones. The layers were overlaid with the GIS, and the resulting sectors of the intersecting zones were interpreted through a set of decision matrices which formed the basis of the predictive model. The ultimate decision for each segment of land was the intensity of field survey to be applied in the segment.

Systematic field surveys were then carried out, and identified sites were classified according to type and were then mapped and inventoried. These sites form the basis for evaluating the utility of the predictive model. This evaluation included two assessments: (1) the comparative frequency of occurrence of sites in areas of predicted high and low sensitivity; and (2) the ability of the survey to identify sites which appeared on historic maps.

Finally, an attempt was made to identify flaws in the model, so that future model implementations could be improved. This was done by detecting patterns in the actual distributions of the different site types encountered. The resulting patterns were compared with the patterns generated by the original predictive model. Discrepancies between actual locational patterns and those patterns assumed by the model were identified.

Sources of model data

Environmental data

Three key environmental variables were considered primarily for development of the prehistoric predictive model, and were used for the historic model as well. These were surface slope, proximity to streams and soil drainage. Slope was derived from the 7.5 minute topographic quadrangle map (United States Geological Survey, 1982). Zones of slope, based on distance between contour lines, were delineated on the topographic map according to three discrete categories: (1) slopes under 5 per cent (gentle); (2) slopes between 5 and 15 per cent (moderate); and (3) slopes over 15 per cent (steep). The resulting map of slope–class zones was digitized as a polygon coverage (see 'GIS map encoding' below). Stream locations were also taken from the topographic quadrangle, and were supplemented with information on intermittent stream locations and springs, as indicated on maps provided by the US Army including the Fort Drum Terrain Analysis document (US Army, 1977). Stream courses were digitized as vector coverages (see 'GIS map encoding' below). Soil classes were derived from available soil maps. These were first traced onto USGS quadrangles by photo-rescaling and overlay and were then digitized as a polygon coverage. Roughly two thirds of the project area was mapped by modern soil survey (Soil Conservation Service, n.d.); the remaining third was mapped by an early, low-resolution survey (Carr *et al.*, 1913). Soil drainage classes were derived from soil descriptions, and included eight classes ranging from very poorly drained through excessively drained. The digitized soil zone maps were ultimately recoded into drainage classes.

Historic map data

Data pertinent to the historic sensitivity model consisted of mapped site locations and courses of historic roads. Locations of mapped structures were to be field-examined for evidence of archaeological remains. Roadways were deemed to be the most significant variable accounting for historic site locations based on previous studies conducted in the northeastern United States (eg., Louis Berger and Associates, Inc., 1984). The primary data sources for historic sites and roadways were the Beers Atlas of 1864 (Beers and Beers, 1865) and the Robinson Atlas (1888). Data from these maps were transferred to the USGS quadrangle. From this base map the site distribution was digitized as a point coverage and the road network was digitized as a vector coverage (see 'GIS map encoding' below). In all, 128 historic site locations, mostly farmsteads, were inventoried (Figure 23.2). Those sites within the historic town of LeRaysville represent a special case, that is, a rural village, and were not included in any of the analyses.

Ground disturbance areas (GDAs)

Since the establishment of Pine Camp in 1909, the project area has been subject to a variety of ground disturbances resulting from erosion and military activities. The ultimate integrity of archaeological sites, and hence their significance, is dependent on the level of disturbance. This is measured along two axes: (1) the degree of destruction of the natural soil profile within disturbed areas; and (2) the proportion of a given area that has been subjected to disturbance. Together these two measures yield a probability that a given archaeological site has remained intact. A professional pedologist, Dr Daniel Wagner of Geo-Sci Consultants, was hired to define ground disturbance areas. Through field sampling, 14 different disturbance types were identified, ranging from historic ploughing to mine fields, and including trails, roads, foxholes and bivouac areas. Disturbance types were classified into two groups: slight and severe. The latter class consists of disturbances in which the original soil profile has been disturbed to depths exceeding one foot or where the original surface horizon has been completely removed. Slight disturbances typically involved historic ploughing, which occurred throughout most of the project area.

Through a field survey by pedestrian reconnaissance and subsurface testing, the entire project area was mapped according to four classes of ground disturbance. These classes were based on the percentage of each zone subjected to severe disturbance, as follows: (1) slightly disturbed areas contained less than 20 per cent severe disturbance; (2) moderately disturbed areas contained between 20 and 75 per cent severe disturbance, but were generally less than 50 per cent disturbed; (3) severely disturbed areas contained more than 75 per cent severe disturbance; and (4) totally disturbed areas, which included such areas as bivouac locations, eroded and graded areas, borrow pits and paved developments. A final, miscellaneous zone was incorporated into the ground disturbance scheme as a fifth class, excessively wet areas. These included the floodplains of major streams in the project area, which were inaccessible during field survey due to flooded conditions. A map of the distribution of ground disturbance zones was drafted, and was overlaid onto the base map, the USGS topographic quadrangle. The map was then digitized as a polygon coverage discussed below.

GIS map encoding

All geographic source maps were digitized into a raster-based GIS. This involved subdividing the project area base map into a high resolution grid of pixels, or picture elements. A pixel size was selected such that a standard computer printout, where each print character represents a pixel, could be overlaid directly onto the 1:800 project area field maps. Accordingly, each pixel represented 80×100 feet on the ground, or 0.18 acre.

Each environmental source map (for example, soil zones) was then digitized into the GIS grid to form a gridded coverage, or digital image of the environmental map.

All gridded environmental maps were then combined into a three-dimensional, layered computer database. Each grid cell of the project area was represented by a collection of environmental variables, one variable from each layer of the database. The grid cells became the units of analysis, or records of a data file, and the layer values became the variables (see Kvamme, in Chapter 10 of this volume, for further explanation). The resulting data file consisted of 37,582 pixel records.

Maps were digitized at the Digital Image Analysis Lab (DIAL) of the University Computing Center (UCC), University of Massachusetts at Amherst. One of two techniques was used depending on the type of data. Vector data such as stream and road courses were digitized on an electronic digitizing table using MAPMAKER, a UCC-specific software package (Anderson, 1985). Polygon data such as soil and ground disturbance maps were optically scanned with an Optronics C4500 drum scanner. Zone centroids were encoded on a digitizing table and the zones were filled with their appropriate values using a computer polygon-fill routine. Raw digital data were encoded into the final GIS using software written by the senior author in ANSII standard FORTRAN V. The resulting maps, or digital images, were raster images containing data in each cell, either feature presence/absence (eg., road course) or zone value (eg., soil type).

Primary data images were then filtered through a variety of functions to produce secondary or synthesized coverages. Soil zones were re-classified into drainage categories. Stream and historic road courses were processed through a proximity function to produce proximity maps. Such a coverage represents the distance from a given pixel to the nearest feature of a given kind, measured in pixel units, or 100 foot intervals. A 1 km buffer surrounding the project area containing streams and roads outside the project area was included when the proximity coverages were calculated so that valid values would be assigned along the edges of the project area.

The basic predictive variables comprising the GIS included: slope, drainage, proximity to streams, proximity to historic roads, proximity to mapped historic sites, and ground disturbance areas. The proximity coverages were reclassified into discrete buffer zones to facilitate implementation of the predictive models.

Development of the predictive models

Overview

Archaeological sensitivity maps were generated through a series of GIS overlays according to the illustration in Figure 23.3. Models for both prehistoric and historic sensitivity were derived from the raw GIS variables and then overlaid with ground disturbance to determine the probability that predicted archaeological resources would still be intact. The outcome of this final overlay was a prescription for a given intensity of field survey for each parcel.

Model definitions

The rules governing each overlay were a set of logical decisions made for each pixel in the GIS based on the values of the input variables. Each pixel was assigned a discrete output value according to a predefined formula. The formulae used constituted the sensitivity models and were derived by Dr Michael Alterman and Robert Foss of LBA based on previous studies of the patterning of prehistoric and historic site locations (Foss, 1987).

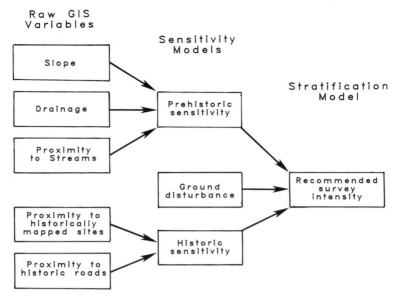

Figure 23.3. Logic of the predictive modelling procedure. Arrows indicate polygon overlay and reclassification decisions.

The basis of the prehistoric model was outlined above; high sensitivity areas were defined as those within 250 m of water, on moderately or better drained soils, and on slopes less than five per cent. Low sensitivity areas, on the other hand, were any areas greater than 750 m from water, on poorly drained soils, or on steep slopes greater than 15 per cent. All other areas were classified as being of moderate sensitivity.

The historic sensitivity model was based primarily on the locations of documented historic farmsteads and related rural sites and secondarily on historic roadways. Critical buffer zones surrounding these features were defined as the basis for the model. High sensitivity areas were defined simply as those within 400 feet of a mapped historic site location. Moderate sensitivity areas ranged between 400 and 700 feet from a mapped site or within 300 feet of an historic road. All other areas were classified as having a low potential for containing historic resources. The resulting distributions of archaeological sensitivity throughout the project areas are illustrated in Plates IV (prehistoric) and V (historic).

The final, and most complex, overlay operation was that combining the prehistoric and historic sensitivity models with the model of ground disturbance areas. The impact of disturbance and the prescribed field survey methods were defined separately for the prehistoric and historic models, as the nature of anticipated remains differed; historic sites would have included structural remains which would have greater preservation potential and visibility than most prehistoric sites. Given the three sensitivity classes in each of the archaeological models, and the five classes of ground disturbance, the simultaneous overlay of the three coverages resulted in 45 discrete possible combinations. Each of these combinations was assigned a survey strategy, chosen from among six separate levels of prescribed survey intensity. The decision matrix defining the overlay of the three models is shown in Table 23.1. Survey strategies are numbered 1 to 6 according to level of intensity.

Table 23.1. Decision matrix for the assignment of survey strategies to strata.

Archaeological sensitivity		Ground disturbance area				
Prehistoric	Historic	Slight	Moderate	Severe	Total	(Wet)
High	High	6	6	5	2	1
High	Medium	4	4	3	2	1
High	Low	4	4	3	2	1
Medium	High	6	5	5	2	1
Medium	Medium	4	3	2	1	1
Medium	Low	3	3	2	1	1
Low	High	5	5	5	2	1
Low	Medium	3	3	2	1	1
Low	Low	2	2	1	1	1

Notes: Recommended survey intensity:
(1), No survey;   (4), Systematic survey;
(2), Judgemental survey;   (5), Mapped site identification;
(3), Sample survey;   (6), Systematic survey and mapped site identification

Survey strategies were chosen on the basis of ground disturbance and the resource types anticipated, whether prehistoric or historic. Different survey techniques were anticipated for identifying the two site types, minimally, surface reconnaissance for locating historic cellar holes and subsurface shovel testing for identifying prehistoric habitation sites. The following are definitions of the six levels of survey intensity prescribed:

–No survey. Areas felt to have no potential for containing cultural resources were excluded from survey. Such areas fell into two categories: (1) totally disturbed areas with other than high potential for either prehistoric or historic resources; and (2) areas deemed to be excessively wet. The latter category included the entire floodplain of all the major streams in the project area.
–Judgemental survey. Zones with either low cultural resource sensitivity or total disturbance were assigned this strategy. Through pedestrian reconnaissance, unique features within the low probability zones would be identified and further surveyed. An example of such a feature might be a prehistoric rock shelter.
– Sample survey. Areas anticipated to contain a low density, but definite distribution of cultural resources were to be sampled. These included areas of moderate cultural resource potential and areas of moderate disturbance. Sampling designs were to be stratified, systematic, or random, depending on the size and composition of each survey segment.
–Systematic survey. This strategy was generally prescribed for high prehistoric sensitivity areas with partial or greater integrity. The strategy was to include subsurface testing to identify prehistoric sites. The entirety of each parcel in this class was to be surveyed, provided that test areas were undisturbed and possessed ground conditions suitable for prehistoric habitation.
–Mapped site identification. This strategy was assigned specifically to all zones surrounding mapped historic sites, unless the survey area was totally disturbed or on a floodplain. The intent of the strategy was to locate the documented site. At the minimum, surface inspection was to be used to identify structural remains. In addition, subsurface testing would be employed if structural remains could not be identified.
–Combined systematic survey and site identification. Areas with both a high

prehistoric potential and proximity to documented historic sites fell into this class. Presumably, surface inspection for historic structural remains would be made in the course of conducting subsurface testing for prehistoric sites. Historic artifacts recovered from subsurface testing would supplement structural archaeological information.

Implementation of field surveys

Plots of the distribution of prescribed survey strategies were produced as colour-coded overlays of the USGS quad, laminated in plastic for field use. Field survey initially commenced according to the prescribed strategies. Shortly into the survey two problems became apparent. One was the difficulty of locating survey segments on the ground. After all GIS polygons were overlaid, the resulting slivers often measured only a pixel or two in width, a couple of hundred feet. At the scale of a USGS quad, the resulting map was unrealistically detailed and it became difficult to identify slivers in the field. The other problem was that historic sites soon became recognized in areas of predicted low sensitivity. For these reasons, the modelled survey maps were abandoned and actual field strategies shifted to include at minimum a systematic surface reconnaissance of the entire project area. This strategy resulted in the identification of numerous historic sites which provide a critical, unbiased database for evaluating the validity of the original model.

Methods for model evaluation

As stated in the overview, the cultural resource sensitivity models were evaluated by comparing the relative frequencies of sites encountered in various modelled sensitivity zones, and by accounting for the number of historically documented sites that were identified. Finally, the locational patterning of site types identified was ascertained, and these patterns were compared with the original model assumptions.

All analyses were performed using a 100 per cent sample of the 37,000 project area GIS pixels as the spatial universe for potentially occurring cultural resources. Compared with this universe were the locations of the identified sites. Sites were classified into three predominant types; farmsteads, trash dumps and maple sugar processing sites. To each site record was appended the collection of GIS variables characterizing the pixel within which the site was located. A file of combined project area pixel records and site records constituted the database used in all analyses. One exception in the database was the exclusion of the settlement of LeRaysville from all analyses (Figure 23.2). This was done because the settlement consisted of a dense cluster of residences and support facilities which were not representative of the remainder of the project area. Accordingly, all historically documented sites in LeRaysville and the project area pixels immediately surrounding them were excluded so as not to confound analytical results. Below are descriptions of analyses performed.

Analysis of site frequencies

This analysis was simplified by reclassifying the project area into zones where historic sites were predicted to occur and those where they were not. Generally, zones of predicted high historic potential were to have received intensive, systematic survey, whereas low potential zones were subject to marginal survey coverage, consisting of sampling, judgemental inspection or no survey at all. Because prescribed survey included intensive prehistoric survey in some low historic sensitivity zones, this variable was not used as a basis for reclassifying historic potential. Instead, zones were reclassified explicitly on the basis of modelled historic sensitivity and ground disturbance according to the survey decision matrix (Table 23.1). Specifically, areas were considered to have low historic potential if one of the following conditions held: (1) the zone was deemed to have low sensitivity according to the historic model; (2) the zone was rated as moderately sensitive, and ground disturbance was severe or worse; or (3) the zone was rated highly sensitive, but ground disturbance was total or the area was deemed excessively wet.

Following this reclassification scheme, both project area pixels and sites identified by type were tabulated into either low or high potential categories. Total pixels tabulated represent the area of zones classified and sites tabulated represent the number of sites identified within these areas. Site frequency, or density, was calculated by dividing the latter by the former. If the model were valid, site frequencies would be significantly higher in high potential zones than in low potential zones.

Assessment of historic site identification

This analysis had several purposes: (1) to evaluate the effectiveness of field survey strategies in identifying historically documented sites; (2) to assess the impact of ground disturbance on the population of historic sites; (3) to evaluate the relative accuracy of the original historic maps; and (4) to evaluate the accuracy of transferral of site locations to the USGS base map. The analysis was conducted by comparing the location of each historically mapped site with the nearest farmstead identified in the field. This was done arbitrarily, without any attempt to match site for site. Despite this limitation, the analysis was productive (see 'results of analysis' below), as many historically documented sites were mapped in isolated locations, far from any identified site. The analysis was simplified by classifying mapped sites into three categories based on a bimodal distribution of distance to the nearest identified site: (1) mapped site verified, or archaeological site identified within 400 feet of mapped location; (2) site verification uncertain, or identified site between 400 and 800 feet distant; and (3) site not identified, or the nearest identified site greater than 800 feet distant.The three classes of site verification were then compared with ground disturbance rating, to ascertain the role of site destruction in the failure to identify sites.

Analysis of site locational patterns

The chief goal of this analysis was to detect non-random tendencies in the location of the three site types identified in the field, farmsteads, trash dumps and maple sugar processing sites, in addition to predicting the locational patterns of documented historic sites. This was done by executing a discriminant analysis between the locations of each site type and the potentially inhabitable universe which was the file of 37,000 GIS pixels. The latter data set constitutes a control data set with which to compare the site locations, and is a key requirement of pattern recognition (see Kvamme, Chapter 10 in this volume, for further discussion). The discriminant analysis procedure defines the difference between site locations and random locations across the landscape in terms of a collection of critical independent environmental variables. The independent variables used in this analysis were all those which may have affected the location of sites identified. These included slope, drainage, modern ground disturbance and proximity to a variety of locations features, viz., streams, historic roads, historically mapped sites, mills and town centres. For the analysis of trash dumps and maple sugar processing sites, proximity to farmsteads included those that were historically mapped as well as those identified in the field. Proximity to historically mapped sites was not included in the analysis of mapped historic sites themselves, for obvious reasons.

Statistical procedure employed

The procedure employed was a step-wise discriminant analysis. In particular, the subprogram DISCRIMINANT provided within the SPSS statistical package was used, and the Wilks step-wise method within the subprogram was selected (Klecka, 1975, p. 434–467). On the first step, the procedure isolates the single environmental variable that best explains the difference in location between the sites analyzed and the corresponding random control points. This may be interpreted as the key factor in site location, although the role of the variable may be indirect. The second step of the procedure isolates the next variable which best explains the difference in location between sites and random control points beyond the difference explained by the first variable. The procedure continues until all patterned differences between sites and control points have been accounted for. Beyond the terminating step, site locations are essentially random with respect to the remaining environmental variables.

At each step, the subprogram provides a statistical measure of the relative strength of each variable included in the discriminant function up to that step. This is the *F-to-Remove* statistic, which is a measure of the ratio of variance within each group (for example, sites versus control points) as compared to the variance between groups, with respect to the given variable. The statistic measures the relative proportion of the total variability in site location accounted for by each variable. The absolute value of the statistic decreases for each variable as more variables are included in the discriminating formula. That is to say, as more variables are combined to account for site location, the relative strength of each individual variable diminishes, as each variable accounts for some of the total variability.

After the final step has been executed, the procedure generates the formula for a discriminant function used to differentiate sites from control points on the basis of the discriminating variables. The formula contains coefficients for each variable included. The coefficients are weighting factors for each standardized variable, that is, each variable as measured in standard deviations. Hence the absolute values of the coefficients indicate the relative weight of each variable in the discriminating equation. The sign of each coefficient indicates whether the variable is directly or indirectly associated with site location. A negative sign for proximity variables indicates that sites are closer to the given variable than random control points.

Statistical significance of the discriminant function

The SPSS DISCRIMINANT procedure calculates the significance of the difference between the two groups being analyzed on the basis of the number of cases in the groups. For most analyses, this usually involves two sample groups, such as male students in a class versus female students. As the present GIS analysis represents an unconventional application of the discriminant procedure, the total number of cases analyzed is artificially high, as a substantial sample of control pixels was selected (numbering in the thousands). Hence, the significance statistics generated are all artificially significant, and are essentially meaningless.

An alternative significance test which does not rely on number of cases, the one used here, is the Jackknife test technique. This test was selected as OPTION 5 of the SPSS DISCRIMINANT procedure. According to this test, the discriminant function generated by the procedure is applied to each case in the analysis, and an overall discriminant function value is calculated for the case. The value is given in terms of standard deviations of the discriminant function value, according to its variation across all cases. Mean discriminant function values are calculated for the two separate groups (sites versus control points here). Because the control data set is random, its mean value is zero, or close to it. The mean value for the site sample is some number of standard deviations from zero; the greater the figure, the clearer the separation between sites and random points.

In the Jackknife test, each case is assigned to one of the groups being analyzed, on the basis of the discriminant function value calculated for it, and its comparison with the two group mean values. Hence, the assignment of a case to a group is independent of its original group assignment. The accuracy of the discriminant procedure is thereby measured in terms of the numbers of cases that are correctly re-classified into their proper group assignments on the basis of the discriminant function. If both groups are randomly distributed with respect to each other, then fifty per cent of each group will be re-classified into the other group. If the groups are well separated, then, ideally, less than five per cent of each group will be incorrectly re-classified. The Jackknife test is further discussed below.

Results of analyses

Project area stratification

Acreage breakdowns of the project area according to prescribed survey strategy were calculated by tabulating the frequencies of GIS pixels in each strategy class, then dividing each total frequency by pixel area (0.18 acre). Roughly a quarter of the entire project area was to be intensively, systematically surveyed for either prehistoric sites, historic sites or both (Figure 23.4). A substantial proportion, 13 per cent, was exempted from survey altogether on the grounds of being either totally disturbed or excessively wet. Finally, roughly two thirds of the project area was to be surveyed in part, either by sampling or through judgemental selection.

As stated above, the stratification model was abandoned in the field when it was realized that sites were occurring in areas predicted by the model to have low site potential. Following this decision, most of the project area was subjected to walkover survey, at minimum. As a result, over 100 historic sites were identified. These consisted of 63 farmstead sites, 25 trash dumps and 22 maple sugar processing sites (Figure 23.2). The trash dumps inventoried do not include midden areas associated with farmstead sites. The maple sugar processing sites typically consisted of stone ramps and abandoned equipment. Most of the maple sugar processing sites, and many of the trash dumps, were situated in the back lots of farms, away from houses and roads. As such, many fell within zones predicted to have low historic sensitivity by the model.

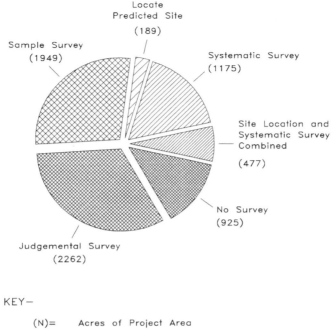

Figure 23.4. Breakdown of the project area by recommended survey strategy.

Site frequency by zones of historic potential

As described in 'methodology', the project area was reclassified for this analysis into zones of high and low potential for containing historic sites, based on the model's assumptions regarding proximity to roads, proximity to mapped sites and modern ground disturbance. Numbers of sites of the three different types identified within each of these zones are tabulated in Table 23.2.

Table 23.2. Area surveyed and sites identified, by historical potential.

Area surveyed/ sites identified	Anticipated historic potential	
	Low	High
Acres of project area	5,298 (78%)	1,500 (22%)
Farmstead sites	10 (16%)	53 (84%)
Trash dumps	16 (64%)	9 (36%)
Maple sugar processing sites	20 (91%)	2 (9%)

The distribution of farmstead sites illustrates a typical example of the '80–20' rule, viz., 84 per cent of the sites occurred within 22 per cent of the project area, and 16 per cent occurred within the remaining 78 per cent of the project area. Whereas this distribution may seem managerially reasonable, it was scientifically unjustifiable to 'write off' the 10 farmsteads in the low potential areas; hence, the areas were surveyed. Further examination of the 10 site locations reveals the problems with the original stratification model.

Of the 10 sites, seven occurred in zones that were deemed to be excessively wet, eg., on stream floodplains. These were areas that were exempted from survey by the stratification model. Because these sites represent 11 per cent of the population of farmsteads identified, and because the floodplain zone constitutes only 7.6 per cent of the project area, it appears that floodplains are actually actually *high* potential areas, rather than being of low potential as originally assumed.

Of the remaining three farmsteads not conforming to the model, two were located in severely disturbed zones. Because they were both in moderate sensitivity historic zones, they fell within areas exempted from intensive survey by the stratification decision matrix. The problem, therefore, was not with the historic sensitivity model, *per se*, but with the assumptions about ground disturbance areas, which will be discussed further below.

Were it not for disturbance and perceived excessive wetness, the 10 'aberrant' site locations would have fallen within the parameters specified for the historic sensitivity model alone. Six of the 10 sites fell within 300 feet of a mapped historic site, or within the high-sensitivity zone. All of these were also within 400 feet of an historic road. Of the remaining four sites, three fell within 300 feet of an historic road, which would have been the moderate sensitivity zone. Only two of the sites exhibited locational parameters falling outside the historic model limits; one was situated 800–900 feet from the nearest mapped historic site, and the other was situated 600–700 feet from an historic road.

Unlike farmsteads, trash dumps and maple sugar processing sites failed to conform to the historic sensitivity model. Nearly two thirds of the trash dumps and almost all of the maple sugar processing sites were identified within zones of predicted low site potential. This was not because of disturbance or wetness; sites were distributed rather randomly with respect to these variables and, in fact, no maple sugar processing sites occurred in floodplain zones. The chief reasons for their non-conformity were their distance away from houses and roads alike. As stated above, they tended to be located in back lots (Figure 23.2). Their particular locational patterns will be discussed in the following section.

Verification of mapped historic sites

As stated in 'methodology', an attempt was made to assess the success of the field survey in relocating farmsteads that were historically mapped. Excluding the sites in the town cluster of LeRaysville, 94 sites were mapped throughout the project area (Figure 23.2). These were classified for verification on the basis of the nearest field-identified farmstead. If the latter were within 400 feet, the site was considered relocated, if greater than 800 feet, it was considered not.

Of the 94 mapped sites, one fifth (19) were not identified on the ground within a reasonable distance of the mapped location; roughly sixty per cent (55) were. To examine the possibility that the 19 missing sites were destroyed, the sites were tabulated by ground disturbance zone (Table 23.3). This revealed that disturbance probably had an effect on site survival and, hence, site identification. Two thirds of the sites in the slightly disturbed zone were identified, whereas two thirds of the sites in the severely disturbed zone were not identified. The latter sites may have been destroyed by military activities (eg., tank manoeuvres) within the severely disturbed zones.

Results of this analysis show that the site locations indicated on the historic

Table 23.3. Verification of historically mapped sites by disturbance zone.

Ground disturbance area	Verification status		
	Site located	Status uncertain	Not located
Slightly disturbed	45 (65%)	15 (22%)	9 (13%)
Moderately disturbed	2 (100%)	0 (—)	0 (—)
Severely disturbed	2 (22%)	1 (11%)	6 (67%)
Totally disturbed	2 (50%)	0 (—)	2 (50%)
Excessively wet	4 (40%)	4 (40%)	2 (20%)
All areas combined	55 (59%)	20 (21%)	19 (20%)

maps are reasonably accurate, and can be used confidently for modelling purposes. Failure to locate some of the mapped sites may be due to the lack of surviving foundations on some of the mapped structures. Another factor may involve varying degrees of cartographic error, which will be discussed further below.

Patterning of site locations

Overview

As an independent control on the stratification model, actual site locational patterns were ascertained through discriminant analysis, as outlined in 'methodology'. The procedure was intended to identify environmental variables to which sites were non-randomly patterned. The procedure was executed for each of four different site types; mapped historic sites, field-identified farmsteads, trash dumps and maple sugar processing sites. Results of these four analyses are presented in Table 23.4. For each of the analyses the variables are listed that were selected as significantly accounting for differences between the locations of sites and locations of all pixels in the GIS. The variables are listed in order of importance, according to the step in which each variable entered the function. The statistics listed are for the final discriminant function; these are: (1) the F-to-Remove statistic, which measures the relative strength of the variable in the final function; and (2) the variable coefficient, which indicates the sign of the association and—along with the F statistic—the relative strength of the variable.

The statistical significance of the site locational patterns were tested by the Jack-knife method, as explained in 'methodology'. The purpose of this test was to ascertain how well the discriminant function could classify an actual site or an actual random control point into the proper category. The results of the test reflect how distinctive site locations are with respect to the random landscape. Sites that are located randomly across the landscape will have a fifty per cent chance of being re-classified as a site by the Jackknife test.

The Jackknife test was run for each of the four analyses executed; results are

Table 23.4. Results of discriminant analyses on four different site classes.

Step entered	Variable entered into the function	F-to-Remove	Function coefficient
(A) Historically mapped sites			
1	Distance to the nearest historic road	109.1	−0.89
2	Distance to the nearest mill/factory	6.8	−0.21
3	Soil drainage	6.6	0.23
4	Distance to the nearest stream	4.9	−0.23
5	Topographic slope	2.2	−0.13
(B) Field-identified farmsteads			
1	Distance to the nearest historic road	21.6	−0.62
2	Distance to the nearest historically mapped site	5.4	−0.34
3	Severity of ground disturbance	4.0	−0.20
4	Distance to the nearest stream	3.4	−0.21
5	Topographic slope	2.3	−0.15
6	Soil drainage	2.4	0.15
7	Distance to the nearest mill/factory	2.2	−0.15
(C) Trash dumps			
1	Topographic slope	5.7	0.65
2	Distance to the nearest historic road	2.5	0.44
3	Soil drainage	2.7	−0.46
4	Distance to the nearest stream	1.9	0.40
5	Distance to the nearest mill/factory	1.6	−0.36
(D) Maple sugar processing sites			
1	Distance to the nearest stream	24.2	0.90
2	Distance to the nearest historic road	4.9	0.36
3	Topographic slope	6.6	0.42
4	Distance to the nearest mill/factory	2.3	−0.26
5	Soil drainage	1.8	−0.22
6	Severity of ground disturbance	1.3	0.19

presented in Table 23.5. For each run test statistics are listed for the two groups analyzed, sites and control points. Statistics listed by column are: (1) the mean overall discriminant function score for the group; (2) the number of cases analyzed in the group; (3) the number and percentage of the group cases classified by the Jackknife test as site locations; and (4) the number and percentage of the group cases classified by the Jackknife test as random control points. Actual sites classified as control points represent aberrant sites, or those not conforming to the modal pattern of location. Random control points classified as site locations represent potential locations of unreported sites, eg., blocks of land sharing environmental character-istics typical of known sites, but within which no sites have been reported.

The following are summaries of the locational patterns of the four site types and their significance.

Farmsteads

Results of the analysis show farmstead sites to be predictably patterned, and generally confirm the original model of historic sensitivity developed by LBA. Mapped historic sites, most of which are farmsteads, are by far most patterned toward roadways (Table 23.4A). Field-identified farmsteads show a similar tendency,

Table 23.5. Significance of discriminant analyses: the Jackknife test.

Sample analyzed: site type	Actual group membership	Mean function value	Cases in group	N of cases (and %) reclassified as			
				Site loc's		Proj. area	
(A) Historically-mapped sites	Site loc's:	−1.24	94	89	(95%)	5	(5%)
	Proj. area:	−0.00	36799	11487	(31%)	25312	(69%)
(B) Field-identified farmsteads	Site loc's:	−1.40	59	55	(93%)	4	(7%)
	Proj. area:	−0.00	36799	10111	(27%)	26688	(73%)
(C) Trash dumps	Site loc's:	0.77	24	14	(58%)	10	(42%)
	Proj. area:	−0.00	36799	11551	(31%)	25248	(69%)
(D) Maple sugar processing sites	Site loc's:	1.34	22	15	(68%)	7	(32%)
	Proj. area:	−0.00	36799	8808	(24%)	27991	(76%)

though proximity to mapped historic sites is secondary (Table 23.4B). Figure 23.5 illustrates the distribution of identified farmsteads with respect to proximity to historic roads and mapped sites. Sites are distributed very closely to roads, nearly all lying within 400 feet. This is partly a result of the stability of the road courses within the project area since the early nineteenth century, which permitted their accurate mapping onto the USGS base map. This may not be the case in other project areas.

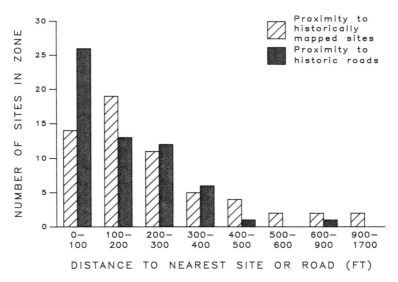

Figure 23.5. The distribution of identified farmsteads according to proximity to historically mapped sites and roadways.

Proximity to mapped sites is less well defined than proximity to roads. The several outlier sites could represent cartographic error, or else they could be sites that did not appear on the historic maps (Beers and Beers, 1865; Robinson 1888). They could represent eighteenth century sites that were abandoned by the time the maps were compiled, or they could be sites post-dating the maps—possibly through 1941 when the military reservation was established. It is also possible that the sites existed at the time the historic maps were compiled, but were not entered on the maps. This would be a typical problem in areas remote from town centres.

The patterning of farmstead sites according to the discriminant analysis is rather significant, as indicated by the Jackknife test (Table 23.5). Generally 95 per cent of the sites, mapped or field identified, were correctly classified as such. Of the project area, generally 30 per cent was classified as a location potentially containing an historic farmstead. Apparently, the discriminant function stratified the project area somewhat more conservatively than did Alterman on the basis of intuition. It included roughly 10 per cent more project acreage and sites in the pattern than did the LBA model, which estimated roughly 20 per cent of the project area and 80 per cent of the sites to fall within the high sensitivity zone. If the model's buffer zones were extended by another layer of pixels, or 100 feet, then the resulting stratification would closely mirror that of the discriminant procedure.

Trash dumps

Results of the discriminant analysis show an association of trash dumps with steep slopes (Table 23.4C). This variable accounts for nine of the 25 dumps, or 36 per cent, which occur on moderate or steep slopes. This is significant since 81 per cent of the project area is level or only gently sloping. The association is probably related to the practice of disposing refuse in ravines and over hillsides. Other locational factors are remoteness from roads and association with poorly drained areas. These probably relate to dumps at the rear end of farm lots and to the filling of low areas. The overall locational pattern of trash dumps is not distinctive, as the Jackknife test classified them close to 50-50 as being either site locations or random points on the landscape (Table 23.5C). Figure 23.2 shows that trashdumps were generally located either in remote areas or else near farmsteads. This may reflect functional differences in the type of dump. Clearly more study needs to be done of dumps before an accurate locational model can be developed.

Maple sugar processing sites

Curiously, the chief variable accounting for the location of this site type is remoteness from streams (Table 23.4D). This association is spurious and simply reflects the fact that most of the maple sugar processing sites are situated in the interior of the project area (Figure 23.2). The project area is generally circumscribed by streams (Figure 23.1), hence the negative association. There was also a weak association with remoteness from roads and with moderate or steep slopes. Many of the sites were situated in woodlots to the rear of cleared farmlands, often in upland areas—accounting for the latter associations. Based on these criteria, the Jackknife test was able to stratify the project area such that only one quarter of it was optimal for maple sugaring activities (Table 23.5D). The sites themselves were not so well sorted; only two thirds of them were classified as being typical locations. This is probably because some of the sites were situated immediately adjacent to farmsteads (Figure 23.2).

Summary and discussion

The goal of this study was to determine the most effective means for surveying the 6,700 acre project area being slated for development at Fort Drum. Appropriate archaeological, historical and environmental data were incorporated into a GIS which was employed as a tool for developing a predictive model of cultural resources. The model was used to divide the project area into five discrete strata in which varying levels of survey intensity were to be applied. The results of the field survey, that is, the distributions of sites identified, were used as the basis for evaluating the predictive model. This evaluation exposed several problems.

First, stream floodplains were initially assumed to have been too wet for human habitation. This assumption proved to be invalid, as several farmsteads were located within floodplain zones. In fact, zones along streams are particularly sensitive to early historic sites which focused on water power, viz., mill sites. Floodplain zones could contain localized areas of high ground, such as levees or terraces, and could have been dry on a seasonal basis, such as during the autumn. In the future, archaeologists should be more careful about discounting the sensitivity of floodplain zones.

Another problem was the assumption about disturbance. Because ground disturbance zones were defined on the basis of the proportion of area actually disturbed, the likelihood of encountering an intact historic site in a given zone became a matter of probability. For example, if a zone were deemed 80 per cent disturbed, it could be expected that 20 per cent of the sites originally in the zone would have survived. Hence, such zones should have been subjected to further refined stratification based on localized patterns of disturbance and intact ground. The model would merely estimate the acreage of intact ground to be surveyed.

A third problem was the difficulty of implementing the final survey distribution on the ground. As is pointed out elsewhere in this volume by Allen, Green and Zubrow (Chapter 29), the more layers that are overlaid in a GIS, the more the accuracy of the resulting map coverage diminishes geometrically. The result of overlaying many polygons is a mosaic of discrete slivers. As sliver size diminishes with overlay, the slivers become increasingly difficult to locate on the ground. One means of resolving this problem may be to abandon the polygon overlay methodology, and use a strictly raster-based modelling procedure. Instead of creating buffer zones from proximity maps, or pigeonholing pixels into sensitivity strata, pixels could be assigned sensitivity indices on a continuous scale, using one of several statistical procedures (see Kvamme, Chapter 10 in this volume). The resulting sensitivity maps would be probability surfaces continuous over space. These would be much easier to project onto the ground than a polygon overlay. Another possibility for improving the usefulness of GIS on the ground is to adopt multiple levels of resolution. For example, proximity to town centres could be derived from regional-scale maps, whereas stream terraces could be derived from project-specific topographic survey maps.

Another problem, related to map accuracy, was the reliability of historic maps in locating former structures on the ground. In this study, the maps were found to be reasonably accurate, owing to the stability of road and town locations through time. This is not always the case in other areas. A common problem is cartographic accuracy, which could be controlled by computer geocorrection, except that the content of the map may be erroneous. Sites predating the earliest maps may have been occupied and abandoned by the time the maps were compiled. Sites existing at the time of map compilation may have been overlooked; particularly in areas remote from town centres. Often, placement of sites on maps was done by relative distance rather than absolute, for example, a sequence of neighbouring farmsteads along a rural route. The pinpointing of mapped historic site locations on the ground could be facilitated by a number of measures. Land deeds could be consulted to ascertain permanent landmarks. Early USGS topographic maps, which are fairly accurate, could be examined for structures that may still have been standing at the time of mapping. Aerial photographs could be inspected for traces of old roadbeds, or for stone walls which would have marked property boundaries. Such efforts would be cost-effective especially when stratifying large project areas.

Perhaps the most significant problem encountered in this study was that of predictive modelling of historic site locations. The primary problem with the historic model was its emphasis on historically mapped farmsteads and roads. This resulted in an underemphasis on the rear lots of farm properties. Subsequent survey indicated that potentially significant historic resources, including refuse deposits and resource processing locations (eg., maple sugar sites), were often situated well outside the high and moderate historic sensitivity areas. This problem exemplifies a common flaw of archaeological predictive models: they are typically designed to predict the location of a generic site, whether historic or prehistoric. In fact, there are many different functional classes of site, as well as differing settlement patterns during different time periods. What is truly needed is a composite of many different predictive models, for different site classes, which can be overlaid onto any given project area to yield a composite distribution of archaeological sensitivity. To achieve such diversity in predictive modelling, GIS will need to incorporate new and different variables. For example, the present study could have benefited from an examination of soil suitability for maple trees, to predict where maple sugar processing sites could occur. New site types need to be identified, and their distributions studied, so that variables pertinent to their location can be built into archaeological GIS. Only then can a broad spectrum of site types within a given project area be assessed.

Conclusion

GIS are a useful tool for cultural resource management. They can be used to estimate acreages of and produce maps for areas to be surveyed. Planners can use GIS to model the impact of alternative construction designs, to implement sampling strategies, whether random or systematic, and thereafter to estimate from the sampling results the population of cultural resources present.

Perhaps the most common use of GIS in cultural resource management is in predictive modelling. This may also be the most common misuse. Predictive models are often designed to model the 'typical' site, and to predict the locations of other sites of its kind. Optimization of survey strategies is often biased towards that 20 per cent of a project area likely to yield 80 per cent of the sites. Yet, those 80 per cent are likely to be the least anthropologically interesting. It is the tails of the pattern distribution, the 'aberrant' sites, that are the most unique, and, according to National Register criteria, the most significant.

Although GIS can be used to define conformity, they can also be used to examine diversity. GIS permit models—or hypotheses regarding human settlement patterns—to be formulated and tested rigorously in the field, as has been done in this study. Flaws in a model can be defined, and the model can be refined and retested. Tails in the distribution of sites can be defined, and those tails can become the focus of future modelling effort. By continually refining our models, we can hope to advance our knowledge of history and human behaviour.

Acknowledgements

This study was carried out under contract by Louis Berger and Associates, Inc. (LBA), for the National Park Service, Mid-Atlantic Office and The US Army. The research was facilitated with the cooperation of: Dr John Hotopp, Director, Cultural Resource Group, LBA; Lloyd Chapman of the National Park Service Mid-Atlantic Regional Office, Philadelphia; Curt Williams and Dr David Guldenzopf (Archaeologist) of the 10th Mountain Division, Fort Drum, New York; and Bruce Fullem and Robert Ewing of the New York State Office of Parks, Recreation and Historic Preservation. Dr Daniel Wagner of Geo-Sci Consultants, Inc., along with Dr Gary Shaffer and Henry Holt of LBA, were responsible for surveying ground disturbance zones within the project area. The GIS was developed at the Digital Image Analysis Lab, University Computing Center, University of Massachusetts at Amherst, with the support of David Oliver and Dr Duncan Chesley. The bulk of the project research and analysis was conducted by the staff of LBA. Dr Amy Friedlander was responsible for the historic background research. The predictive model was formulated by Dr Michael Alterman and implemented by Robert Foss. Archaeological field survey and analyses were conducted by Carolyn Pierce, Victor Stolberg, Marcus Grant, Henry Holt, Alain Outlaw, Cait Shaddock, and Phillip Waite. Finally, several individuals at University of Massachusetts at Amherst assisted in the study. Paul Abbott digitized the raw GIS data. Dave Lacy shared his insights in historic site identification in the Green Mountain and Finger Lakes National Forests. Mark Goodberlet and Elizabeth Chilton formatted the final text for the manuscript.

References

Anderson, D., 1985, *MAPMAKER Digitizing System User's Guide.* Version 1.0. Information Systems Group, University Computing Center, University of Massachusetts, Amherst

Beers, S.W. and Beers, D.G., 1865, *New Topographical Atlas of Jefferson County, New York* (Philadelphia, PA: C.K. Stone)

Carr, M.E., Gilbert, B.D., Morrison, T.M. and Maxon, E.T., 1913, *Soil Survey of Jefferson County, New York* (Washington, DC: US Department of Agriculture, Bureau of Soils)

Foss, R., 1987 *Cultural Resource Inventory Evaluation, Recording and Management Planning; Fort Drum, New York: A Report on Task Order No. 2, Archaeological Resource Preservation Potential,* (East Orange, NJ: Louis Berger and Associates, Inc.)

Friedlander, A., LeeDecker, C. and Foss, R., 1986, *Re-Evaluation of Rural Historic Context for Fort Drum, New York Vicinity.* Fort Drum Cultural Resources Project Report No. 2 (East Orange, NJ: Louis Berger and Associates, Inc.)

Klecka, W.R., 1975, Discriminant Analysis. In *Statistical Package for the Social Sciences*, 2nd edition, edited by N.H. Nie, C.H. Hull, J.G. Jenkins, K. Steinbrenner, and D.H. Bent, (New York: McGraw-Hill), pp 434–467

Klein, J.I., Wise, C., Schaffer, M. and Marshall, S.B., 1985, *An Archaeological Overview and Management Plan for Fort Drum.* Report submitted to the National Park Service and the US Army by Envirosphere Company

Louis Berger and Associates, Inc., 1984, *Route 92: Technical Environmental Study,*

Cultural Resources. Report Submitted to the New Jersey Department of Transportation and the Federal Highway Administration

Powell, T.F., 1976, *Penet Square: Episode in Early History of Northern New York,* (Lakemont, NY: North Country Books)

Robinson, E., 1888, *Robinson's Atlas of Jefferson County, New York,* (New York: E. Robinson)

Schlebecker, J.T., 1975, *Whereby We Thrive: A History of American Farming, 1697–1902,* (Ames: Iowa State University Press)

Soil Conservation Service, 1990, *Preliminary Soil Survey of Jefferson County, New York,* (Washington, DC: US Department of Agriculture)

United States Department of the Army, 1977, *Fort Drum, New York Terrain Analysis* (Fort Belvoir, VA: The Terrain Analysis Center, US Army Engineer Topographic Analysis Center)

United States Geological Survey, 1982, Black River, New York sheet, 7.5 minute topographic quadrangle series, 1:24,000. Revision of 1958 original.

24

Modelling and prediction with geographic information systems: a demographic example from prehistoric and historic New York

Ezra B.W. Zubrow

Introduction

Geographic information systems (GIS) are beginning to be used in anthropology and archaeology (Zubrow, 1987; Kvamme, 1989). They are being applied to a wide variety of contexts and for a diversity of purposes. GIS are spatial record keeping systems. Thus, they are tailor-made for archaeology. They allow one to store, query, display and manipulate spatial data which are the bread and butter of every archaeological problem. One may create a regional database for examining site distributions. These can be used to assess the archaeological sensitivity of a region, to predict dynamically new site locations or to retrieve comprehensible patterns of coherent information from large amounts of spatial data.

GIS are being used as tools for reconstruction. They make archaeologists into chefs who can 'rebake' the 'layer cake' of an archaeological site. Since most GIS are based upon an overlay geometry, it is easy to utilize the GIS layers as stratigraphic layers. The artifacts can be placed in each geological and cultural layer and the cake can be built. Each layer is added until the icing of the present topographic surface is completed. Once finished the cake can be cut in as many different ways as the archaeologist can imagine and new relationships discovered. *Bon appetit.*

Sophisticated GIS are also useful for terrain representation. One can reconstruct present terrains or palimpsest past terrains. Thus, one may be able to reconstruct archaeological sites and place them in a valid interpretation of their past environment. Combined with their reconstructive abilities noted above, GIS make true 'landscape' interpretation possible. GIS can be used at all levels of resolution. From the macroscopic scale of archaeological regions and sites to the microscopic scales of palynology and spectroscopy, they are revolutionizing how evidence from archaeology is being marshalled.

However, this paper focuses on the use of GIS in a very different manner. Rather than highlighting their ability to store, query, display and manipulate spatial data, it emphasizes GIS as a tool for modelling. Modelling requires a cognitive shift. Some might even say a lateral lobotomy. There is a shift from 'left-sided' to 'right-sided' thinking. One shunts aside the empiricism of most archaeological information gathering and places one's preconceived expectations front and centre. The result is seldom the confirmation of ones expectations but the development or intuition of new ones.

Predicting the past

There is something incongruous about predicting the past. It strikes the reader as being vaguely amusing and slightly disconcerting. One is, after all, supposed to know the past. One experiences the present, and predicts the future. If part of the past is not known, one should 'find it' or 'discover it', not predict it. There is an awkward juxtaposition of tense and time which is a reflection of our concepts of prediction.

Prediction is making suppositions about future phenomena on the basis of well-grounded knowledge of the present. Analogously, predicting the past is making suppositions about past phenomena on the basis of well-grounded knowledge of the present. However, there are two important types of past prediction. They are based upon time moving in opposite directions. The first is very familiar. It is predicting backward from the present. It is well known to anthropologists and archaeologists under the rubric 'ethnographic analogy'. More formally, one builds a model based upon present knowledge and runs time backwards until one reaches the past moment one wishes to study. The second type of past prediction is to make a model based upon present knowledge and set it at a date in the past prior to the time one wishes to study. Then, one allows time to run forward until one reaches the period one wishes to examine. One makes predictions from an earlier past to a later past.

Purpose

This study uses GIS techniques to simulate various models of demographic migration and settlement in order to achieve three goals. These goals are:

(1) To determine more accurately the processes of settlement of a frontier;
(2) To learn heuristically the advantages and disadvantages of different models and their consequent scenarios;
(3) To determine the limiting factors of growth and migration.

More specifically, this paper concerns the spread of European populations through the state of New York between 1608 and 1810 and their interaction with the native, prehistoric and ethnohistoric populations. In this study, predictions are made of the rates and routes of the contacting populations.

Archaeological substance: the end of prehistory

From an archaeological perspective, one could call this subject the 'end of prehistory'. For as the European populations settled New York, the native populations and traditions were systematically destroyed through intentional and inadvertent genocide. The European colonization of New York took approximately two centuries. The population moved from the Hudson River to Lake Erie and from the Delaware River to Lake Ontario. These are linear distances of roughly four hundred miles which meant an average spread of two miles per year. Of course, the actual movement was complicated as towns were settled and farmland cleared. The exact sequence of settlement was determined by a number of factors including land availability, Native American politics and warfare, modes of transportation, protection and luck. However, once the European settlers had occupied the land the traditional forms of native culture ended and new forms based upon economic dependency and reservation became the new native culture.

Overview

In order to model the migration and settlement of the Europeans in New York as well as their interaction with the native population, I have constructed models which have four aspects. First, they are dynamic allowing populations to move along the hydrological routes of the state. Second, they have variable parameters including the rates of growth, the location and the initial size of the populations. Third, they include a mechanism by which one may trace the subsequent growth and spread of the population. Fourth, they allow one to test the simulated data from the models with the historical records of the actual occurrences of people moving and settling the land.

Methodology

The methodology for testing these models is relatively complex and consists of several stages. I constructed a network corresponding to the hydrology of New York State. Onto this network, I grafted a series of demographic growth models. I initialized the populations at different sizes and different locations and then ran a series of simulations to determine the settlement outcomes. These initializations corresponded to the six major drainages and directions towards which colonization took place. These models of growth are the Hudson model (to the west), the St. Lawrence (to the south), the Lake Ontario model (to the south), the Susquehanna (to the north), the Delaware (to the northeast) and the Mohawk (to the north and south). I tested these against the real data.

Perhaps the clearest presentation of my methodology is to draw an analogy to the methodology of constructing a house. One begins with the foundation. In my case the foundation is the hydrology of New York State which was digitized. This is a computerized version of a standard hydrology map. It will be the backbone of my model.

Next, one frames the walls. The analogous process for the model is to construct the network. Networks, like walls, serve a number of functions. Walls separate rooms, they provide infra-structural support, and they are surfaces upon which to hang pictures. Networks are systems of linear phenomena between junctions through which something flows—water, people, electricity. The junctions are usually abstracted as nodes and the linear phenomena as arcs. Their abstraction is a subset of graph theory which is a branch of mathematics (Price, 1977). For the purposes of my test of the model, I constructed a network using the spatial structure of the hydrology. In other words I have simplified reality. I claim that during this colonial period people travelled generally along the same routes as the water flowed. This assumption is based on considerable historical evidence. At the time of contact, there was heavy boreal forest across the state of New York making overland transportation difficult. The native populations used the canoe as a traditional means of transport. Many archaeological sites are along the waterways and along the *portages*. Early explorers such as Hudson followed the waterways. Early trade was based upon water transportation. First, the *voyageurs* extending the fur trade, then small agricultural villages and towns, and even industry used water at the primary means of transport. Of course, this is the state whose agricultural and industrial production in the nineteenth century was dependent upon the Erie Canal. In short, this simplifying assumption is based on a long tradition of water transport.

Flow networks may be used for a variety of functions including routing and allocation of resources. There are a number of algorithms and equations which define these functions. The reader will be relieved that I shall forego a detailed discussion. For example, if one wishes to determine the maximum flow between two nodes of a network, one may use the process based upon the 'maximum flow, minimum cut' theorem (which states that the value of the maximum feasible flow is equal to the capacity of the minimum cut separating the nodes) or the more dynamic Ford Fulkerson maximum flow equation (Haggett, 1967; Price, 1977). Not only may one determine and compare flows but one may compare the degree of connectivity of networks. Similarly one may compare parts of the same network. Connectivity is the relationship of extant routes with the maximum number of routes possible. Thus, one can create a scale of connectivity from 0 to 1 where 0 is absolute nonconnectivity, minimum connectivity is $2/m$, intermediate connectivity is $n/(0.5(m^2 - m))$ and 1.0 is the maximum connectivity. For these equations m is the observed number of places and n the observed number of routes (Abler *et al.*, 1971).

By allocation I mean finding how people will be distributed among centres and their surrounding routes or arcs. Each town or centre has a potential capacity for receiving or distributing people. Based on the capacity of the centre, people along the arcs will move to the centre or vice versa. People are allocated to the centres or arcs according to a variety of criteria such as time or distance or some other function. The networks are dynamic and can be initialized at different locations with different values. Every house has multiple entrances and exits. Thus, I have constructed a series of doors which can be opened and closed. Each door may be opened or closed individually or in groups. In my case the door is an initial population at a particular location in time. For example an individual door might be the twenty people who originally settled at Fort Nassau near Albany in 1614. From there people may have moved in numerous ways following the hydrology. For example they may have moved west from the Hudson along the Mohawk or along more minor tributaries into a variety of places. Furthermore, I recognize that the boundaries of frontiers are permeable. People passed back and forth and entered the system in more than one place. For example, I have a relatively old house and people not only enter through the door you expect but often through other doors. My eight-year-old daughter demonstrated this phenomenon this week when she entered through the old coal chute.

I conceived of each test as a line or a plane from which colonization could begin. In other words, I wanted to open all the doors on the east side of the house at once, then all the doors on the north side of the house, then the doors on the south side, etc. After opening the doors, I watch how the European population colonizes New York.

I test my hydrological network model with six alternative simulations based on different initializations. First, I open the three doors—the very first settlements on the Hudson. Then, for the second test, I open three early doors on the St. Lawrence, and then for each of the drainages. In other words I can reconstruct different versions of history. If the colonization by the Europeans was the result of spread from the Hudson, there will be one type of settlement pattern. If colonization was the result of spread from the north and St. Lawrence, there would be a different type of settlement pattern. Next, I take the 83 early settlements for which I know the dates of first settlement and compare them with the various results.

For this study I use a series of programs and subroutines which determine the allocation. They are part of the ARC/INFO Network package (ESRI, 1988a,b). ARC/INFO is a vector-based GIS which uses points, nodes and arcs to construct the spatial representations.

Results

Plate VII is the digitized hydrology of the state of New York. It includes Lake Ontario, part of Lake Erie, as well as smaller lakes. The data has been massaged to improve accuracy. First, the location of each node in the coverage was compared to every other node and if they were beneath a small threshold distance about 0.01 inches they were snapped together and made equal. This was done in order to correct small errors in the digitizing when arcs do not meet exactly at the nodes. The second correction was splining. The spline smooths and generalizes the arcs (ESRI, 1988a).

The network was created from the digitized hydrology coverage with a standardized format. The resistance to travel was set at a constant impedance which was arbitrarily chosen at 10 units. Every arc could be traversed in both directions moving from node *a* to node *b* and back again. However, after I experimented with bidirectional movement, I rejected it. Instead, unidirectional outward movement from the centres was chosen. The size of the potential capacities of the nodes ranged from 50 to 1000.

The resulting network had 80 arcs and 84 nodes. For each drainage, centres were selected for initial settlements. These centres were chosen on two ranked criteria. First, they were chosen so that the settlements were more or less evenly distributed along the main channel of the drainage. Second, they were chosen to correspond to the earliest settlements in the drainage. The first criterion had priority over the second. However, there were sufficient settlements so that both could be optimized.

Figure 24.1 is a visual presentation ot the type of growth which occurs. First, there are the initial settlements or centres (Figure 24.1a). Second, there is the initial spread of population from each of the centres along the nearest rivers and streams. The general rule to determine how this process occurs is to examine each possible route and determine section by section, arc by arc, which routes have the highest 'demand' for new population and the least 'resistance' to the migration (Figure 24.1b). This process continues until either the number of migrating people is completely settled, which is another way to say that all the demand is met, or migration will end if the resistance to settlement is so great that no more migration may occur (Figure 24.1c).

Since the written word is not dynamic, nor can I put a moving picture or a video into a book, there is no way that I can actually show the reader the actual process of migration. However, I will try to describe the process. I opened the flood gates of migration on the Hudson River by initializing three centres—one near Albany, one near Fort Nassau and one near the Canadian border. The growth is initially from south to north. Above Albany it starts to move to the north and to the west and simultaneously begins to fill in populations onto the secondary rivers. Thus, it creates a more packed and dense settlement pattern. The settling process

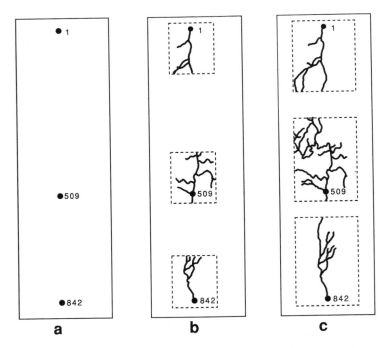

Figure 24.1. A visual presentation of the type of migration which occurs dynamically using a GIS: (a) the initial settlements; (b) part of the way through the migratory growth process, as populations reach new locations and create new settlements along the waterways; (c) the end of the process showing the furthest extent the migration reached.

in the northern part of the state persists in moving populations to the west. Secondarily, there are several more northern tributaries which are settled. Plate VIII renders the completed process. The initial centres are represented by circles and numbers. The coloured lines represent the rivers and streams along which settlement took place. The white lines represent the waterways which were not settled. As one can see the entire state is not settled. Indeed settlement ends with less than one third of the state settled. The eastern counties (Westchester, Putnam, Columbia, Rensselaer, Washington, Essex and Clinton) are completely settled. The northern counties are only partially occupied. No European populations are in Jefferson or St. Lawrence. None are west of the eastern boundary of Lake Ontario. Only a few populations penetrate into the Finger Lake region.

Similar analyses may be done for each of the other models. However, for the sake of brevity their discussions will be reduced. The second model suggests that migration and European settlement begins from the north. There are three initialized northern settlements along the St. Lawrence. The growth of settlements is divided into two separate patterns. The first pattern is south along the Hudson. The second is southwest through St. Lawrence county towards Lake Ontario. As the process continues, the entire Hudson area is settled. The two groups of settlements reach toward each other from St. Lawrence county towards Oneida county. Figure 24.2 shows the conclusion where the two groups of settlements do meet forming one northeastern settlement pattern. The darker lines represent the waterways along which predicted migration takes place. The lighter lines are the unoccupied rivers and streams.

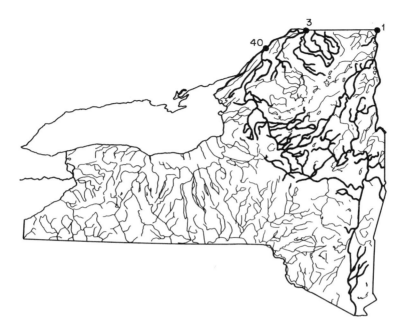

Figure 24.2. The Saint Lawrence model. The darker lines show the greatest extent of European migration based upon three initial centres along the Saint Lawrence River. The lighter lines are the unoccupied drainages of the New York waterways.

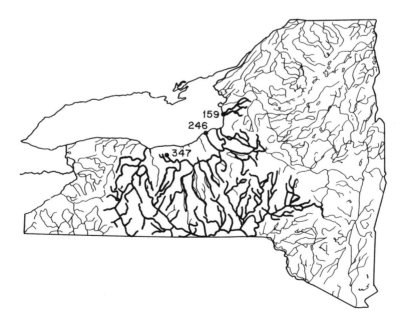

Figure 24.3. The Lake Ontario model.

The third model may be called the Lake Ontario model for the initial three settlements are found along the southern shore of Lake Ontario. The result is again two settlement expansions. One elaboration originates in the Oswego and Rochester areas and is quite large and densely connected. The other which is further north, near Watertown, is much smaller. Figure 24.3 shows the end result of this growth process. The western settlement area has grown quite rapidly occupying many of the western central counties and including Allegany, Steuben and Corning. This growth reaches south to the Pennsylvania line. However, it should be noted that neither the far western counties such as Erie or Chatauqua nor any of the counties east of Hamilton are settled.

The fourth model is the Susquehanna model. After settlements are initialized the growth is primarily northwards. The migration pattern is one single connected composite. It grows north to Lake Ontario and then eastwards almost to the Hudson as seen in Figure 24.4, the completed growth process.

The fifth model is the Delaware model. It can be characterized as the 'growth that was not'. Three settlements were initialized along the Delaware River, at the same values as the others. There was some growth in the north but, as can be seen, when the growth is completed there was almost no growth (Figure 24.5).

The sixth model is the Mohawk model, exhibited in Figure 24.6. Growth moves south along the Hudson and simultaneously northeast from the Mohawk towards the Hudson. It is the converse of the Hudson pattern. The reader should compare Figure 24.6, the completed growth, with Plate VIII, the terminated growth of the Hudson River.

The next stage in the research design is to compare these models with the real data. Figure 24.7 shows the actual location of 83 settlements for which I know the date of earliest European settlement or the first known European settler. I have recreated the hydrologically-based network and allowed growth to take place from all 83 settlements. However, I have time-lagged them to correspond to temporal reality. Almost the entire state is settled. Since there is no way to recreate the dynamic spread, and a figure would simply show most of the state's hydrological system in darkened lines, the reader will have to be satisfied with a verbal description. There appear to be four major settlement systems operating. One corresponds to the Hudson and the Mohawk, one to the Susquehanna and Lake Ontario, and one to St. Lawrence. The fourth growth pattern is a western system which is not similar to any of the models.

The results can be augmented by examination of Table 24.1. It shows each model and the number of arcs which are filled by the complete growth of the system. The Lake Ontario and the Susquehanna models create the largest settlement area (276 and 278 arcs) and are the migratory converse of each other. Each covers approximately the same area. One moves from north to south and the other from south to north. Of course, the reconstruction in which the 83 settlements were used spanned an even larger settlement area. A careful examination of the maximum impedances shows a clear relationship between resistance and the spread of population. In most cases, the greater the spread of the population the greater the resistance. The major exception to this rule is the case of the 83 settlements. There is a very different type of migration taking place. In the former three settlement models, the growth tends to be extended along the waterways and each new settlement is part of the extensive network. In the case of the 83 settlements, there are numerous small networks, none of which is very extensive.

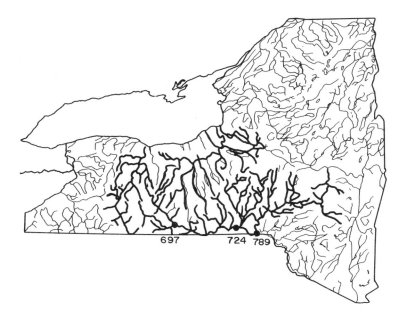

Figure 24.4. The Susquehanna model.

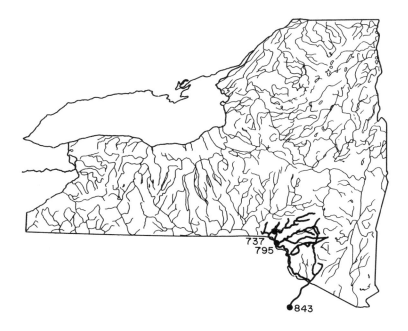

Figure 24.5. The Delaware model.

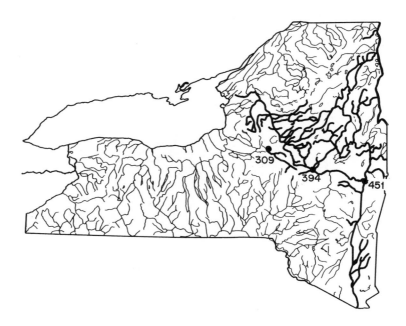

Figure 24.6. The Mohawk model.

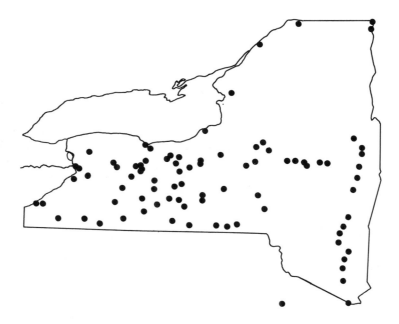

Figure 24.7. The actual location of 83 settlements where the date of the first European settler is known.

One of the advantages of such a model-building and predictive exercise is that one may try to produce better results on the basis of what one has learned. In some sense, one can produce a 'second generation model'. Thus, *ex post facto*, I tried to create a model which came closer to reality by combining the hypothetical models. An example is noted in Table 24.1 as the 'Combined Triangle' model. It was initialized as two settlements along the Hudson and one on the Susquehanna. The population settles 460 arcs which is quite close to the time-delayed 83 settlements.

Table 24.1. *A comparison of predicted migration and settlement patterns based upon different models of initial donor settlements.*

Models	Arcs	Maximum impedance
Hudson	103	240
Saint Lawrence	231	340
Lake Ontario	276	410
Susquehanna	278	350
Delaware	60	180
Mohawk	183	200
83 Settlements	442	90
Combined Triangle	460	370

Conclusions

This paper had one purpose and three goals. The purpose was to show that GIS could be used to model spatial processes and develop interesting, testable predictions. Specifically, there were three goals; to use GIS modelling for learning about settlement processes on the frontier, for evaluating alternative models of historical migration, and for determining the limiting factors of growth and migration. How well were these met?

It would appear that GIS are powerful tools for modelling. I was able effectively to model different migration patterns following a complicated hydrological system given different initial situations. In other words, each model began with populations at different locations. Did I learn about settlement processes on the frontier? Yes, I learned that there were good competitive reasons why certain settlement patterns were successful and why others were not. This frequently had to do with the comparative connectivity of the networks as well as the particular structure of the settlement system. I learned that there were numerous reasons why populations would settle secondary tributaries prior to completely settling the primary channel areas. The most frequent reason was that the accumulative resistance or accumulative cost was less.

Was I able to evaluate alternative models of historical migration? The answer to this question is also affirmative. The traditional historical description of migration in New York is to suggest that the population spread up the Hudson and then west along the Mohawk. My models and my tests with real data indicate that it is far more likely that the population spread simultaneously up the Hudson and along the Susquehanna. Thus, I have gained new knowledge spatially and temporally.

Finally, have I learned what are actually the limiting factors of growth and migration? Not yet, nor do I know whether or not it will be possible. That is the challenge.

Acknowledgements

The digitizing of the hydrology of the state of New York was done in ARC/INFO 3.2 by members of the Anthropological Laboratory for Geographic Information Systems including Jill Pacillo, Kenneth Kaiser, Judith Sutton, Cindy Schiavitti, Donna Belz and Kathleen M. Allen. I wish to gratefully acknowledge my debt to Jack Dangermond and the ESRI corporation for initially providing ARC/INFO to Duane Marble and myself as well as providing considerable professional and background support. In addition, I wish to thank the Department of Anthropology for spatial, financial and temporal support while working on this project. The State University of New York at Buffalo Computing Center in the persons of Frank Rens, Reb Carter, Roger Campbell and Michael Sher provided resources and consultation for using a VAX cluster version of ARC/INFO while my colleagues of the National Center of Geographic Information and Analysis provided equipment, advice and more importantly large amounts of technical assistance through James Smith, the local ARC/INFO system manager.

References

Abler, R., Adams, J. S. and Gould, P., 1971, *Spatial Organization: The Geographer's View of the World*, (Englewood Cliffs, NJ: Prentice-Hall) pp. 259

ESRI, 1988a, *Users Guide: ARCEDIT. Release Notes and Installation Guide, version 4.0.* (Redlands, CA: Environmental Systems Research Institute)

ESRI, 1988b, *Users Guide: NETWORK. Release Notes and Installation Guide, version 4.0.* (Redlands, CA: Environmental Systems Research Institute)

Haggett, P., 1967, Network Models in Geography. In *Models in Geography*, edited by R. J. Chorley and P. Haggett (London: Methuen and Company) pp. 609–668

Kvamme, K. L., 1989, Geographical Information Systems in Regional Archaeological Research and Data Management. In *Archaeological Method and Theory, Volume 1*, edited by M. Schiffer (Tucson, AZ: University of Arizona) pp. 139–203

Price, W. L., 1977, *Graphs and Networks: An Introduction* (Princeton, NJ: Auerbach)

Zubrow, E.B.W., 1987, The Application of Computer-Aided G.I.S. to Archaeological Problems. In *Proceedings of the First Latin American Conference on Computers in Geography,* edited by D.F. Marble (San Jose: Editorial Universidad Estatal a Distancia, Costa Rica) pp. 647–676

25

Modelling early historic trade in the eastern Great Lakes using geographic information systems

Kathleen M. Sydoriak Allen

Introduction

Geographic information systems (GIS) have been used in a variety of contexts in archaeology. Although most of the applications have been in predictive modelling of site location, some attempts have been made to use GIS in a dynamic way to model change in prehistoric cultures. This temporal dimension is at the heart of archaeological research and the introduction of new techniques for the analysis and interpretation of process is an exciting development.

In this paper I use GIS to model spatial aspects of the development of trade between Native Americans and Europeans in the early historic period (ca. AD 1550 to 1750) in the eastern Great Lakes area of North America. This is done through use of the NETWORK module of the ARC/INFO geographic information system. Trade is examined at three periods of time from the late 1500s to the mid 1700s and the flow of goods among native populations and between natives and Europeans is identified.

My model of trade simplifies reality in order to illuminate those areas where knowledge is clear as well as to point out the regions where our understanding lies in the shadows and is most limited. Toward this end, the model begins with the assumption that the natural hydrology of the state was the most important communication and transport route in prehistoric as well as historic times.

Early historic trade

At the time of first European contact, the eastern Great Lakes area was occupied by a number of tribal groups. The most well known of these was the Five Nations Iroquois in New York State. Other Iroquoian groups, including the Huron, Neutral and Petun were located in Ontario, while another Iroquoian group was along the St. Lawrence River (part of the present border between Canada and the United States). Each of these tribal groups was localized within a restricted area. In New York State, the Iroquois were spaced at regular intervals across the state.

Although the exact dates for the first trading between Native Americans and Europeans is not known, a significant amount of trade was taking place by the mid 1500s (Trigger, 1976). This trade originated along the coast of North America but it reached inland along existing native trade routes such that interior groups (like the Iroquois) had acquired a large amount of European trade goods by the mid to late 1500s (Bradley, 1987).

The traditional view of the beginnings of trade among the Iroquois sees trade and European contact originating in the St. Lawrence River valley (Trigger, 1978; Bradley, 1987). However, trade goods may also have arrived by way of river travel along the Delaware and Susquehanna Rivers. In the mid 1500s, events to the south of the Iroquois among native groups along the Susquehanna and Delaware Rivers in Pennsylvania suggest that European goods were available at the outlets of these rivers into the Atlantic Ocean. Native settlements in these areas moved closer to the source of European goods, evidently in an attempt to serve as middlemen between the source of goods and their ultimate destination, the native settlements (Bradley, 1987). By the early 1600s, the establishment of trading posts and forts changed trading patterns away from native distribution to more direct contol over the process by Europeans. Prior to this time, face to face contact between inland natives and Europeans was virtually non-existent. In 1609, Hudson sailed up the river that now bears his name and a new era of increased contact began. Fort Nassau (later known as Fort Orange) was established as a Dutch trading post in 1614 on the present site of Albany. Shortly thereafter, the Mohawk attacked the former residents of the area around Albany to restrict access to the post and maintain control over it (Fenton and Tooker, 1978; Trigger, 1978).

European settlement profoundly altered native exchange systems. Competition among groups for control of the fur trade increased conflict. Among the effects was the realignment of trade from its former north-south axis to an east-west axis (Bradley, 1987). By the early and mid 1700s, several additional forts were established further in the interior and the European presence was more strongly felt.

The model

The model presented flows out of the known situation of trade in the early historic period. I model this early historic trade process first by placing native populations in their respective locations across the state. Second, European trading posts and forts are added as they developed over time. The movement of trade goods is very simply modelled as moving out from these centres over hydrological routes. It is of particular importance to note that the Iroquois were located at the headwaters of four major waterways: the St. Lawrence, the Mohawk/Hudson, the Delaware and the Susquehanna. Their location made them well suited for receiving European goods both through native exchange routes in the early historic period and from European trading posts and forts in later times.

The alteration in native exchange networks as a result of European demands for furs is evaluated using the NETWORK module of the ARC/INFO GIS. This system allows one to watch the flow of goods over short time periods as the network develops as well as to model the evolution of trading networks as European posts are established at several different periods of time.

The network

The network used in this study was created from a digitized map of the hydrology of the state of New York. This coverage was taken through several transformation processes in order to correct errors and make it suitable for the analysis. This included re-digitizing and editing several arcs, matching nodes in order to eliminate extraneous ones, and splining the arcs for the final maps (ESRI, 1988a).

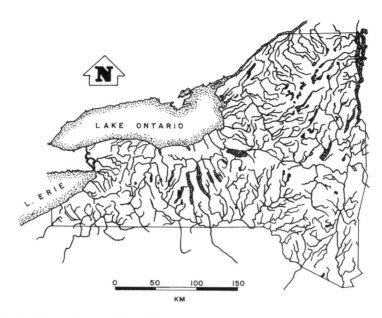

Figure 25.1. Hydrological map of New York State digitized for use as the network for modelling trade.

The final map of the hydrology was used to build the network (Figure 25.1). The arcs are transportation links whose attributes affect resource movement. In the NET-WORK subprogram of ARC/INFO (ESRI, 1988b), there are several characteristics of the arcs and flow which the user assigns at the time of network creation. These include impedances and demand (both associated with arcs) and turning directions. Impedance is the resistance associated with travelling across an arc; demand refers to the number or amount of resources associated with a feature such as an arc. The purpose of impedance is to simulate the variable conditions on lines and turns in real networks that affect the flow of resources; when there is a choice of paths that resources may flow along, the optimal path is the path of least resistance. In the network, I used constants for impedance (10) and demand (25). The net effect of this was to set up a network where the demand at each of the centres was equal (i.e., each centre of native population and European settlement had an equal need for goods) and the flow of goods was as likely to move along one arc as another. The NETWORK program also allows one to issue rules for turns when a choice of arcs is available. In my model, no rules were issued and, therefore, the path chosen was directly related to demand at the centres and not controlled by the model or my expectations.

The centres

Centres were added to the network based on known locations of contact era native populations and European trading posts and forts. Three categories of centres were added: (1) ten centres represented native populations; (2) three centres represented posts from the early 1600s; and (3) seven centres represented trading posts and forts from the first half of the 1700s (Figure 25.2). Delays were built into each of these centres corresponding to known site succession and chronology from the early historic period.

In addition to delay, centres could have two other characteristics: resource capacity

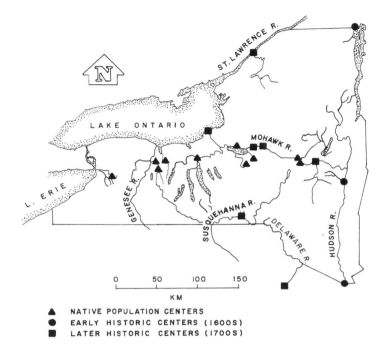

Figure 25.2. Map of New York State showing locations of all centres: native populations (triangles), early historic centres (circles), and late historic centres (squares).

and impedance limit (ESRI, 1988b). Resource capacity is the capacity of the centre to receive or supply resources to the surrounding arcs; the impedance limit is the maximum allowable impedance from the centre to any arc allocated to the centre. These values were assigned so as to mimic the historic trade patterns. The capacity was set at 5000 for native populations, 10,000 for the early 1600s trading posts, and 5000 for the later historic forts. The maximum impedance was set at 1000 for the native populations and the early European trading centres and at 500 for the later European outposts. I gave higher values to the capacity of the early historic trading sites in recognition of the high demand associated with this early period in the European fur trade. I gave lower impedance values to the later European forts due to the increased number of settlements at that time, the greater competition and the greater ease of transportation.

In the network program, arcs, with their associated demands, are assigned to centres until the capacity of a centre is reached. The sequence in which these arcs are assigned is based on the demand and impedance associated with the arcs. Arcs with high demand and low impedance are assigned faster than arcs with the opposite characteristics. No additional arcs can be assigned once a centre's capacity and/or maximum impedance limit are reached.

One may think of this process as occurring in two directions. In one case, goods flow out from the centre along paths determined by demand and impedance, while in the other, goods flow in to the centre. Trade may occur in a number of ways and includes both of these processes. In this paper, I have simplified the situation such that I treat all centres as if goods only flow outward from them.

States of flow

Four examples are presented to illustrate this pattern of network flow.

Native populations—the Iroquois

Ten centres of native population are shown across the state of New York. These correspond to known prehistoric and early historic Iroquois site locations and include six tribal locations: these are the Mohawk, the Oneida, the Onondaga, the Cayuga, the Seneca and the Erie (a tribe located in western New York). Figure 25.3 shows the locations of these tribes across the state. Note the east-west orientation of the tribal groups and their location at the headwaters of several major waterways. Several of the tribes are represented by several centre locations. These reflect the number of villages occupied by members of particular tribes at the time of historic contact. So, for instance, there were three villages occupied by the Mohawk and the Seneca, two by the Onondaga, and one by the Oneida.

The figure also shows the initial allocation of the network with the flow of resources along hydrological routes. Routes are indicated on the figure by the patterned lines moving out from the centres. Note the north-south movement along the Hudson River and in the Finger Lakes region. As noted, this corresponds to known patterns of indigenous trade. Figure 25.4 shows the full extent of arc allocation through the hydrology network of the state. Clearly, there is a strong horizontal component to the trade patterns corresponding to the horizontal distribution of native centres and also a clear extension of trade north towards the St. Lawrence River and to the south along the Hudson and the Susquehanna Rivers.

In the next step of the model we look at the effect of early European settlement on the native trade pattern. As noted before, there is some indication that patterns of trade shifted from an east-west axis to a north-south one. Can we identify this pattern in our model based solely on hydrology of the state?

Native/early European contact—early 1600s

To the first map of the native populations, three centres representing the first more or less direct contact between these interior natives and Europeans are added (Figure 25.5). These centres are represented by circles and include one at the top of the map representing Montreal (settled in 1608), one near the confluence of the Mohawk and Hudson Rivers, Fort Nassau (founded in 1614), and one at the outlet of the Hudson River with the Atlantic Ocean. In this Figure we can see encroachment of the Europeans along the eastern edge of the state from north and south and the resultant trading pattern. The Mohawk extend east to Albany but the Europeans clearly control trade on the eastern waterways.

Although the European centres are given twice the capacity of the native centres, the latter still control much of the state. Among the native populations, the network of resource flow is more evenly divided between the two halves of the state than formerly. At this point, the Europeans control movement from north to south; the natives control it from east to west. This corresponds to the model suggesting that the orientation of trade changes from north-south to east-west.

Europeans—1700s

In the next step, we examine all of the European centres in existence by the mid 1700s and none of the native centres (Figure 25.6). In this case we are interested in the flow

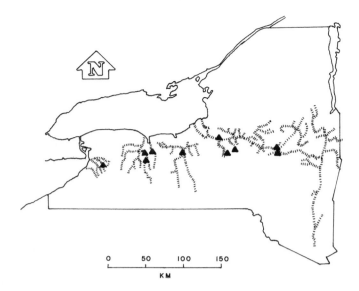

Figure 25.3. Native population centres at an early stage in network formation. The patterned lines moving out from the centres represent the movement of goods.

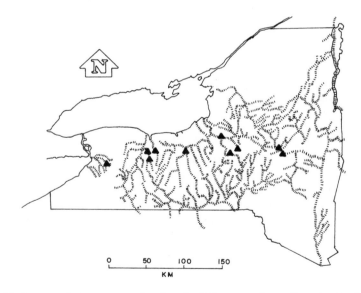

Figure 25.4. Native population centres showing the full extent of arc allocation in the network.

of goods out of the European centres. This represents the situation as if the natives were not participating in the exchange network. On the basis of hydrology alone, how do the goods distribute across the state? The three centres from the early 1600s are represented (again as circles) along the east side of the state as are seven later sites (represented as squares). As noted above, delays in the movement of goods are built into the centres that correspond to the known dates of establishment of the sites.

Two centres provide access to the interior of the state from centres along the Atlantic Coast south of New York. The centre in the south central part of the state on the Sus-

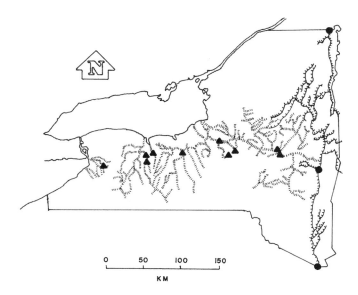

Figure 25.5. Early stage in the development of the trade network between native populations (triangles) and early historic centres (circles).

quehanna River represents access from centres (including Baltimore) on the Chesapeake Bay; the centre located further to the east is on the Delaware River and represents access from the Delaware Bay and Philadelphia. Although these coastal centres were established at an early time period, direct contact with natives of the interior took longer to accomplish.

Two centres are from the early 1700s. These include the trading post at Oswego on Lake Ontario and Fort Hunter along the Mohawk. Two others are from the mid 1700s; Ogdensburg in the north along the St. Lawrence and Fort Stanwix at Rome, New York. This latter site sits between the western extent of the Mohawk River and the eastern extent of the Seneca River. Each of these is the major east-west waterway for their respective parts of the state.

Figure 25.6 shows the flow of goods out from these European centres. Notice the filling-out of the hydrology. More of this occurs along the north-south axis than the east-west and this occurs primarily in the eastern part of the state. As the network allocation proceeds, more filling-out occurs to the west (Figure 25.7). Note the large area accessible from Fort Oswego at the eastern end of Lake Ontario. Several key areas are apparent in the interior. These include Fort Nassau, Fort Hunter and Fort Oswego. The Europeans had direct access to all Iroquoian areas from those centres.

Combined native and European—1700s

All of the centres previously noted are combined for this run. There are a total of 20 centres. This model is interactive again as it models the flow of both natives and European goods. This models the flow of goods out of European centres and their interaction with native centres.

Resource flow from the centres at an early stage is shown in Figure 25.8. The native populations are spread out horizontally across the state and from the earliest time we can see the movement of goods out from each of the centres.

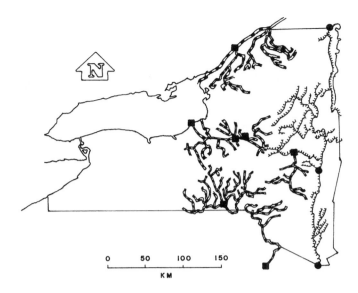

Figure 25.6. Early historic (circles) and late historic (squares) centres and their associated trade networks.

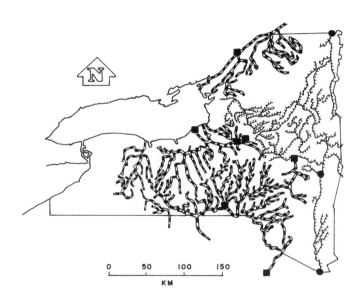

Figure 25.7. Early historic (circles) and late historic (squares) centres and the full allocation of arcs in the network.

As the network continues to run, the central area arcs are allocated first, followed by the peripheral ones (Figure 25.9). At the end of the allocation, the portion of the network controlled by native populations is mostly confined to the western part of the state while the main trade with Europeans occurs to the east. In this last model we can see that there is less native expansion of any sort. It appears that we are seeing not only the trading situation but also a reflection of European settlement.

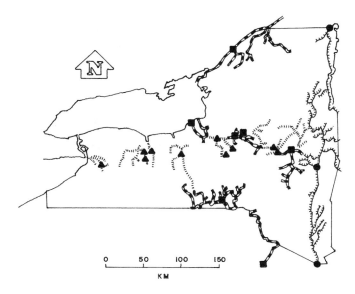

Figure 25.8. Native (triangles) and European (circles and squares) centres in the early stage of network development.

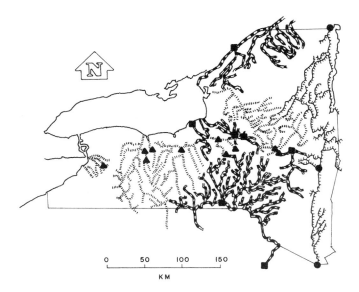

Figure 25.9. Full allocation of arcs for all population centres. Note the European monopoly of trade in the eastern part of the state.

Results

The modelling of trade patterns based on hydrology and the locations of native and early historic settlements has illuminated several points. Recall that this model has been based on known settlements and historic forts and trading posts.

(1) The horizontal orientation of the native populations is readily apparent although the early trading pattern was in a north-south direction along the major waterways.

(2) The distribution of European goods followed that same east-west pattern once forts were established in the interior.

(3) The hydrology of the state played a major role in the development of trade relations with the native populations. In other words, it appears that much of the trading took place along these communication arteries. Hydrology was more important to the early patterns of European contact and trade than originally thought. We can clearly see the movement from the east along the Hudson and Mohawk Rivers into central New York.

(4) I attempted to model the movement of goods from European centres into the interior. The model ends up showing (in some respects) the movement of those doing the trading as well. For example, the native population and early European stage of the model showed the Mohawk moving east to Albany. In actuality, the Mohawk did launch an attack on the Mahican (the former native inhabitants of the area around Fort Orange) to get control over the flow of people and goods in and out of Fort Orange.

(5) The model is a spatial representation of trading relationships and is successful in that respect. Further analysis is necessary if we want to identify the actual mechanisms of trade. At this point we cannot differentiate between market trade and reciprocal exchange.

Although the model has been successful in replicating the early historic trade situation, further work is needed in two major areas:

(1) The results from this application have not been compared with the archaeological record in any real quantitative sense. While the visual patterns correspond well with what we know was occurring on the basis of historic evidence, we have not yet been able to compare the allocations within the networks with archaeological evidence. There are several complications that arise as soon as one attempts to do this. The different categories of European goods that were traded have different kinds of distribution patterns. Also, during the time period that is considered here, the native American acceptance and use of European items changed substantially. European trade items were first altered and used in native ways, such as when copper kettles were cut into pieces and used as projectile points.

(2) This model is very simple in that it examines trade as if the goods were equally likely to flow along any of the routes. In reality this is not the case. The next step is to make the model more complex by varying network and centre variables.

The use of a geographic information system has been very useful in this exercise. However, more work is needed to test and refine this model of early historic trade.

Acknowledgements

I would like to acknowledge the efforts of the following individuals who aided in the development of this paper. First, Ezra Zubrow has encouraged me throughout the course of this study. He has assisted with the design and implementation of this research and his assistance is gratefully acknowledged. Second, James Smith, the technical assistant in the Department of Geography and the National Center for Geographic Information and Analysis, has provided assistance in the form of advice and problem-solving as well as space. Third, I would like to thank J. Stuart Speaker for his illustrations and Stan Green for his comments on this chapter. Finally, I would like to thank the students who worked in the Anthropology GIS lab and who so ably digitized the New York State map. They include Jill Pacillo, Ken Kaiser, Cindy Schiavitti, Judy Sutton and Donna Belz.

References

Bradley, J.W., 1987, *Evolution of the Onondaga Iroquois: Accomodating Change, 1500–1655.* (Syracuse: Syracuse University Press)

ESRI, 1988a, *Users Guide: ARCEDIT. Release Notes and Installation Guide, Version 4.0,* Environmental Systems Research Institute, Redlands, CA

ESRI, 1988b, *Users Guide: NETWORK. Release Notes and Installation Guide, Version 4.0,* Environmental Systems Research Institute, Redlands, CA

Fenton, W.N. and Tooker, E., 1978, Mohawk. In *Handbook of North American Indians, Volume 15: The Northeast,* edited by B.G. Trigger (Washington, DC: Smithsonian Institution Press) pp. 466–480

Trigger, B., 1976, *The Children of Aataensic: A History of the Huron People to 1660,* 2 Volumes (Montreal: McGill-Queen's University Press)

Trigger, B., 1978, Early Iroquoian Contacts with Europeans. In *Handbook of North American Indians, Volume 15: The Northeast,* edited by B.G. Trigger (Washington, DC: Smithsonian Institution Press) pp. 344–356

26

Modelling the Late Archaic social landscape

Stephen H. Savage

Introduction

Archaeologists traditionally study the past in terms of Albert Spaulding's (1960) three archaeological axes of space, time and form. The third leg of this archaeological triad, space, has been studied in terms of settlement patterns, artifact distributions and the like, but has generally not been as widely used as the other two. This is in spite of what Conkey (1988) has termed 'the overwhelming spatiality of our data'. The failure to develop and explore the spatial nature of archaeology has been in part a methodological problem. Archaeologists have not been able to manage effectively all three axes of space, time and form simultaneously. In this chapter I propose to develop the third archaeological axis, space, through the paradigm of landscape archaeology. Landscape archaeology is proposed as an integrative paradigm for studying past cultural systems that incorporates many traditional approaches. It offers a holistic view that places individuals and individual behaviour at the centre of research. Past human culture is studied in terms of the physical, material and cognitive ways humans inherit, transform and bequeath their natural and cultural environments. Models of locational choice, for example, consider site location in terms of information and action, and by recognizing the possible contradictions between the two, and among the perceptions of individual members of a society. Models of social organization can be conceptualized in terms of the social, cognitive and physical landscapes within which people live. Similarly, models of differing modes of subsistence, of different settlement and site types, and of different collecting strategies can be studied under the general umbrella of landscape archaeology. Each activity associated with such models produces patterned remains in the archaeological record that can, if studied as a landscape, tell us much about the ways people in the past viewed and used their world. Virtually all of human behaviour results in patterning in the physical, cultural or cognitive landscapes, and is, therefore, amenable to studies informed by landscape archaeology. Landscape archaeology can in this way subsume the other two archaeological axes of form and time.

This chapter uses landscape archaeology to bring several diverse approaches to the past together in a study of social organization in the Late Archaic in the Savannah River Valley of Georgia and South Carolina (Figure 26.1). The approaches of geographic location theory, and models of social space and subsistence are brought together to model the landscape of the Late Archaic.

Geographic information systems

Geographic information systems present a powerful method for developing and testing theories related to landscape archaeology. Traditional approaches to the study of the archaeological axes of form, space and time have not been able to cope successfully with all three simultaneously. Analytical methods such as spatial autocorrelation and the mapping of principal components have not been able to provide understandable answers to the questions asked of them (Paynter *et al.* 1974).

The development of GIS promises to provide much more effective means to control all three archaeological axes. In GIS, all data are spatially referenced. Data representing form and time may thus be analyzed through mathematical or Boolean techniques in a manner that preserves their location. The interplay of GIS methods and landscape archaeology presented in this study clearly shows the power of this new combination for future archaeological research.

The fifty-one Late Archaic sites used in this study were first reported as a part of the archaeological survey of the proposed Richard B. Russell Reservoir (Taylor and Smith, 1978). This work was carried out in order to assure compliance with the Section 106 process, prior to and during construction of the Russell Dam and Reservoir by the US Army Corps of Engineers. The survey and testing reports, and site data, are on file at the Institute of Archaeology and Anthropology in Columbia, South Carolina.

The data set and hypotheses

Work by Wobst (1974), Clark (1975) and Dennell (1983) in the European Palaeolithic has suggested that hunter/gatherer groups were organized into minimum band subsistence groups and maximum band social groups. Based on these studies, I have developed the hypothesis that the Late Archaic landscape was divided into social territories based on minimum group subsistence, or resource collection, bands and maximum group socialization bands. Subsistence groups are organized around such things as resource procurement, household maintenance and child-rearing. Maximum band social groups, on the other hand, tend to be organized around the maintenance of a viable mating network, ritual, and cooperation between the smaller minimum bands. Each minimum band is associated with a particular territory which is theirs by habitual use. Maximum band territories represent the sum of the habitual use areas of a given maximum band's constituent minimum bands. These territories may have been defended against other maximum bands, considering the evidence for Late Archaic warfare at places like Eva (Lewis and Kneberg-Lewis, 1961) and Indian Knoll (Webb, 1946).

This chapter has two primary goals. First, I am concerned with people and society, and cultural change. My primary goal is to develop an understanding of Late Archaic social organization that will provide a framework for studying the social and technological changes attending the period. Second, I am interested in the development of a GIS-based methodology that can work hand-in-hand with a powerful body of anthropological theory to provide new avenues of research into past cultural systems. The geographic information systems approach to the Late Archaic landscape has been extremely productive in reaching both of these goals.

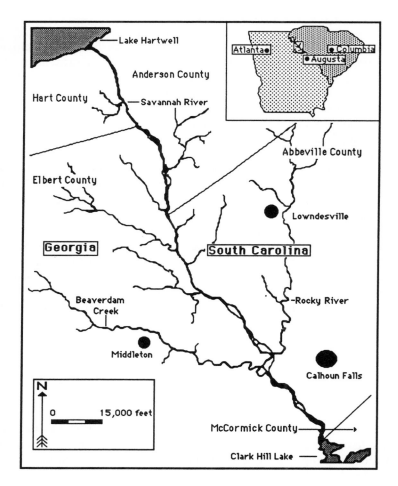

Figure 26.1. Map of the project area prior to inundation.

Models of social grouping

Three primary models need to be considered at this point: (1) Dennell's (1983) subsistence and reproductive group model; (2) Wobst's (1974) study of minimal/maximal bands; and (3) Clark's (1975) model of social territories.

Subsistence/reproductive groups

Dennell has noted that two demographic groups need to be considered in hunter/gatherer societies: the subsistence group and the reproductive group (Dennell, 1983). The nature and activities of the subsistence group are defined as follows:

'(The subsistence group) can be defined as a group of people habitually associated with each other throughout at least part of the year for the procurement of those resources necessary for their physiological well-being, and for the rearing of young and caring for the old and sick.... The size of subsistence groups does not have to remain constant throughout the year, but may change as members form smaller sub-units; for example, a hunting band might split into smaller groups at some times of the year to exploit dispersed resources. . . . .Nor does its membership have to remain constant, since members may leave to join another group and be replaced by others. However, at any given time of the year, its size should remain roughly the same from year to year. A subsistence group should also be associated with an annual territory: that is to say, with an area that it and its neighbouring groups will recognize as containing its food resources' (Dennell, 1983, p. 12)

Dennell presents six different types of subsistence group land use patterns, two of which, the forager pattern and the logistic pattern, apply to hunter/gatherer groups. Foraging groups, such as the !Kung San (Lee, 1979), display a pattern involving location at several base camp locations in a given year. A group will stay in one location until the resources are exhausted in that area. Daily food collecting, or foraging trips, will be made from the base camp. There is little storage of food. In contrast, logistic collection involves splitting the subsistence group into smaller resource collection groups, whose task is to move to a location other than the base camp and there collect or extract specific resources at specific times of the year. When a sufficient amount has been collected, the workers return to the base camp. This strategy involves, therefore, a planned seasonal dispersion, collection of specified resources, a central base camp, and logistical camps occupied by the collecting groups while they are away from the central camp. The use of such a system involves considerably more organization and planning, with its attendant opportunities for information management and task direction. Most of the subsistence models developed in the Southeast involve this kind of resource procurement strategy (Taylor and Smith, 1978; Sassaman, 1983). Note, though, that in both examples, Dennell stresses that resource procurement takes place within a defined territory, understood by both the endogamous group and its neighbours. The reproductive group in Dennell's (1983) model is the mating pool:

'As a demographic unit, the subsistence group is usually too small to provide its members with an adequate range of potential mates. For this reason, we need to recognize a larger unit which can be called the *reproductive group*. This comprises a set of subsistence groups within which the members of any one unit will tend to find a mating partner; it is, in effect, the regional breeding population that ensures the long term viability of each subsistence group. Since it functions both by encounters between and within groups, it also serves as an information network that can provide each subsistence group with knowledge about their neighbours and their regional—as opposed to local—environment' (1983, p.14)

During the Late Archaic in the Southeast subsistence groups were probably dispersed into the hinterland for much of the yearly round, while the reproductive group came together at floodplain agglomeration sites (Green and Sassaman, 1983), probably in late winter/early spring. Depending on the size of the agglomerative groupings, more than one reproductive group may be represented in an 'information management group', or band.

The subsistence group, reproductive group and information management group are three examples whose interrelationships are part of a broader dynamic political economy. Other age/gender groups which may be envisaged are also a part of this larger system (Conkey, 1988). The operations among these interrelationships provide substantial opportunities for information managers to manipulate relations between and among groups for political advantage (Root, 1983), and some of that manipulation can be seen in the spatial distribution of sites on the landscape, with respect to each other and to environmental factors.

Minimum/maximum bands

Wobst's (1974) study of Palaeolithic social systems also emphasized a two-tiered social system. In his system, minimum and maximum bands take approximately the same positions as Dennell's subsistence and reproductive groups. The minimum band is defined as:

> '...the most permanent and strongly integrated unit in hunting and gathering society. Its size is large enough so that it will survive prolonged periods of isolation through the cultural practices of cooperation among its members, division of labor according to age and sex, and mutual food sharing. On the other hand, it is sufficiently small to not place an undue strain on the local food resources. Such minimum bands tend to consist of several families of consanguine and/or affinal relatives who, at least part of the year, share a common settlement and participate in a given range of cultural activities. The size of these units allows the unimpaired transmission of the cultural system from generation to generation' (Wobst, 1974, p.152)

Wobst conducted a simulation study based on an average assumed minimum band size of twenty-five people, derived primarily from ethnographically observed hunter/gatherer groups. Other studies suggest, though, that minimum band populations may have ranged from fifteen to fifty (Hassan, 1981). Perlman indicates that in the temperate Southeast, during the Late Archaic, minimum bands may have been considerably larger, perhaps between 100 and 300 people (based on one person per square km) (Perlman, 1985).

Leaving aside the issue of minimum band size for the moment, Wobst (1974) points out, as does Dennell, that the minimum band would not have contained enough people to maintain the reproductive viability of the group, thus a larger group is required.

The larger, maximum band, provides the 'glue' that holds hunter/gatherer society together by providing (1) a larger mating pool, (2) an informal exchange of raw materials, (3) the means for cooperative resource extraction and (4) the means for mitigating the effects of micro-environmental perturbations. The maximum band size is related, according to Wobst, to the various rules governing the selection of mates. A completely open system, with no incest taboos, for example, would require about 175 people to insure reproductive viability. The more restricted the mate selection rules become, the more people are required to operate the mating network and, hence, the maximum band size increases. Hassan (1981) estimates a required size range of between 200 and 500 people.

Like Dennell, Wobst assumes a territorial organization for the minimum bands:

> 'The movement of entire maximum bands, or their components, beyond the area which their cultural system permitted them to exploit, and with which they were familiar is...effectively blocked by social boundaries. A given society was not located in a vacuum but in a social environment, that is, in a network of

neighboring maximal bands. The territoriality of hunters and gatherers is determined at the organizational level of the minimum band. The 'territory' of these groups is usually not maintained through an exclusive claim but through habitual use. It is delineated by the proximity of other minimum bands, by distance, by familiarity with the environment and by natural obstacles' (Wobst, 1974, p.153)

These territories, within which minimum bands operate, and defined by other minimum bands, distance and the physical landscape, are what I choose to call 'habitual use areas', since they are maintained not through claim but by use. One of the aims of this chapter is to delineate such habitual use areas within the Richard B. Russell project area.

Social territories

Both Dennell and Wobst assume some sort of territoriality associated with subsistence, or minimum bands (that which I call habitual use areas). Clark (1975) assumes four levels of social territories. The annual and the social territory are roughly analogous to the territories exploited by subsistence or minimum bands, and reproductive or maximum bands.

At the lower end of Clark's territorial scale is the home base, which is analogous to the catchment area (Higgs and Webley, 1971) of a minimum band's base camp. At the larger end of Clark's scale is the technological territory, archaeologically defined by common tool types, flaking methods and the like. An example can be drawn from the entire southeastern United States during the Late Archaic, when the region shared a common lithic technology and produced a common tool type, the Savannah River point. It seems, then, that we can think of the Late Archaic in terms of social territories occupied by distinct minimum bands, organized at a higher, though looser, level into maximum bands. Such groups are believed to have occupied distinct physical territories on the basis of habitual use, especially at the minimum band level. The hypothesis that will be advanced later in this chapter is developed from this notion of social territory and social grouping.

Models of frontiers and boundaries

Once we begin to think in terms of social groups in habitual use areas, or territories, then questions of group boundaries arise. Marquardt and Crumley (1987) address the notions of the boundary as an edge, that is, as a division between two areas or groups, and that of the boundary as a centre, where different activities, some related to boundary maintenance and others related to exchange, cooperation, communication, and the like, take place.

For us the dual nature of boundaries is of primary concern. Boundaries are dual in that they are artificial divisions of the physical landscape; by virtue of their continuity, they effect discontinuity. But beyond this conception of boundary as barrier or as dividing line, boundaries themselves are worthy of study because they often serve simultaneously as *edges* and *centres* within the landscape under investigation. For example, the quantity of information and/or goods moving *along* a boundary may often be significantly greater than the quantity moving *across* that boundary. From the standpoint of the groups divided by the boundary, that boundary is an edge, a periphery. From the point of view of participants in commerce and communication, the boundary is in fact an important kind of functional centre (Marquardt and Crumley, 1987).

The concept of the boundary as a centre becomes powerful for the Late Archaic when coupled with Wobst's notions of interaction between minimum bands. As I have noted above, Wobst pictures a number of integrative processes, including mate selection, exchange, sharing of extractive tasks and mitigating environmental fluctuations, that occur between minimum bands. Although these bands are defined as residing in territories, there is a significant amount of interaction across the boundaries of the various habitual use areas. For those engaged in such interaction, the boundary thus becomes a centre. Green and Perlman (1985) have noted the importance of studying boundaries as part of open-system research:

'Frontier and boundary studies recognize that societies are open. By so doing, they can contribute insights into the processes that produce the spatial, temporal, and organizational variability observed in the archaeological record. First, they open prehistoric and historic archaeology to a systematic study of noncentral places and the links between these and the traditionally studied central place sites.....Second, broad historical patterns have taught us that social change often is most visible, and in some cases most active, on the peripheries of social systems... Finally, frontier studies are a natural and perhaps necessary element for the study of long distance spatial process' (Green and Perlman, 1985, pp. 9–11)

Once we begin to think of the Late Archaic in terms of a social system delineated by minimum and maximum bands, occupying habitual use areas, then we open the archaeological record to the study of both the areas themselves, and the boundaries between them. By recognizing that the boundary can serve both as an edge and a centre, we can begin to consider those processes that run along the boundary, and those that cut across it, and begin to ask questions related to their functions in the Late Archaic cultural system.

Applying geographic location theory to the Late Archaic

A geographic approach to the past should address issues related to individual responses, the role of information as either a shared commodity or a 'currency', and variations in either individual or group ability to utilize resources for individual or group ends. 'Such a body of theory would embellish existing location theory by taking into account nonoptimal behaviour, imperfect knowledge, other psychological variables, socially dictated constraints, and the impact of existing patterns on subsequent patterns (processes)' (Pred, 1967, p.16).

The archaeological record is seen as the result of human decision-making activities, whether as individuals or as groups of people. Locational decisions are the result of the interplay of numerous factors related to foraging, as well as other constraints operating on groups and individuals—many of which are related not to subsistence but to social relationships. Pred (1967) has created a 'behavioural matrix' that models decision makers and their ability to act upon information. There are two issues here: (1) social factors are important to location decisions and (2) knowledge or information needs to be considered. 'Every locational decision is viewed as occurring under conditions of varying information ability, ranging, at least theoretically, from null to perfect knowledge of all alternatives, and as being governed by the varying abilities (as well as objectives) of the decision makers' (Pred, 1967, p.24).

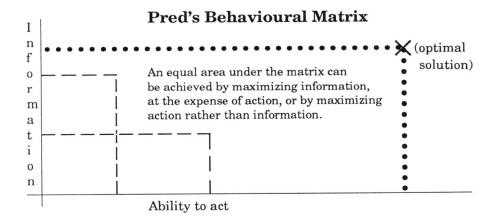

Pred's Behavioural Matrix

An equal area under the matrix can be achieved by maximizing information, at the expense of action, or by maximizing action rather than information.

✗ (optimal solution)

Information

Ability to act

Pred's matrix takes the form of information on the vertical axis, and the ability to use that information on the horizontal axis. What has been described as an 'optimal solution' may be thought of as occupying the far upper-right position of the matrix—it is dependent on perfect knowledge and perfect ability to exercise that knowledge.

Since we can assume that no individual or group can control perfect knowledge, it follows that each contextual particular must fall at some point below and to the left of the optimal solution point. Most significantly, Pred's offers spatial implications for this model:

'Hence, in any given situation, each locational decision making unit or actor, be it a single person or a firm (or band, tribe, or chiefdom), can be thought of as jointly having a real spatial attribute (site and situation, land use or path of movement) that is reproducible on a map, and behavioural qualities that can be hypothetically located in the behavioural matrix' (Pred, 1967, p.24)

The role of political economy in the creation of settlement patterns can be brought into this model when we consider the use of information as a currency to maintain control over individuals or groups who do not have access to that information. This can impact, for example, upon their position along the vertical axis of the locational matrix.

'Marshall (1959:337) notes…limitations on locational behaviour: '!Kung of the Nyae-Nyae region almost never went outside their region because in strange places they cannot depend upon food reciprocity and either do not know where wild foods grow or might not be allowed to gather them'. Note that food reciprocity and information sharing about the distribution of food resources are implied to be coterminous, and that withholding information about the distribution of resources can be an effective mechanism for controlling the use of resources' (Moore, 1981, p.202)

Differential

knowledge may well have played an important role in prehistoric foraging societies. The presence of age- and gender-differentiated grave goods at Indian Knoll and Eva, and the existence of long-distance trade, implies that the Late Archaic is not as egalitarian as we once thought. The presence of emerging elites in an incipient ranked society should be considered, and their role in society explored. Restricting knowledge about resource 'hot spots' could have meant selecting for non-survival, and hence could have been a powerful coercive tool in the hands of incipient elites. Furthermore, since dispersed populations require a greater expenditure of effort to maintain communication and share intelligence, it becomes easy to see how an emerging elite might capitalize on this difficulty by encouraging dispersal during some times of the seasonal round (cf. Sassaman, 1983, and Root, 1983).

The information managed in this manner may be either environmental or social. Environmental information is coded *into* the environment, and includes such factors as the location of good quality raw material sources, or the distance to water sources. Social information is coded *into* people, through ritual and tradition, and *onto* the environment. An example would be the location of a site with respect to its nearest neighbours, or the centrality of a site with respect to both its environment and its function in a social setting.

Equally exciting possibilities accrue when we interpret the horizontal, 'ability to act' axis of the locational matrix. If an emerging elite is skimming off the surplus production in a society in order to maintain power, there may have been times when the ability of others to exploit a given situation was hampered by obligations to the incipient elites. In addition, since part of the role of information and its managers may have been to promote alliances, the ability to act on a given piece of information may well have been influenced by the nature of an agreement made between regional elites.

Another example of the 'ability to act' matrix may be seen in the location of particular site types with respect to their surroundings. Since environmental information is coded into sites, we would expect it to vary depending on the function of a particular site. For example, in hunting camps much of the information coding may have to do with the view of the surrounding territory, thus affording better game visibility. Such sites may compensate for an increased difficulty in actually capturing the prey. That is, the information axis is manipulated in order to compensate for a relatively low position on the 'ability to act' axis. With the application of a locational matrix of this sort, it becomes possible to interpret the Late Archaic settlement pattern in the Savannah River Valley as a series of locational decisions whose efficiency was directly influenced by the 'management' of an emerging elite class. We are able to consider the roles of individual actors and groups of actors. People are active participants in creating and transforming the rules and expectations of the society within which they live; culture is 'meaningfully constituted' (Hodder, 1986).

Landscape archaeology

People live in a world that is partly a product of their natural surroundings, partly a product of their inherited cultural surroundings (subject to individual interpretations), and partly a product of their own actions. If 'All the world's a stage', then the actors on that stage are engaged in not only the action of the drama of life, but also engaged as set builders and playwrights.

'Societies form and are formed by their natural and constructed environments. . . . how a group adjusts to a geographic area reflects much of the group's
history, organization, and values, and in turn such adjustments influence that
group's perception of the physical and the constructed environment. The *landscape* is the spatial manifestation of the relations between humans and their
environment' (Marquardt and Crumley, 1987, p.1)

Models of social organization, such as Dennell's subsistence/reproductive
groups, or Wobst's minimum/maximum bands, can be joined to models of
subsistence under the general umbrella of landscape archaeology. Social and
subsistence activities actively transform the physical landscape and create a cultural
landscape. 'In all these instances, spatial structure is both the medium and the
outcome of social practices. It is neither ideology nor social reality but it integrates
both in the moments of daily life' (Hodder, 1987, p.143).

Because it recognizes the role of the cognized environment, both physical and
cultural, landscape archaeology is able, through such constructs as Pred's geographic
location theory, to take an actor-centred approach to the past. Such an approach creates
a dynamic understanding of past cultural systems, since it allows conflict and resolution,
control and manipulation of information, and various abilities of the actors involved,
to change the landscape actively. Change thus becomes a product of the normal functioning of past cultural systems, not a phenomenon which cannot be explained. Landscape models have not been applied to the Late Archaic in the Southeast. Research
in this period has tended to focus on the relative importance of various items in the
Late Archaic diet, and on exploitative mechanisms for obtaining those resources. What
gets lost in traditional research is the emphasis on people, the emphasis on within-
and between-group conflict and resolution, and the emphasis on change mechanisms.
By introducing the human element, through an hypothesis related to social groups
in the Late Archaic, I hope to redirect the discussion to the more interesting issues.

The Late Archaic social landscape model (the hypothesis)

Based on the works of Clark, Wobst and Dennell in European prehistory, the
hypothesis which will be considered in this chapter may be stated as follows: **The
Late Archaic social landscape consisted of maximum band social territories, divided
into minimum band subsistence territories.**

This hypothesis requires seven assumptions:

(1) The Late Archaic landscape reflects the patterned behaviour of Late
Archaic social groups.

(2) The archaeological record preserves a sufficient amount of that patterning,
in material and spatial relationships among sites, artifacts and features, that
an understanding of the patterned behaviour may be approached.

(3) The function of archaeological sites in the past cultural system may be
understood by analyzing them along the archaeological axes of space, time
and form.

(4) The Late Archaic landscape included people using sites in 'habitual use
areas' (Wobst, 1974; Dennell, 1983).

(5) Sites identified as base camps (based on analysis of space, time and form)
form the centres of 'habitual use areas'.

(6) The boundaries between habitual use areas may be either edges or centres but, in either case, least-cost movement across the physical landscape must be considered during boundary formation.

(7) The population density of Late Archaic hunter/gatherers is assumed to be in the range of 0.39 to 1.2 people per square km (Hassan, 1981, Table 2.1). Minimum band size is assumed to range from 20 to 120 (Perlman, 1985). Maximum band size is assumed to range from 200 to 600 (Wobst, 1974; Perlman, 1985).

The assumption of 'habitual use areas' follows Wobst and Dennell and requires some discussion.

'Paleolithic social groups are territorial. 'Territorial' implies that the members of a given social group moved within an area which was more or less delineated by social factors, by the proximity of other such groups, by considerations of distance, by familiarity with the environment, and by natural obstacles.....The 'territory' of these groups is usually not maintained through an exclusive claim but by habitual use' (Wobst, 1974, pp.151–153; emphasis mine).

The placement of a base camp at the centre of an habitual use area is based on principles developed from site catchment analysis. Vita-Finzi and Higgs (1970, p.5) define this term as 'the study of the relationships between technology and those natural resources lying within economic range of individual sites'. Roper notes that

'The study of Higgs *et al.* (1967) and Vita-Finzi and Higgs (1970) exemplify the two techniques most commonly used for delimiting the territory to be examined in site catchment analysis—namely, the use of circular territories of fixed radii and the use of time contours' (Roper,1979, p.123).

The actual distance travelled in search of resources will depend on topographic features which enhance mobility, such as a flat, treeless plain, or reduce mobility, such as a Piedmont environment with many small streams and rivers to cross. The fact that the environment, especially topography, vegetation and hydrology, influences the distance that can be travelled has caused the one- or two-hour walk method of catchment analysis to be preferred. This is simply a technique for compensating for environmental variation. The principle of least-cost movement is thus factored into catchment analysis.With the advent of GIS, a more direct method is available. This method measures distance from certain features, such as base camps, by calculating for the effects of moving over topography and through hydrology. The GIS method insures that least cost measures are included in the consideration of boundary formation.

The boundaries between habitual use areas, following Marquardt and Crumley (1987), may be either edges or centres, depending on the people looking at the boundary. For example, on the one hand, individuals or groups that are not involved in inter-group communication or exchange are likely to view the boundary as an edge that separates their group from another. On the other hand, people who are involved with inter-group activities are likely to view the boundary as a centre for those activities. A potential source of conflict is thus presented between the two differing cognized views of the boundary zone.

Test implications

Four test implications may be derived from the hypothesis and assumptions. They are general in character, having to do with the nature of habitual use areas, including the types of sites likely to be found in them, and the groups that lived in them, and include the following:

(1) The distribution of Late Archaic sites in the project area is clustered, rather than random or regular.

(2) A variety of site types, reflecting different temporal, spatial and functional uses, will occur in each cluster.

(3) The habitual use areas in the project area support populations in the range of minimum bands, and thus reflect the activities of minimum bands, rather than maximum bands.

(4) The boundaries between habitual use areas will reflect uses both as edges and centres. Small clusters of sites occurring on both sides of a boundary between different habitual use areas can be interpreted as centres. Edges are boundary areas where no such clusters exist.

Assessing the test implications

Each of the four test implications will be assessed in turn, using a variety of statistical techniques and geographic information systems methods, as appropriate. In this section I will include only a summary discussion of the GIS methods used to generate the maps which I will present. The statistical techniques involved are straightforward enough that a summary discussion of their application in this part of the text is all that is required. A more detailed description of the GIS and statistical methods used can be found in Savage (1989).

Test implication 1: the distribution of Late Archaic sites in the project area is clustered, rather than random or regular

The nearest-neighbour statistic (Johnston, 1984) will be used to test the nature of the site distribution in the project area (Figure 26.2). This statistic is calculated by dividing the actual mean distance between sites by the expected distance if the sites were located randomly in the project area. Johnston notes that the value of the nearest-neighbour statistic 'ranges between 0.0 and 2.1491: 0.0 indicates a totally clustered pattern of points; 1.0 indicates a random distribution...and 2.1491 a uniform distribution' (Johnston, 1984, p.220). To operationalize the nearest-neighbour statistic, the mean Euclidean distance from the fifty-one site location centres was calculated by averaging the distance from each site to its nearest neighbour. The mean distance is 855.71 m for fifty-one sites. The size of the project area is 20×31 km, or 620 square km. The expected value equals 1.75 km, or 1750 m. The value of the nearest-neighbour statistic equals 0.489, indicating a distribution that tends to be clustered.

A problem with the use of the nearest-neighbour statistic is related to the size of the area in which the sites are located. The same set of sites will tend to be clustered if the area including the sites is large, or will tend to be random if the area is small. Gettis and Boots (1977) suggest that one way to eliminate this problem is to position the sample area within the total cluster of sites, that is to create, for the purposes of the test, a smaller area.

To address this problem it is, therefore, necessary to define a smaller window within the project area. The window was calculated by eliminating two site locations on each of the four sides of the project area. These locations were the sites with the two most extreme UTM coordinates in each direction.

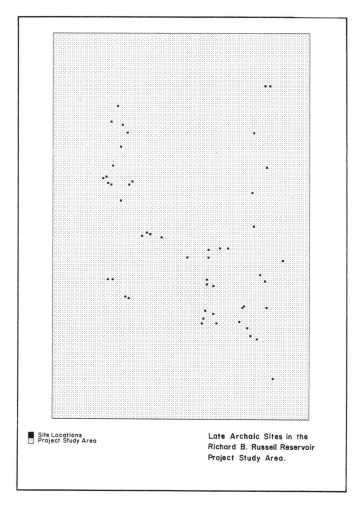

Site Locations
Project Study Area

Late Archaic Sites in the
Richard B. Russell Reservoir
Project Study Area.

Figure 26.2. Late Archaic sites in the Richard B. Russell Reservoir Project study area.

In this second test the nearest-neighbour statistic equals 0.674, indicating a distribution that is tending towards random, but is still somewhat clustered. Thus, regardless of which sized project area is chosen, the distribution of sites in the project area is clustered, rather than random or regular. The fact that the entire project area will be used for further GIS analysis indicates that the 0.489 figure is a more accurate reflection of the site distribution than the 0.674 figure calculated on a truncated window of the entire project area. Test implication 1, therefore, has been demonstrated.

Test implication 2: a variety of site types, reflecting different temporal, spatial and functional uses, will occur in each cluster

There are several steps involved in assessing this test implication: (1) the sites and artifacts need to analyzed, and site functions assigned based on the archaeological axes of space, form and time; (2) least cost movement ranges, centred on base camp locations, need to be calculated taking into account the project area's natural terrain; (3) once these distances have been determined, the project area may be divided into Thiessen Polygons (Gettis and Boots, 1977, pp.126–128, 135–142), representing habitual use areas; and (4) a site-type distribution map overlaid on the polygon map will allow this test implication to be assessed by inspection.

Step 1: determining site function based on form, space and time

Published artifact inventories were consulted (Taylor and Smith, 1978; White, 1982; Goodyear *et al.*, 1983; Sassaman, 1983). For reasons of artifact-type comparability the inventories published in Taylor and Smith (1978), supplemented by the testing results from Goodyear *et al.* (1983) are used to conduct the basic analysis of tool type and lithic raw material variation.

The analysis of tool-type variability in the site inventories provides control over the archaeological axis of form, through the use of an index of variability and function (I.V.). Using such an index it is possible '...to infer site function or intrasite activity areas by the use of functional categories defined from specified artifact traits' (Watson, in Garrow *et al.*, 1979, p.103). The index of variability and function is used to develop functional tool categories. The specific index of variability for a given site is simply a percentage of the tool types present with '...a higher I.V. indicating more diverse activity and a presumed longer or more diverse site usage' (Watson, in Garrow *et al.*, 1979, p.103). For example, small numbers of different tool types present tend to indicate an extractive site, while a wide range of types is more indicative of a maintenance camp (Binford and Binford, 1966).

For this data set an index of variability is based on the presence or absence of nine tool types (Taylor and Smith, 1978). For the purposes of further analysis, extraction sites are judged to be those with five or fewer tool types present. Maintenance sites, in contrast, have six or more tool types (Table 26.1, variable *form*). These cut-off points, and those described below, were selected to facilitate statistical analysis.

Site longevity (the temporal axis) was addressed in a similar manner, although instead of considering tool types, lithic raw material variability was used to create an 'Index of Connectivity'. Five 'exotic' lithic materials occur in the project area: (1) Coastal Plain chert; (2) slate; (3) Ridge and Valley chert; (4) steatite; and (5) other. Sites with more than two types of lithic raw material were assigned a 'good' connectivity rating, two or fewer a 'poor' rating (Table 26.1, variable *time*).

This measurement is valuable because it serves as an indicator of long-range spatial processes that occur over long time periods. Sites with many different types of exotic lithic material, for example, demonstrate a 'connectedness' to outside raw material sources that other sites do not. The measurement also serves as an indicator of long-term processes occurring at a site, since connectedness is more likely to develop at sites used over long periods of time, whether permanently occupied or used repeatedly, year after year, as seasonal base camps (Goodyear *et al.*, 1983).

Site size forms the spatial, or third, dimension of the data set. Sites were grouped into large and small categories (Table 26.1, variable *space*). Large sites included those of 7,500 square m or greater, small sites are under 7,500 square m.

Eight possible combinations of the three variables measuring space, form and time are possible. Table 26.1 summarizes the different site types developed from the eight permutations. For a more detailed description of the procedure described above, see Savage (1989).

Table 26.1 Site types based on analysis of of space, time and form

Archaeological dimensions			Site type
(Space)	(Time)	(Form)	
Large	Good	Extraction	Long-term extraction area
Large	Poor	Extraction	Short-term extraction area
Large	Good	Maintenance	Long-term base camp
Large	Poor	Maintenance	Short-term base camp
Small	Good	Extraction	Long-term extraction locus
Small	Poor	Extraction	Short-term extraction locus
Small	Good	Maintenance	Long-term logistical camp
Small	Poor	Maintenance	Short-term logistical camp

Step 2: deriving least-cost movement ranges from base camps

Once the types listed above have been assigned to the sites in the project area, it is possible to create movement ranges, based on principles of least-cost movement. In the past, site catchment analysis has approached this procedure by looking at either simple concentric rings of increasing radius around some centre of an isotropic plain (based on Von Thunen's (1966) model), or by creating one- or two-hour 'walks' (Roper, 1979). In contrast to these methods, the geographic information system approach creates distance rings of increasing radius from some centre as in the Von Thunen method; however, the method is able to consider topographic and hydrological features directly when distance is calculated. Instead of creating perfectly symmetrical distance rings, as would exist on an isotropic plain, the GIS method creates irregular distance rings, based on real-world conditions.

To implement this method, hydrology and 'terrain roughness' map overlays are required. The hydrology layer (Figure 26.3) was digitized from USGS 1:100,000 quad sheets containing the project area. The terrain roughness overlay was created by calculating the first derivative of an elevation map layer (Figure 26.4), producing an overlay of slope values (Figure 26.5). The first derivative of slope was calculated, and the result is an overlay of 'change in slope', or terrain roughness (Figure 26.6). Once these two map overlays are available, the distance from base camps may be determined in the GIS with the 'Spread' command. The base camp sites are isolated from the overall map of site types (Figure 26.7), and distance is spread from them, over the 'roughness' map and through the 'hydrology' map, to create the 'basecamp' map layer (Plate IX). This overlay represents the cost of movement from known base camps across the physical landscape, expressed in terms of variable distances. Reference to Plate IX shows the irregular nature of the distance rings spread from the base camp locations, reflecting the relative ease or difficulty of movement from various locations in the project area.

Step 3: creating habitual use areas with Thiessen Polygons

The Thiessen Polygon approach has been used by a number of archaeologists to study territoriality in hierarchical societies (Cunliffe, 1971; Renfrew, 1975; Hodder

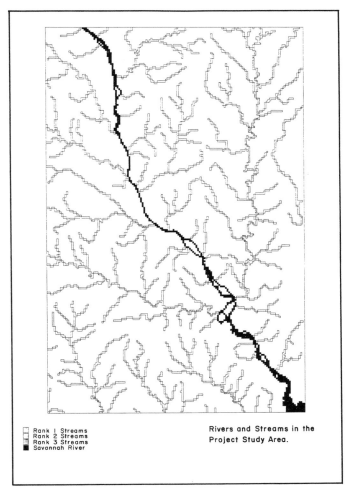

Figure 26.3. Rivers and streams in the project study area.

and Orton, 1976). In the absence of evidence for a political hierarchy, such as in the Late Archaic, Cherry advocates drawing weighted polygons around each highest order site (1987). In my case, these 'highest order' sites would have been the base camps. The Thiessen Polygon boundaries (Figure 26.8) were drawn by dividing the landscape along straight lines which ran between areas of greatest equal distance between centres (base camps). Exceptions to this rule were made when a number of base camps occurred in a very small area (within one to two km of each other). In these cases, it was felt that what the site placement reflects is the use of an area as a base camp location, rather than a specific site. If the sizes of the sites are considered (on the maps only their centre points are plotted), then some of the base camps become very close to each other. This situation may reflect different seasonal occupations by groups who essentially had a different idea of 'site', or 'place', than the archaeologists who recorded the remains. The Thiessen Polygons created thus reflect the existence of six different habitual use areas (Figure 26.9) within the project area. Four had base camp locations situated within a relatively small area, while the other two are represented by a single base camp location.

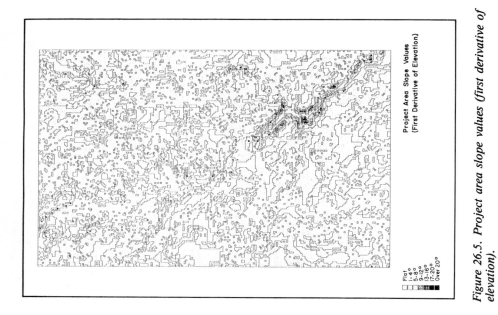

Figure 26.5. *Project area slope values (first derivative of elevation).*

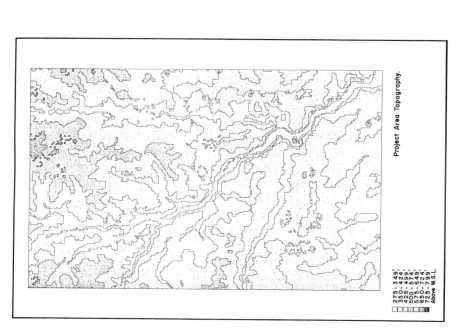

Figure 26.4. *Project area topography.*

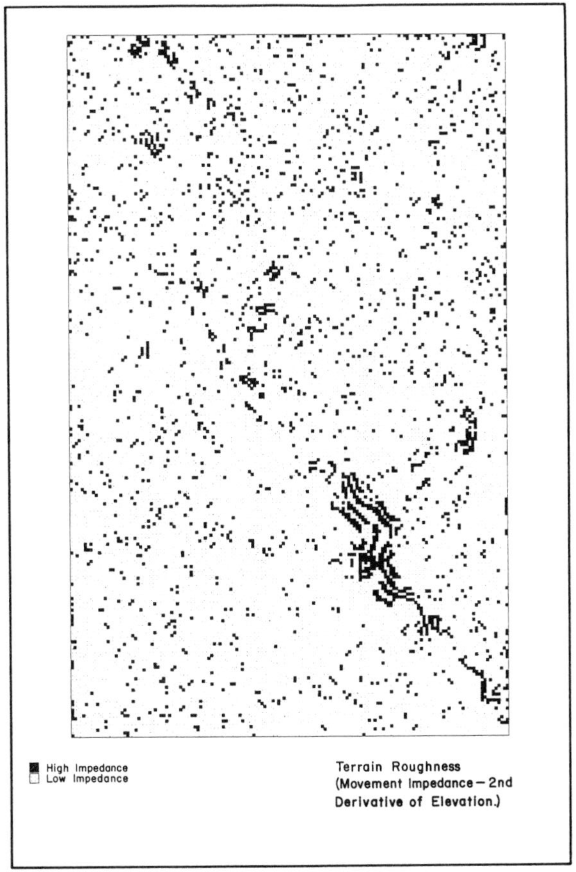

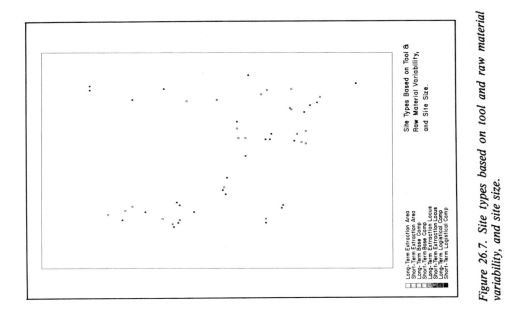

Site Types Based on Tool &
Raw Material Variability,
and Site Size.

Long-Term Extraction Area
Short-Term Extraction Area
Long-Term Base Camp
Short-Term Base Camp
Long-Term Extraction Locus
Short-Term Extraction Locus
Long-Term Logistical Camp
Short-Term Logistical Camp

Figure 26.7. Site types based on tool and raw material variability, and site size.

Terrain Roughness
(Movement Impedance — 2nd
Derivative of Elevation.)

High Impedance
Low Impedance

Figure 26.6. Terrain roughness (movement impedance:—2nd derivative of elevation).

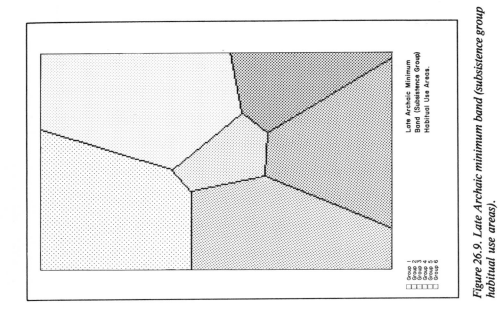

Figure 26.9. Late Archaic minimum band (subsistence group habitual use areas).

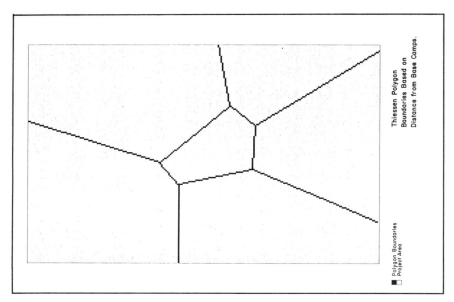

Figure 26.8. Thiessen Polygon boundaries based on distance from base camps.

Step 4: overlaying site types on Thiessen Polygons

The final step in assessing site clusters is accomplished simply by overlaying a map of the different site types (Figure 26.7) on a map of Thiessen Polygon boundaries (Figure 26.8), thus creating a map of site types within habitual use areas (Figure 26.10). Since the boundaries of the Thiessen Polygons were created without reference to any site types except the base camps required to create the distance measurements, the distribution reflected in Figure 26.10 may be judged to be free from bias. When we do this we see that each habitual use area contains a number of different site types.

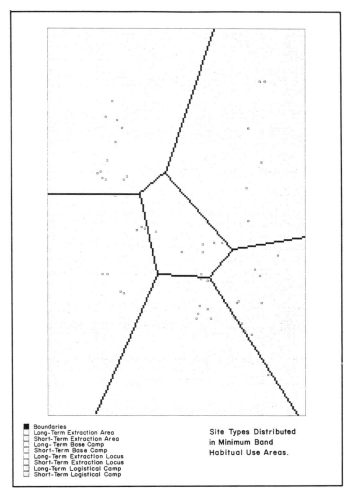

Boundaries
Long-Term Extraction Area
Short-Term Extraction Area
Long-Term Base Camp
Short-Term Base Camp
Long-Term Extraction Locus
Short-Term Extraction Locus
Long-Term Logistical Camp
Short-Term Logistical Camp

Site Types Distributed
in Minimum Band
Habitual Use Areas.

Figure 26.10. Site types distributed in minimum band habitual use areas.

Test implication 3: the habitual use areas in the project area support populations in the range of minimum bands, and thus reflect the activities of minimum bands, rather than maximum bands

This test implication may be assessed by estimating the population size for each habitual use area. This estimate is calculated by multiplying the size of each area by an assumed population density. Minimum bands would be expected to range from 20 to 120 people (Wobst, 1974; Perlman, 1985). A range of population densities has been provided in Hassan (1981, Table 2.1) for acorn-gatherer/hunter/fisher societies. The number of grid cells in each polygon can be found via the GIS. Grid cells are 127 m square; the area of each polygon in km^2 is obtained by multiplying the number of grid cells for each area by 16,129 square m (127^2) and dividing by one million (one million m^2 per km^2); see Table 26.2. Population estimates are then calculated (Table 26.3).

Table 26.2 Size of habitual use areas in square km

Group One	143.61 km^2
Group Two	152.84 km^2
Group Three	118.23 km^2
Group Four	36.66 km^2
Group Five	110.24 km^2
Group Six	62.77 km^2

Table 26.3 Estimated population ranges of habitual use areas

	Population densities		
	0.39	1.2	Average
Group One	56.0	172.3	114.2
Group Two	59.6	183.4	122.0
Group Three	46.1	141.9	94.0
Group Four	14.3	44.0	29.2
Group Five	43.0	132.3	87.7
Group Six	24.5	75.3	49.9
Total	243.5	749.2	497.0

The results calculated with the averaged population density show that groups ranged between 29 and 122 people, suggesting minimum band organization. The population estimate for the entire project area indicates the presence of at least one maximum band, assuming that those groups ranged from 200 to 600 people.

Test implication 4: the boundaries between habitual use areas will reflect uses both as edges and centres. Small clusters of sites occurring on both sides of a boundary between different habitual use areas can be interpreted as centres. Edges are boundary areas where no such clusters exist

In order to consider a boundary as a centre there should be material evidence for some kind of interaction across it. When no such evidence exists, the boundary may be considered to have functioned as an edge, adding a 'whereness' to the understanding of boundary edges and centres.

In the Late Archaic there is evidence for long distance trade; Marquardt (1985) and Bender (1978) discuss the exchange of Great Lakes copper for southeastern marine shells. Such trade would have cut across many minimum and maximum band areas, but the chances of its touching any given minimum band must be considered very low, given the apparent low volume of materials being exchanged.

Interactions among minimum bands are more likely, it seems, to have been related to subsistence-related exchange or cooperation. This notion seems all the more probable when we consider that a lot of people in any given minimum band were probably related to people in neighbouring minimum bands. It should not surprise us to find evidence for cooperative hunting/gathering episodes among such groups of related people.The evidence for this kind of interband cooperation across minimum band boundaries can be expected to exist in the form of logistical camps and extractive areas near the boundaries between minimum bands.

The boundaries between habitual use areas were drawn with reference only to the base camp sites in the project area. Two areas of potential cross-boundary cooperation between minimum bands are shown in the project area (Figure 26.11). This map overlay shows, in each of the small boxes, a long-term logistical camp, and one or more long-term extraction areas along the boundaries between minimum band habitual use areas.There is, in addition, a short-term extraction locus (a kill site) associated with each group. Therefore, the boundaries serve as both edges and centres.

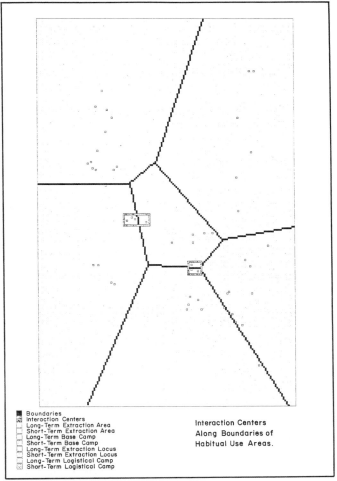

Boundaries
Interaction Centers
Long-Term Extraction Area
Short-Term Extraction Area
Long-Term Base Camp
Short-Term Base Camp
Long-Term Extraction Locus
Short-Term Extraction Locus
Long-Term Logistical Camp
Short-Term Logistical Camp

Interaction Centers
Along Boundaries of
Habitual Use Areas.

Figure 26.11. Interaction centres along boundaries of habitual use areas.

These site groups demonstrate evidence for cooperative hunting/gathering between different minimum bands, and that the boundaries between the bands involved may be interpreted as centres for this intergroup cooperation at these places. Conversely, where no such site groupings exist, the boundary may be interpreted as an edge.

We can summarize our findings as follows:

(1) The distribution of Late Archaic sites in the project area is clustered.
(2) A variety of site types, reflecting different temporal, spatial and functional uses, occurs in each cluster.
(3) The habitual use areas in the project area reflect the activities of minimum bands that may have comprised one maximum band.
(4) The boundaries between habitual use areas serve as both edges and centres.

The landscape analysis presented the hypothesis that *the Late Archaic social landscape was comprised of maximum band social territories divided into minimum band subsistence territories*. This analysis has produced six habitual use areas within the project area, ranging in size from 36.66 to 152.84 square km. Based on a population density ranging from 0.39 to 1.2 persons per square km, the landscape in the project area would have supported between 245 and 749 people, average 497, well within the range of one maximum band. Thus, the project area appears to have supported six minimum bands associated in one maximum band.

Conclusion: theory and method in landscape archaeology

The concept of landscape archaeology unites many of the traditional approaches to archaeological research. In particular, this study has demonstrated how such traditionally divergent themes as geographic location theory, models of social organization, boundary studies, site function, demographics and subsistence models may be shown to converge on questions examined under a broad theoretical perspective illuminated by landscape archaeology. The issues that make the Late Archaic period exciting are the social changes that accompany the period. By using the landscape archaeology paradigm it has been possible to go beyond subsistence-related studies, to get at these social issues.

Geographic information systems provide the means to create and interpret cultural landscapes. Most significant in this study is the ability to consider natural features (hydrology and topography) in conjunction with archaeological evidence in order to examine behaviour over real space. The social landscape model presented here (Plate X) considers the natural terrain in place of an isotropic plain.

Possibilities for future work

With these beginnings, other questions may be asked; questions relating to inter- and intra-group dynamics, as some minimum band groups demonstrate greater successes in exploiting their surroundings, and greater reproductive success. We might expect minimum band group boundaries to shift over time in ways that reflect the various successes and failures of the maximum band's constituent groups—as one group expands and another contracts. Inter-group cooperation and conflict may be examined in response to physical and social environmental perturbations.

Issues of this sort may be examined at the larger, maximum band level as well. There is evidence for increasing hostility in the Late Archaic. It seems possible, given the minimum/maximum band model demonstrated here, that such hostility may have existed primarily between maximum bands, where fewer personal interrelationships are likely to have existed than in a single maximum band's constituents.

Another avenue of research in this area would take advantage of the ability of the GIS to model long-term cultural processes. By conducting a similar study with Middle Archaic sites in the Russell Reservoir, a diachronic perspective of changing social organization may be achieved. It will be possible to explain the complex technological and subsistence changes observed between these two periods in terms of social dynamics, rather than just changing point types and new food resources. The combination of landscape archaeology and geographic information systems, therefore, offers a truly new way of looking at old things.

References

Bender, B., 1978, Gatherer-Hunter to Farmer: A Social Perspective. *World Archaeology*, **10**(2), 204–221

Berry, B. J., 1964, Approaches to Regional Analysis: A Synthesis. *Annals of the Association of American Geographers*, **54**, 2–11

Binford, L. R. and Binford, S. R., 1966, A Preliminary Analysis of Functional Variability in the Mousterian of Levallois Facies. In *Recent Studies in Paleoanthropology*, edited by J. D. Clark and F. C. Howell, *American Anthropologist*, **68**(2), pp. 238–295

Cherry, J. F., 1987, Power in Space: Archaeological and Geographical Studies of the State. In *Landscape and Culture: Geographical and Archaeological Perspectives*, edited by J. M. Wagstaff. (Oxford: Basil Blackwell) pp. 146–172

Clark, J. G. D., 1975, *The Earlier Stone Age Settlement of Scandinavia.* (Cambridge: Cambridge University Press)

Conkey, M. W., 1988, Contexts of Action, Contexts for Power: Material Culture and Gender in the Magdalenian. Paper presented at the *Conference on Women in Prehistory*, held at the University of South Carolina, March 1988

Cunliffe, B. W., 1971, Some Aspects of Hill Forts and Their Cultural Environments. In *The Iron Age and its Hill-Forts*, edited by M. Jesson and D. Hill. (Southampton: Department of Archaeology, Southampton University) pp. 53–69

Dennell, R., 1983, *European Economic Prehistory: A New Approach.* (New York: Academic Press)

Garrow, P. H., White, M. E., Watson, G. M., Nicklas, S. D., Savage, S. H. and Muse, J., 1979, Second Draft Report: Archaeological Survey of the John H. Kerr Reservoir, Virginia and North Carolina. Soil Systems, Inc. Project ES-1202. Marietta, GA

Gettis, A. and Boots, B., 1977, *Models of Spatial Processes: An Approach to the Study of Point, Line, and Area Patterns.* (Cambridge: Cambridge University Press)

Goodyear, A. C., Monteith, W. and Harmon, M., 1983, *Testing and Evaluation of the 84 Sites and Reconnaissance of the Islands and Cleveland Property, Richard B. Russell Dam and Lake, Savannah River, Georgia and South Carolina.* Research Manuscript Series 189, Institute of Archaeology and Anthropology, University of South Carolina, Columbia

Green, S. W. and Perlman, S. M., 1985, Frontiers, Boundaries, and Open Social Systems. In *The Archaeology of Frontiers and Boundaries,* edited by S.W.

Green and S.M. Perlman. (New York: Academic Press) pp. 3–14

Green, S. and Sassaman, S., 1983, The Political Economy of Resource Management: A General Model and Application to Foraging Societies in the Carolina Piedmont. In *Ecological Models of Economic Prehistory,* edited by G. Bronitsky. Arizona State University Anthropological Research Papers, Number 29, Tempe, AZ, pp. 261–290

Harris, M., 1979, *Cultural Materialism: The Struggle for a Science of Culture.* (New York: Random House)

Hassan, F. A., 1981, *Demographic Archaeology.* (New York: Academic Press)

Higgs, E. S. and Webley, D., 1971, Further Information Concerning the Environment of Paleolithic Man in Epirus. *Proceedings of the Prehistoric Society,* **37** (part II), pp. 367–380

Higgs, E. S., Vita-Finzi, C., Harriss, D.R. and Fagg, A.E, 1967, The climate, environment and industries of Stone-Age Greece, Part III. *Proceedings of the Prehistoric Society,* **33**, 1–29

Hodder, I., 1986, *Reading the Past: Current Approaches to Interpretation in Archaeology.* (Cambridge: Cambridge University Press)

Hodder, I., 1987, Converging Traditions: The Search for Symbolic Meanings in Archaeology and Geography. In *Landscape and Culture: Geographic and Archaeological Perspectives,* edited by J. M. Wagstaff. (Oxford: Basil Blackwell) pp. 134–145

Hodder, I. and Orton C., 1976, *Spatial Analysis in Archaeology.* (Cambridge: Cambridge University Press)

Johnston, R. J., 1984, *Multivariate Statistical Analysis in Geography.* (New York: Longman)

Kohl, P. L., 1981, Materialist Approaches in Prehistory. In *Annual Review of Anthropology,* edited by B. J. Siegel, A.R. Beals and S. A. Tyler. (Palo Alto, CA: Annual Reviews) pp. 89–110

Lee, R. B., 1968, What Hunters do for a Living, or, How to Make out on Scarce Resources. In *Man the Hunter,* edited by R. B. Lee and I. DeVore. (Chicago: Aldine Press) pp. 30–48

Lee, R. B., 1979, *The !Kung San: Men, Women and Work in a Foraging Society.* (Cambridge: Cambridge University Press)

Lewis, T. M. N. and Kneberg-Lewis, M., 1961, *Eva: An Archaic Site.* (Knoxville: University of Tennessee Press)

Marquardt, W. H., 1985, Complexity and Scale in the Study of Fisher-Hunter-Gatherers: An Example from the Eastern United States. In *Prehistoric Hunter Gatherers: The Emergence of Cultural Complexity,* edited by T.D. Price and J.A. Brown.(New York: Academic Press) pp. 59–98

Marquardt, W. H. and Crumley, C. L., editors, 1987, *Regional Dynamics: Burgundian Landscapes in Historical Perspective.* (New York: Academic Press)

Marshall, L., 1959, Marriage among !Kung Bushmen. *Africa,* **29**, 335–365

Moore, J. A., 1981, The Effects of Information Networks in Hunter-Gatherer Societies. In *Hunter-Gatherer Foraging Strategies,* edited by B. Winterhalder and E. A. Smith. (Chicago: University of Chicago Press) pp. 194–217

Paynter, R. W., Green, S. W. and Wobst, H. M., 1974, Spatial Clustering: Techniques of Discrimination. Paper presented at the *39th Annual Meeting of the Society for American Archaeology*

Perlman, S. M., 1985, Group Size and Mobility Costs. In *The Archaeology of Frontiers and Boundaries,* edited by S. W. Green and S. M. Perlman. (New York:

Academic Press) pp. 33–50

Pred, A., 1967, Behavior and Location: Foundations for a Geographic and Dynamic Location Theory. *Lund Studies in Geography, Series B: Human Geography,* Number 27. Lund, Sweden

Price, B. J., 1982, Cultural Materialism: A Theoretical Review. *American Antiquity,* **47**, 709–741

Renfrew, C., 1975, Trade as Action at a Distance. In *Ancient Civilisation and Trade,* edited by J. A. Sabloff and C. C. Lamberg-Karlovsky. (Albuquerque: University of New Mexico Press) pp. 3–59

Roberts, B. K., 1987, Landscape Archaeology. In *Landscape and Culture: Geographical and Archaeological Perspectives,* edited by J. M. Wagstaff. (Oxford: Basil Blackwell) pp. 77–95

Root, D., 1983, Information Exchange and the Spatial Configurations of Egalitarian Societies. In *Archaeological Hammers and Theories,* edited by J. A. Moore and A. S. Keene. (New York: Academic Press) pp. 193–200

Roper, D. C., 1979, The Method and Theory of Site Catchment Analysis: A Review. In *Advances in Archaeological Method and Theory,* Volume 2, edited by M. B. Schiffer, (New York: Academic Press) pp. 119–140

Sassaman, K. E., 1983, Middle and Late Archaic Settlement in the South Carolina Piedmont. Unpublished M.A. thesis, University of South Carolina, Columbia

Savage, S. H., 1989, *Late Archaic Landscapes: A Geographic Information Systems Approach to the Late Archaic Landscape in the Savannah River Valley, Georgia and South Carolina.* Anthropological Studies 8, Occasional Papers of the South Carolina Institute of Archaeology and Anthropology, The University of South Carolina, Columbia

Spaulding, A. C., 1960, The Dimensions of Archaeology. In *Essays in the Science of Culture: in Honor of Leslie A. White,* edited by G. Dole and R. Carneiro. (New York: Thomas Crowell)

Steward, J., 1969, Postscript to Bands: on Taxonomy, Processes, and Causes. In *Contributions to Anthropology: Band Societies,* edited by D. Damas. National Museums of Canada, Bulletin **228**, pp. 288–295

Taylor, R. L. and Smith, M. F., 1978, The Report of the Intensive Survey of the Richard B. Russell Dam and Lake, Savannah River, Georgia and South Carolina, Research Manuscript Series, Number 142, Institute of Archaeology and Anthropology, University of South Carolina, Columbia

Vita-Finzi, C. and Higgs, E. S., 1970, Prehistoric Economy in the Mount Carmel Area of Palestine: Site Catchment Analysis. *Proceedings of the Prehistoric Society* (Cambridge), **36**, 1–37

Von Thunen, J. H., 1966, *Von Thunen's Isolated State.* (Translated and edited from the 1842 original by C. M. Wartenberg and P. Hall). (London: Pergamon)

Webb, W. S., 1946, Indian Knoll, Site Oh 2, Ohio County, Kentucky. *University of Kentucky Reports in Anthropology,* **7**(4), Lexington

White, J. W., 1982, An Integration of Late Archaic Settlement Patterns for the South Carolina Piedmont. Unpublished MA thesis, University of Arkansas

Wobst, H. M., 1974, Boundary Conditions for Paleolithic Social Systems: A Simulation Approach. *American Antiquity,* **39**(2), 147–178

27

Sorting out settlement in southeastern Ireland: landscape archaeology and geographic information systems

Stanton W. Green

Introduction

In this chapter I would like to introduce a general discussion of landscape archaeology and geographic information systems, followed by a brief discussion of our work on the early prehistory of Ireland. I will begin with an assertion that underlies my enthusiasm for the use of GIS; geographic information systems are much more than a set of cartographic techniques. They are a range of methodologies that facilitate problem solving through the description and interpretation of data as well as the development of theory. GIS make available cartographic theory, spatial theory and computer mapping and analysis to the archaeologist (and social scientists in general) interested in the spatial aspects of human behaviour. They make possible the true integration of natural and cultural factors in modelling and recreating past cultural landscapes.

The three dimensions of archaeology and GIS

Archaeologists often speak of the three dimensions of archaeology: space, time and form. However, we have never really been very good at coping with all three simultaneously—not because of the nature of our archaeological data but because of limitations of our theory and methodology. For example, the study of cultural landscapes often comes down to the need to overlay, compare and correlate multi-dimensional maps. Sometimes these maps represent different cultural and natural variables; sometimes different time periods or their stratigraphic representatives. Difficulties arise when we attempt to combine these maps for analysis and display, or perform mathematical manipulations on the variables we mapped.

We start off with the problem of variable classification. For nominal variables we must decide upon an effective number of categories. Land cover, for example, can be broken down into general categories, such as forest and arable, or more specific ones such as deciduous, coniferous, wheat, barley and oats. For interval variables, the question becomes one of intervals. Should one use equal or non-equal intervals? How many intervals should be used? How should these intervals be distributed (eg. logarithmically)? This type of data manipulation is central to both data analysis and display. For manual map-making classification is extremely time consuming; for GIS they are routine. All GIS provide reclassification functions that allow for the easy and fast reclassification of single or multiple variables. Variables can be combined in various ways (adding them, multiplying them, giving them different weightings or priorities, etc.) and then classified into any number of categories and intervals. This aspect of GIS methodology provides much more than just the technical advantage of saving time; it provides a tool for thinking about and interpreting space in ways previously unavailable.

Things get even more complex as we combine variables in spatial analysis. Common methods such as scatter and triangular graphs can often yield reasonably interpretable displays of relationships between two or three variables (Dickenson, 1973). Classic among archaeologists are the triangular graphs that are used to describe soils according to their sand/silt/clay composition. However, the mapping of these derived categories requires rather broad arbitrary categories if the map is to be interpretable.

Finally, we move onto the comparison of distribution maps. The visual comparison of maps has been shown to be unreliable and inaccurate except in cases where distributions are highly associated and therefore obviously similar (McCarty and Salisbury, 1961). Classic statistical models are not designed for three-dimensional spatial analysis and therefore do not provide relief for the problem of map comparison. As I will discuss below, the use of traditional descriptive statistics on spatial data usually forces one to answer relatively simple three-dimensional questions with very complex multi-variate answers. Because GIS is designed to model space mathematically it offers simple solutions to these problems.

The reason for this is that GIS include in their database a graphic component that manages *spatially referenced data*. This allows a GIS to:

(1) Interrelate spatially referenced data as map overlays;
(2) Describe, manipulate and analyze these data-maps; and
(3) Very importantly, create new data in the form of new map overlays.

Applying GIS

My GIS project has three goals. First and most general, I use GIS to solve the methodological problem of analyzing archaeological data without losing its three-dimensional quality. This is the ultimate goal of the application of GIS to archaeology. I will discuss this within the realm of landscape archaeology, the goal of which is to model or recreate past cultural and natural landscapes within a problem-oriented framework. GIS is a powerful methodology for the application of landscape theory.

Second, I am developing flexible software for the field computerization of archaeological data that can be used for GIS analysis. This is based on Stephen Savage's Archaeological Research Management System (ARMS) that is designed to manage spatially referenced archaeological data so that it can be interfaced with the GIS system MAPCGI2 (a derivative of Map Analysis Package, MAP). Field computerization of site and artifact data provides two primary benefits. First, it provides a quick, comprehensive and accurate way to process site and artifact information in the field. This is essential in cases where artifacts cannot be removed from their country of origin (as is the case for Ireland). However, even in cases where artifacts can be brought back to laboratories, field computerization eliminates the necessity of having to double code the data in the field. The objective is to develop field and laboratory forms that are identical (or at least very similar) to the input screens of the database management system. The result is quick and accurate entry of site data, including site location (especially important for, the site survey), and excavation type information such as the horizontal and vertical provenance of artifacts, soil, pollen and flotation samples, and the existence of features.

These data are spatially referenced so that all of the information for a particular site, excavation unit or stratum can be retrieved and output for display or analysis. Moreover, the database is available for output to statistical analysis programs (for example SAS, SYSTAT) and, most pertinent to this volume, it can be hooked to a GIS system such as MAP. In this latter case, any variable or combination of variables can be brought into MAP for GIS description or analysis. Such a set-up can be used to monitor surveys and excavations as they are proceeding. In the case of survey this allows one to keep track of sampling design and its results. For excavation a site can be recreated from the data as it is being excavated. This is a great descriptive, analytical and predictive step forward.

My third archaeological goal is to examine the cultural landscape of prehistoric Ireland's original Mesolithic hunter-gatherer colonizers and changes in this landscape as these colonists and perhaps immigrants incorporated agriculture into their society. I discuss this aspect of the project in the second part of the chapter where I focus on the project itself. Let me first, however, move into a more general discussion of GIS and landscape archaeology.

GIS and landscape theory

The combination of three-dimensional data entry and GIS offers a powerful approach for collecting information on and interpreting the structure of cultural landscapes and archaeological sites. Cultural landscapes are composed of space (two dimensions) and the cultural and natural forms that vary over space (the third dimension). As is implied in its definition, a cultural landscape is inherently determined by the place from which it is being viewed. In other words, geographic areas really incorporate infinite landscapes. This has multiple and profound implications as we try to understand past cultural uses and perceptions of the environment.

Landscape theorists consider a variety of elements in defining landscapes; viewpoint, range of vision, direction (Keiji, in Higuchi, 1983), angle of incidence, angle of elevation, and even light (Martens, 1890). I would also include knowledge of the environment in the definition of landscape. This notion of a relative landscape has theoretical and methodological implications for the archaeologist.

At the level of theory, we often frame questions concerning the cultural use and perception of the environment. Land use modelling, such as catchment analysis, defines land use patterns by delimiting landscapes in terms of the distribution of resources as they relate to sites. Questions concerning how the site inhabitants perceived the environment continually plague these types of models. In a sense, catchment analysis is a method without a guiding theory. It requires a theoretical framework for asking questions concerning how people and groups of people view their environment—just the kinds of questions that landscape theory combined with GIS can be designed to address. GIS offers functions that allow one to view (that is to describe) a landscape from a particular place or point on that landscape. Such a view can be blocked by a barrier, or facilitated by a pass (Madry and Crumley, Chapter 28). A view can be a simple line of sight or—more provocatively—a culturally constructed path. A trading pattern via trading partners, middlemen, or perhaps competing groups or raiders, can be modelled over a landscape. This provides the basis for interpreting the pattern of trade of goods between regions, or more generally their social interaction. The exciting possibility of overlaying theoretical models of cultural interaction over archaeological settlement data is now being developed by archaeologists using GIS, some of whom have papers in this volume (Allen, Chapter 25; Zubrow, Chapter 24).

The three dimensions of archaeology

At a more general level, archaeologists have struggled with the problem of capturing the three dimensions of space, time and form simultaneously. I have long had an interest in the problem of capturing the multi-dimensionality of archaeological data mathematically and will use a particularly frustrating example to illustrate the spatial bottleneck that I believe GIS can relieve.

In this case, we are looking for spatial relationships between environmental and land use variables in historic Denmark (Green and Ulrich, 1977). The questions were 'simply': what are the determining factors for historic grain productivity, and how are these spatially associated? We specifically examined historical land use productivity as recorded in hartkorn—a tax measure of how much grain a particular parcel of land produced. This measure was correlated with environmental variables (eg. soils, frost-free days, average winter and summer temperature, precipitation, etc.) in order to determine which variables most affected grain productivity. Our objective was to examine the spatial co-variation of these environmental variables and land productivity (at least as it was reported to the taxman).

We first regressed the variables against hartkorn (grain productivity) in order to look for possible causal relationships and especially to determine which variables most affected grain production. Each of the environmental variables was then mapped and overlayed on the hartkorn map in order to look for spatial patterns. However, these analyses were not really satisfactory for examining the spatial co-variation of grain productivity and environmental factors. In particular the regression and mapping were independent and therefore did not provide the integrated analysis necessary for our needs. The map overlaying was difficult to interpret in any systematic fashion.

Our next tactic was to combine statistically the independent environmental variables through principal components analysis in order to 'simplify our database' and eliminate redundancy. Several components were produced representing soil, precipitation/sunshine and temperature factors. We regressed these factors in order to prioritize them in terms of their effect on grain production. We then mapped these environmental factors using the then state-of-the-art SYMAP program. In order to distill trends from these maps we put them through (what seemed to be countless) trend surface analyses. Our output was confusing, perceptually inconclusive and aesthetically horrid. Although we had then integrated our environmental elements, and affected a form of spatial analysis through trend surface analysis, the results pointed more to the difficulties with solving this sort of spatial problem than to the answers.

This project illustrates the many ironies and frustrations that led me to explore the application of GIS. To begin with, we had to derive much of our data from maps, only to concoct statistical factors to be once again mapped. These statistical factors (in this case principal components) provide mathematical simplification at a very high interpretive cost. Interpretation had to be based on understanding abstract combinations of variables rather than the original down-to-earth measures of temperature, soils, rainfall etc., as they varied over the historic Danish landscape. In other words, the methodology drove us from simple to complex—and therefore to patterns that were more difficult to interpret.

This complex road to spatial solutions is in direct contrast to the simple, elegant solutions offered by the GIS methodology. With GIS, natural and cultural variables can be mapped independently and combined as three-dimensional maps. So, for example, land use and environmental variables are mapped independently, and then combined as three-dimensional data sets. This avoids the need to combine variables mathematically prior to mapping, and thereby sidesteps the difficulties of interpreting maps composed of complex mathematical combinations of variables, such as principal components. A soil map can be superimposed over a site map (in a number of ways) to analyze site/soil associations. Moreover, GIS programs provide a wide range of functions for exploring individual variables as well as combining them. These methods are referred to as 'map-a-matics' or map algebra (Cowen, 1987). As I stated in the introduction to this paper, GIS is suited for spatially referenced data, and therefore relieves one of the pain of having to contrive traditional statistical models toward spatial ends.

GIS and archaeological excavation

Archaeological sites, in a sense, provide a mirror image of the landscape with their stratigraphic (third) dimension. In fact, if we understand the landscape as an historical process, then as Roberts (1988, p. 79) suggests 'any archaeological excavation may be thought of as the micro-dissection of a small piece of landscape'. The application of GIS to excavation data could be most exciting because it has the potential to allow us to maintain the spatial and temporal context of all artifacts, as well as natural and cultural features within the same data set. We often speak of the destructive nature of archaeology and the need to preserve the site through mapping and photography. GIS offers a truly three-dimensional method of record-keeping that mathematically mimics the reality of three-dimensional archaeological landscapes and sites. As I note above, GIS can be used to monitor excavations, both allowing one to keep track of sampling strategy (and where the samples are in the field process), and in actually recreating the site as it is excavated. This kind of information can be used in the ongoing field decision-making process. Let me now turn briefly to a discussion of our project and how we are incorporating GIS.

The archaeological project

In 1982 I began collaborative research with Dr. Marek Zvelebil (University of Sheffield, England) on the earliest settlement of southeast Ireland. The project is designed to recover information systematically about the prehistoric landscape around Waterford Harbour (Figure 27.1) in order to study the initial colonization of southern Ireland and the subsequent development of agricultural society. Until the time of our research project, no study of this area had been done and the prehistory of the area was quite literally unknown. Our systematic survey of the area has identified some 300 prehistoric sites (including quarries, hunter-gatherer campsites and farming settlements). Artifactual evidence indicates a colonization of the area by hunter/gatherer communities by 6000 B.C. (later Mesolithic period), and continued settlement through the early farming (Neolithic, 3500–1500 B.C.), and the Bronze Age periods (1500–300 B.C.). Test excavations over the past two years have identified several undisturbed, stratified sites that are yielding additional information on chronology, technology and village layout (Zvelebil *et al.*, 1987; Green and Zvelebil, 1990).

Figure 27.1. Ireland, showing the project area around Waterford.

The nature of this project, especially in its initial stages, requires the type of spatial database and methodology that GIS offers. At the current stage of our project more questions than answers are available. We use GIS to ask spatial questions and see our way to simple, interpretable solutions. From a practical standpoint, it provides a flexible, spatially referenced database that can be built during the course of the project by adding cases and variables.

GIS is applied to a variety of tasks. First, we use it to outline and describe geographically the project area. Most basic in this case is the geographic positioning of the project area with regard to the Waterford estuary/harbour, the river system and the Atlantic coast. Second, we document and record the fields we have sampled. Our sampling strategy involves systematic surface collecting in ploughed fields within designed geographic two by three mile grids (Zvelebil *et al.*, 1987; Green and Zvelebil, 1990). The project area has been digitized and fields are added as they are walked. If a field yields archaeological evidence, this is added to the database. This allows us continually to describe and analyze the material we are recovering. We can map out sites and any of their attributes (number of artifacts, number of flint artifacts, number of scrapers) by outputting the site database to the MAP GIS. These variables can be classified into categories, or otherwise manipulated mathematically and then mapped out. Moreover, should we wish to derive a variable from MAP we can add this new variable back into the site database. For example, by dividing a map of the number of artifacts found in each field by a map of the area of each field, we derive an artifact density map. Artifact density can then be added (if we wish) to the site database.

We have used our GIS database to begin to explore some basic geographic relationships. We can, for example, look at simple landscape questions such as distance to coast. A simple spread function in the Map Analysis Package allows us to mark sites within one km of the coast (Plate XI). We could, of course, choose any distance. Such an overlay could then be used for comparison and analysis with other overlays, perhaps representing different time periods, or site types, for example. Again, sites can be marked in terms of their location and this can be added to the site database.

Beyond this we are looking at the cultural landscape as it is expressed in the use of various raw materials. Flint and rhyolite are the primary lithic materials available. Here we can see the limited distribution of rhyolite as it is tied closely to its point source quarry despite its apparent superior quality. In contrast, flint pebbles dominate the landscape. Eventually, we will examine the distribution of various categories of lithic manufacture, Mesolithic and Neolithic sites, etc. As we add geographic overlays such as soils, topography and hydrology, we intend to study the co-variation of the cultural landscape with aspects of the natural landscape. A particularly exciting prospect will involve an overlay of the sea-coast topography. We hope to simulate the effect of sea level on the coast and river systems in order to predict the location of submerged sites and understand discovered sites better in terms of their contemporary landscape.

Summary

To conclude, GIS methodology offers a new set of sophisticated tools for the study of archaeological landscapes and sites. In a sense it combines the basic notions of overlaying maps with the mathematical possibilities of computer analysis and database management systems. Its development as a creative methodology can go a long way towards solving problems of space, time and form. When tied to landscape theory it provides a powerful means for exploring past cultural landscapes. Its beauty is in its simplicity.

Acknowledgments

The field research reported is a collaborative effort with Dr. Marek Zvelebil, University of Sheffield. I thank Claudia for her love, support and perseverance; and Harrison and Devin for their cross-cultural adaptability. A 'tip of the ole cap' to the scores of Irish, American and English students and volunteers who have worked with us over the past eight years. I especially thank Chris Judge and Eddy Moth who have been with the project since its inception. Cheers to Melissa Conner Palmer for digitizing the project area, and Stephen Savage for adapting his computer software to my needs. Specific funding for the GIS component of the project came from the University of South Carolina through a Research and Productive Scholarship Award. I thank Dave Cowen, Lynn Shirley, Tim White and Homer Steadly for their user-friendly support of my GIS and computer endeavours. Finally, thanks to Kathy Allen and Ezra Zubrow for their critical comments and encouragement.

References

Cowen, D. J., 1987, GIS vs CAD vs DBMS: What are the differences? Paper presented at the *National GIS Symposium*, San Francisco, CA

Dickenson, G. C., 1973, *Statistical mapping and the presentation of statistics*, (London: Edward Arnold)

Green, S. W. and Ulrich, T., 1977, The statistical description of environmental variables: a mapping approach. Paper presented at the *Annual Meetings of the Society for American Archaeology*, New Orleans, LA

Green, S. W. and Zvelebil, M., 1990, The colonization and agricultural transition of Southeast Ireland. *Proceedings of the Prehistoric Society of London*, October 1990

Higuchi, T., 1983, *The Visual and Spatial Structure of Landscapes*, translated by C.S.Terry. (Cambridge, MA: MIT Press)

Kvamme, K., 1989, Geographic information systems in regional archaeological research design. In *Advances in Archaeological Method and Theory*, edited by M. Schiffer (New York: Academic Press) Volume 1, pp. 139–203

Martens, H., 1890 *Optishes Mass für den Stadtebau* (Bonn: Max Cohn and Son)

McCarty, H. H. and Salisbury, N. E., 1961, Visual comparison of isopleth maps as a means of determining correlations between spatially distributed phenomena. Department of Geography, Iowa State University, Iowa City

Roberts, B. J., 1988, Landscape Archaeology. In *Landscape and Culture: Geographical and Archaeological Perspectives,* edited by J. M. Wagstaff (Oxford: Blackwell), pp. 77–95

Zvelebil, M., Moore, J., Green, S. W. and Henson, D., 1987, Lithic scatters and human behavior: a case from southeast Ireland. In *Mesolithic Northwest Europe: New Trends,* edited by M. Zvelebil, H. P. Blankholm and P. Rowly-Conwy (Sheffield: Sheffield University Press), pp. 108–123

28

An application of remote sensing and GIS in a regional archaeological settlement pattern analysis: the Arroux River valley, Burgundy, France

Scott L. H. Madry and Carole L. Crumley

Introduction

An American multidisciplinary research team, under the general direction of Dr. Carole Crumley of the University of North Carolina at Chapel Hill, has worked in the Arroux River valley region of Burgundy for over fifteen years. The team has been conducting a regional analysis of the interaction of culture and the environment over a time period covering over two thousand years, ranging from the Celtic Iron Age to the modern period. Much of this work has been directed towards developing diachronic settlement models for the various cultural periods and, more generally, to understand the nature of interaction between culture and the environment (Crumley and Marquardt, Chapter 7). A recent volume has been produced detailing the research results compiled by the various team members (Crumley and Marquardt, 1987). One of the primary requirements in conducting this type of analysis, however, is accurate and quantitative data on the physical environment and its relationship to cultural features such as ancient road networks, political boundaries and archaeological sites.

While much work has been done in the project to date in this regard, the results of these extensive efforts exist as tabular data, or on paper maps at different scales, which reside with over ten researchers around the United States and in France. The advent of new computer technologies, specifically the rapidly maturing capabilities of modern remote sensing systems which can now be fully integrated with the spatial analytical capabilities of geographic information systems (GIS), provides us with a powerful new tool with which to expand the scope, utility and capability of such regional analyses. It was the intention of this phase of our continuing research to acquire the most recent and technologically sophisticated remote sensing data currently available, and to begin the integration of these data into a comprehensive GIS for our research area. Existing tabular and paper map data will also be entered into the system. The GIS portion of the project is still ongoing, our results are tentative at best, and much work remains to be done, but the basic modern environmental context of the region has been entered into the GIS, and we are beginning to use this new tool for a broad range of scientific inquiry.

Research area

The general research area is considered to be the valley of the Arroux River, a tributary of the Loire River in France (Figure 28.1). The area has been extensively documented by the research team (Crumley and Marquardt, 1987). The Arroux valley is located in east-central France, within the modern Department of Saône-et-Loire. The region includes the headwaters of the Seine, Loire and Rhône–Saône drainage basins. These three major waterways flow into the English Channel, the Atlantic and the Mediterranean respectively.

The Arroux River valley flows from its headwaters in the southern Morvan mountains (the northernmost extension of the Massif Central) to the Loire River along a northeast-southwest fault line (Figure 28.2). The river valley is surrounded by precambrian igneous and metamorphic rocks. Along these mountain tops is located a series of Celtic hillforts.

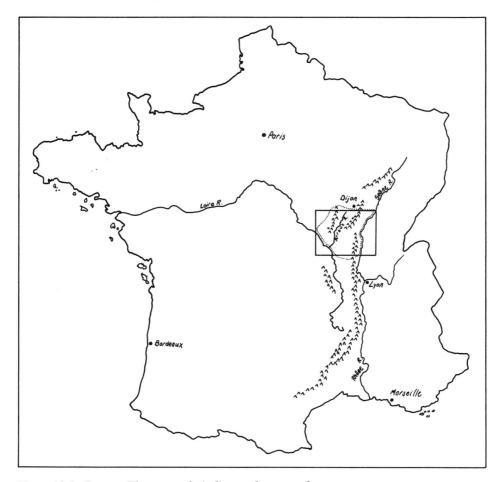

Figure 28.1. France. The rectangle indicates the research area.

For the purposes of this portion of our research, the research area was defined as the area covered by three 1:50,000 topographic maps produced by the Institute Geographique Nationale (I.G.N. numbers 2825, 2826 and 2827). These are the

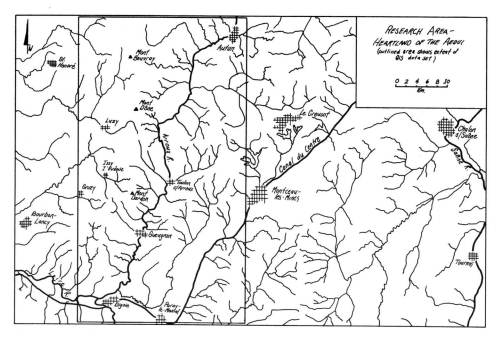

Figure 28.2. Detailed map of the research area.

Autun, Toulon-sur-Arroux and Paray-le-Monial map sheets. This area covers the
Arroux valley from its confluence with the Loire River at the southwest corner, to
just north of Autun at the northeast corner (Figure 28.2). It encompasses an area
of approximately 60 km north-south by 28 km east-west, or 1,680 km square. This
area was chosen because it covers the great majority of the valley, and it was the
area for which digital elevation data were available from the French government.
The French digital elevation data are available by the 1:50,000 map sheet. Since
other important layers were derived from these elevation data, this was decided to
be the general boundary of the GIS database. Some data, such as the SPOT
(Système Probatoire d'Observation de la Terre) satellite data, greatly exceeded this
area, but all GIS layers cover this area completely, with the exception of the geology
layer. This was derived from a 1:80,000 map of the area, which only covers the
northern two thirds of the area. No geological maps are available for the other area
as yet.

Project background

Research in this area has been conducted since the mid 1970s, with an emphasis on
the interplay between regional archaeological settlement patterns and the
environmental make-up of the area. A diverse group of archaeologists, anthropologists, ecologists, climatologists, medievalists, historians and ethnographers have
coordinated its efforts to detail the cultural and environmental components of the
area over a two thousand year time frame (Crumley and Marquardt, 1987).
Particular emphasis has been placed on the Iron Age Celtic component, Iron Age

Gallo-Roman transition period, Early Medieval and modern periods. A series of Celtic hillforts runs along the mountain tops that surround the river valley (Figure 28.2). Foremost among these is Mt. Beuvray, in the northeast corner of the research area. This is the site of Bibracte, once the capital city of the Aedui, the powerful Iron Age Celtic polity whose territory was centred around the Arroux valley. Two other hillforts, Mt. Dône and Mt. Dardon, follow the river to the south of Beuvray, while others are on the other, eastern side of the river. The major city of the valley is Autun, located near the northernmost headwaters of the river. The city was founded by the Romans as Augustodunum, where the Celtic inhabitants of Bibracte were relocated after the Roman conquest. The region has been densely populated since at least the Bronze Age, and is rich in both prehistoric and historic cultural remains. The area at present is generally rural in nature, with much land in pasture for cattle grazing.

The primary threat to cultural resources are modern construction and development, especially a series of gravel mines (sabliers) that run along the Arroux River. These mines have already destroyed significant cultural resources in the area, including the first located Gallo-Roman villa complex, which was located by our research team through ground and aerial survey (Madry and Crumley, 1985).

GIS and settlement pattern modelling in archaeology

The development and use of geographic information systems (GIS) technology is relatively recent, but it is one that is receiving more and more interest from a variety of disciplines interested in regional studies and applications. GIS is finding favour with archaeologists interested in regional settlement pattern analysis and predictive modelling. The basic concept of a GIS is that one creates a computerized 'layer cake' of spatial information, where each layer represents a single, spatial attribute such as roads, streams, soils, elevation, etc. Information in the form of points (such as archaeological sites), lines (such as roads or steams) or polygons (such as soils or geology maps) can be entered into the system through various means. Manual digitizing using a digitizing table is the most common form of data entry from paper maps, but digital data such as digital elevation data, scanned aerial photos or digital satellite data can also be entered into the system. These individual layers are all georeferenced, meaning that they are all entered and stored in the system with the same coordinate system, such as UTM. Once the various 'map layers' are entered into the system, one can conduct a wide variety of display, measurement and analysis operations on the data in ways that are much harder, if not impossible, using traditional methods and paper maps. Useful analysis such as area and distance measurements, coincidence tabulations between layers, and extensive spatial modelling of various layers can be conducted routinely. Hard-copy output in the form of traditional paper maps at various scales can easily be produced. Such maps can represent a single layer, or have various line or point data overlayed (such as roads, streams, UTM coordinates or archaeological sites). Data such as digital elevation data can easily be entered into the system, and new layers, such as slope and aspect, can be derived from the elevation layer. Analysis of the elevation data can include line-of-site, watershed and optimum-route selection processes. Remote sensing data, such as Landsat or SPOT satellite data or digitized aerial photographs, can easily be entered into the system, providing land use/land cover data over very large areas in a most cost-effective manner. In fact, GIS is seen by many as the 'context' within which remote sensing data can truly become useful for regional analysis. This technology has great potential for regional archaeological research and management applications.

There have been several applications of GIS in archaeological research and cultural resources management recently (see the other chapters of this volume). GIS can be seen as the best method available for conducting quantitative regional analysis and archaeological predictive modelling, and great strides are being made in this regard. A volume of over 650 pages on the theory, method and application of archaeological predictive modelling was recently published by the US Bureau of Land Management. This volume devotes a significant portion of its text to GIS (Judge and Sebastian, 1988). GIS provides the researcher with a tool to conduct quantified inquiry into the relationship of environmental and cultural attributes that was unheard of only a few years ago. Much inquiry is currently being made into the proper use (and potential abuse) of GIS. It has the potential of revolutionizing the way that we conduct such regional analyses, and provides us with the ability to model the environmental/cultural interface interactively. It also has the potential for serious abuse, in the form of 'predictive models' based on limited information and insufficient or even no actual ground survey. There is a growing body of such research, and as hardware and software systems become more affordable (Chapter 15 of this volume) this technology will become more and more frequently used.

Remote sensing in archaeology

Research in remote sensing applications in archaeology has been conducted around the world for over fifty years. There are several excellent overviews of the various applications and the technology available in the literature (Ebert, 1984; Sever and Wiseman, 1984; Madry, 1987; Limp et al., 1988). Aerial photographic analysis has been conducted successfully in Western Europe since the end of World War I. Aerial survey and prospecting is a well developed art in Europe, with literally thousands of archaeological sites, roads and other cultural features having been discovered (Goguey, 1968; Agache, 1970; Dassie, 1976). New and more sophisticated digital airborne scanner systems have provided very high spatial resolution data in the Southwest USA, France and elsewhere (Sever and Wiseman, 1984; Tabbagh, 1984). These systems provide digital data like satellite systems, but very high resolution images are obtained, because spatial resolution is defined by the altitude of the aircraft. Data can be acquired on demand, but the cost of data acquisition and processing is high.

Satellite remote sensing can provide enormous amounts of environmental data covering very large areas of land in a very cost-effective manner. Over 63 archaeological projects have used Landsat and other digital data to study prehistoric land use in a variety of different environments around the world, including Mesoamerica, the USA and Europe (Limp et al., 1988). Most of these projects have used the Landsat MSS system, with 80 m spatial resolution, but some have been done using the newer 30 m Landsat Thematic Mapper. These projects go back to the earliest Landsat I MSS projects (Cook and Stringer, 1975), and use Landsat MSS (Madry, 1987), Landsat Thematic Mapper (Johnson et al., 1988), and the new French SPOT system (Inglis et al., 1984). Most of these researchers used the digital data for obtaining synoptic environmental land use/land cover information (Ebert, 1978), although some researchers have attempted correlating site 'signatures' with spectral data recorded by the scanners (Madry, 1987).

The two previous applications using the new SPOT system are of particular interest for this project. The first used the SPOT simulator, an airborne scanner which simulated the spatial and spectral resolution of the satellite system which had not been launched at that time (Inglis *et al.*, 1984). In this study, data were acquired over the Bandelier National Monument.The locations of eighty-six sites were correlated with the imagery values, and a variety of classifications were run on the data. Thematic class maps were produced which were compared with the site locations. One classification process accounted for 45 of the sites, while covering only 25 per cent of the total area of the project.

The second study, and the first to use actual SPOT satellite data for archaeological uses, was conducted over a portion of the Arkansas River Valley in Arkansas (Farley *et al.*, 1988). In this project, the SPOT data were classified and georeferenced using ELAS software, and then entered into the GRASS GIS system. A total of seventy-five archaeological sites in the GIS, which varied widely in size and cultural components, were compared with spectral classes in the SPOT data. The sites were digitized as polygons. High correlations, similar to those of the Inglis *et al.* (1984) study were reported. Three classes in the processed SPOT data accounted for 50 per cent of all sites in only 29.9 per cent of the total area (Farley *et al.*, 1988).

While remote sensing has been used for archaeological applications for some time, it is within the context of GIS that its potential is being realized.

Previous remote sensing research in the region

A significant amount of remote sensing research has been conducted in the research area. This is reported in Madry's chapter of our recent volume (Madry, 1987, pp. 173–236). A multi-scalar method was used, where low-level aerial survey and photography, vertical mapping photographic analysis and Landsat MSS digital satellite data were used to locate sites and old road sections, and to produce a land use/land cover map of the area. This work was successful in several ways. Sites were actually located from the aerial survey, and the first quantification of modern land use/land cover in the region was produced. An attempt was made to correlate known Gallo-Roman villa sites with spectral signatures in the Landsat MSS data but this was generally unsuccessful, although some potentially interesting patterns were seen in the data. It can be reasonably assumed that one reason for the lack of success in locating sites was the 80 m spatial resolution of the MSS data (the only satellite data available at the time). This means that a single pixel of the data covers an area 80 m on a side (or about the size of a major-league baseball field). Clearly the spatial capability of the system was not appropriate for such site location spectral analysis. The technology of remote sensing, both from aircraft and from space, has progressed very rapidly since this phase of our research. Environmental satellite data is now available at 20 and 10 m spatial resolution from the French SPOT system. This new capability should provide us with significantly improved data in two general areas. First of all, we should be able to produce more accurate land use/land cover maps, especially in Western Europe where land use areas are quite small. In fact the SPOT system was specifically designed to be able to conduct such measurements at the scale of European land use. Secondly, we may now actually be able to discriminate spectral differences between individual sites based on anthropomorphic alterations of the land.

Research design and method

The specific purpose of this phase of our long-term research programme was to acquire new SPOT digital satellite data over the research area, and to integrate these data with other environmental and cultural information already available from our research in a geographic information system context. Previous remote sensing research conducted on the project (Madry, 1987) demonstrated the current utility of aerial photograph interpretation and the potential of future satellite digital remote sensing systems. In the rapidly changing field of remote sensing, great advances in technology have been made since our original research, and it was our intention to investigate how these new sources of data could be utilized in our project within a GIS framework, and the general utility of such data for temperate regional archaeological studies.

GRASS GIS database development and analysis

GRASS, the Geographic Resources Analysis Support System, is a general purpose, grid-cell GIS that was developed by the US Army Corps of Engineers Construction Engineering Research Laboratory (USA-CERL) in Champaign, IL (Goran *et al.*, 1988; Madry, 1989). It was originally developed for environmental and land use planning at US Army bases, and was awarded the 'Exemplary Systems in Government' award in 1986 by the Urban and Regional Informations Systems Association (URISA) in 1986, and the 'Army Research and Development Award' in 1988. It operates in the UNIX environment, is written in 'C', and contains about 140,000 lines of code. GRASS is in the public domain, so it is extremely inexpensive (US $ 950 at the time of writing), and source code is distributed along with complete documentation, tutorials and a sample database which includes raster, vector and point data (including archaeological sites). The flexibility of GRASS and the availability of source code provide enormous flexibility for the user, and allow the user to develop new capabilities as required, such as writing new modules or integrating GRASS capabilities with existing databases. GRASS is extremely easy to use and learn, and there is an active user and development community. It has complete digitizing, database development, image analysis and GIS capabilities, as well as extensive graphics, display, statistical and hard copy output. It runs on several common UNIX computer systems, including Masscomp, SUN, AT&T, Tektronix, Apollo, Intergraph, several 386 PCs and the Apple Mac II family (running under the A/UX version of UNIX). Because of its use in military environmental planning activities, it contains several modules which are especially useful in this type of project. Specifically, it has several modules designed for use with point-specific data, which is particularly useful in archaeological site analysis and modelling.

GRASS has been used by a variety of researchers for regional archaeological modelling and management (Farley *et al.*, 1988). GRASS is very appropriate for archaeological researchers for a variety of reasons. One is clearly the low cost of the software and hardware configurations. Users on a budget can start using GIS and remote sensing data in their research with a minimum of cost. It is very easy to use and learn, with both menu-driven and command-line interfaces available. It has modules to read in commonly used digital data from USGS, and Landsat and SPOT digital remote sensing data. Users can develop new applications modules as needed.

Current US archaeological users include the Arkansas Archaeological Survey, Central Washington University, Boston University, Rutgers University, and several governmental agencies with management responsibilities in this area. These include US Army installations and Corps of Engineers districts, the National Guard Bureau, the Soil Conservation Service, and the National Park Service. There is also a growing international user community, with users in France, Spain, the Netherlands, Sweden, Greece, China and Israel.

Project equipment and facilities

The initial work for this project was conducted at the Institute for Technology Development Space Remote Sensing Center, which is located at the NASA John C. Stennis Space Center in Mississippi. The work was done on MASSCOMP 5600 and 5450, and Apple Mac II computer systems. Also used were Calcomp digitizers, Versatek and Tektronix printers, and a 1024 array Megavision digital scanner system.

Additional work was conducted and is continuing at the Cook College Remote Sensing Center of Rutgers University, using a MASSCOMP 5450 computer system, Altek digitizer and Eikonix 4096 array digital scanning device.

GIS database

Various data layers that were considered important to the project research goals were digitized by hand in the GRASS digitizing software package MAP-DEV, at ITD/SRSC and Rutgers. The layers that were developed, in order of completion, were:

(1) geology, taken from the 1:80,000 scale geology maps of the region;
(2) faults, taken from the same geology maps;
(3) ancient road networks, taken from 1:50,000 I.G.N. maps of the region, with the locations of the old roads marked by J. Dowdle of the research team;
(4) modern roads, taken from the same 1:50,000 maps;
(5) modern hydrology (streams and rivers), also taken from the 1:50,000 maps; and
(6) the known Celtic hillforts in the research area.

A second source of data for the database was digital elevation tapes purchased from I.G.N. (the French national mapping agency) covering the three 1:50,000 maps covering the Arroux River valley. These data were provided on digital tape, with a cell size of 97 m. This meant that elevation readings were derived from the maps with a single reading taken every 97 m. These data were read into the ELAS software package, reformatted into GRASS-compatible format, and entered into the system as a digital elevation layer. Aspect (direction of slope) and slope (in degrees) data layers were derived from the original elevation layer using the appropriate GRASS module, so that elevation, slope and aspect layers were created from the original French data tapes.

A third general source of data input into the GIS was scanned map and photograph images. Using a Megavision digital scanning system, we were able to digitize several maps of the area into the GIS. Maps covering the research area dating from 1659, maps of ancient road networks and the three modern 1:50,000 I.G.N. maps were digitally scanned with a 1024 array system. These cell files were then loaded into the GRASS system and georeferenced using the same method as the SPOT images. These maps are of different scales and accuracies, and therefore the registration to the database is only general, but they provide a diachronic aspect to the GIS database that is of interest for such a project, and that is generally lacking in most GIS analysis due to lack of appropriate data.

The fourth source of data input was the SPOT satellite images, which are discussed later.

The current cell layers of the GIS database are:

–elevation
–aspect (derived from elevation)
–slope (derived from elevation)
–SPOT images (20 m false colour infrared)
–land use/land cover map (derived from SPOT image data)
–geology
–faults
–streams
–modern roads
–ancient roads
–known Celtic hillforts
–distances from roads, streams, faults, hillforts, and ancient roads
–1659 and modern maps of the research area

Current vector files include:

–modern roads
–streams
–ancient roads network
–faults
–geology

Current point (sites) files include the locations of the known Celtic hillforts in the area.

Remote sensing data acquisition, analysis and integration

The SPOT system

The SPOT (Système Probatoire d'Observation de la Terre) satellite system is a French commercial earth resources system that was first launched in 1984 (CNES, 1984; Courtois and Weill, 1985). It is in a circular, sun-synchronous, near-polar orbit 832 km in elevation, which has a mean inclination of 98.37 degrees. The satellite orbits the earth just over 14 times per day, and repeats coverage over the same location on the ground every 26 days. The system carries two identical high resolution visible range (HRV) instruments, which use a 'push broom' scanning technique to acquire digital data using charged couple device (CCD) technology. The system acquires data in three multispectral channels with a spatial resolution of 20 m, or in one panchromatic channel with a 10 m spatial resolution, with an 8 bit radiometric resolution (or from 0–255). The spectral bands are 0.5–0.59, 0.61–0.68, and 0.79–0.89 microns for the multispectral, and 0.51–0.73 microns for the panchromatic band. The ground swath is 60 km, which is acquired as $3,000 \times 3,000$ pixels in the multispectral, and $6,000 \times 6,000$ pixels in the panchromatic, mode. The SPOT system also has the ability to acquire data off-nadir by pointing the sensors to the side. This gives the ability to acquire scenes at least once every four days in practice, although the orbital period covers the same location on the Earth's surface every 26 days. This off-nadir capability also makes possible the creation of digital elevation data through traditional photogrammetric parallax methods using two off-nadir data sets. This is particularly useful in developing GIS data sets, and in integrating remote sensing data into a GIS context in areas of the world that do not have digital elevation data. For the price of two SPOT images and some additional data processing, the user can acquire the satellite data for land use/land cover and digital elevation data, from which slope and aspect layers can be derived for a GIS. This can be used as the basis for a combined image processing and GIS database located anywhere in the world. This is a particularly useful capability in areas of the world that do not have digital elevation data available. Unfortunately, at this point, this includes most of the Earth's surface.

All SPOT data for purchase in the United States are marketed and distributed by:

SPOT Image Corp.
1897 Preston White Drive
Reston, VA 22091–4326, USA
Tel. 1–703–620–2200

International marketing and distribution is handled by:

SPOT Image
16 bis, avenue Edouard-Belin
31055 TOULOUSE, CEDEX, FRANCE
Tel. 31–61–53–99–76

SPOT data processing and analysis

Two SPOT digital satellite data tapes were purchased covering the research area. Multispectral data with 20 m spatial resolution were available which were acquired at 10:41 am local time on 9 November, 1986. These data proved to be of limited utility due to thin cloud cover in the centre and southwest corner of the image. There was also a band of cloud shadows across the middle of the research area that was caused by a band of clouds just outside the image. This would have caused significant problems in deriving accurate statistical analysis of the data required for accurate thematic classification for land use/land cover. Therefore, a request was made to SPOT Image to acquire a totally cloud-free image. It was also requested that panchromatic 10 m resolution data be acquired on the same pass if at all possible. These data were successfully acquired on 5 January, 1988, again at 10:41 am local time. These data are contained in three spectral bands in the multispectral mode (20 m resolution), and one band in the 10 m panchromatic mode. Two overlapping 60 × 60 km scenes were acquired, one to the north of the other, thus covering the entire research area and significantly beyond.

These raw data were entered into the GRASS GIS image processing system and a georeferencing process was done using control points, such as road intersections, which could be located visually on the two overlapping images. Multiple control points which were visible on both the SPOT images and the maps were located on the 1:50,000 maps, and their UTM coordinates recorded. The GRASS modules I.POINTS and I.RECTIFY were used to locate the points, compute the RMS error, and then to run the transformation of the raw data to georeference it into the UTM coordinate system used in the GIS. An RMS of less than the 20 m spatial resolution of the system was considered sufficient, and was achieved. The two overlapping SPOT images were combined into a single image using GRASS module 'patch'. This created one single layer from the two overlapping SPOT scenes. A false-colour infrared depiction of the data was made into a GRASS cell map (with colour table) which was entered into the database using the module I.COLORS.

A two-step, unsupervised, maximum likelihood classification was run on the raw multispectral data using GRASS modules I.CLUSTER and I.MAXLIK. The statistics derived were analyzed, and the process was re-run interactively until the final output was considered to be sufficiently accurate when compared with the ground-truthing data acquired in the field (see above). This thematic land use/land cover map was labelled and entered into the GRASS GIS system, along with a false-colour composite image, as additional data layers in the GIS. The data can then be compared with other data layers in the system, display in 3-D, etc.

Plate XII of the scene around Autun shows the detail capable with this 20 m SPOT data. The city itself is clearly visible, along with the small, general aviation airport and the Arroux River. The ancient Roman road network leading into the city is displayed as a vector file in yellow.

Examine the extraordinary feature detail available with these data. Individual field boundaries and hedgerows are clearly visible, as are differentiations in field soils. The Arroux River is unusually high, with low-lying areas along the river clearly being flooded.

Historic map digitization

A copy of a 1659 map of the research area was digitized using a Mega-Vision digital scanning system. This system has a $1,000 \times 1,000$ CCD array device that converts the map or photograph into digital format. This system is useful for digitizing maps, aerial photographs or other analogue data into a format readable by image processing and GIS. The map that was scanned is an original, owned by the first author, which was purchased in Paris in 1983. It was scanned, entered into GRASS, georeferenced and entered into the GIS as a data layer like any other. It was hoped that several maps could be entered into the system to provide us with a means to compare settlement and road patterns in the area over time. Unfortunately, the system was not available for a long enough time to conduct extensive use of this system. There is interesting potential for using such ancillary data, as they can provide the only means of entering such existing information from maps into a GIS. There are obvious problems in referencing these maps to any modern cartographic base such as the UTM system used in this project, but they are the only available data we have of the time and are worth using, even with their associated problems.

Remote sensing ground-truthing field work

Much field work has been conducted over the course of our work in the Arroux River valley, but field research in specific support of this aspect of the work was required and undertaken. Ground-truthing of the research area was conducted over a period from August 3–12, 1987. In preparation for this field work, 8×12 inch hard-copy enlargements of selected areas within the research area were produced from 35mm camera shots of the computer screen image of the SPOT colour infrared images. These were taken into the field, along with 1:50,000 and 1:25,000 maps, vertical aerial photographs and ground-truthing forms. The fields that served as ground-truthing targets were selected for their visibility on the SPOT images, their general distribution throughout the research area, their utility as training sets for specific known land use/land cover types in the region, and their general ease of access by the field crews. Each field plot was located on the imagery, marked on the maps, photographed and the field form was filled out. Plot numbers and Anderson level II classes were assigned to each ground-truthing test plot in the survey area. A total of 63 ground-truthing plots were recorded for this phase of the project. This information was then used in the classification of the digital imagery to produce a land use/land cover layer in the GIS.

GIS analysis and results

While data layers are still being added to the GIS database, some preliminary analysis of the data has begun. Much information has already been gathered within the research area, including descriptive statistical analysis of archaeological sites. What we did not have was quantitative information concerning the relative percentages of the research area that were made up of these attributes such as slope, aspect, elevation, modern land use/land cover, distance from streams or faults, etc. The GRASS GIS analysis provided us with these data through the modules CELL.STATS and REPORT. This provided us with the first truly quantified measurement of the modern environment in the research area, and the relationship between these modern environmental zones and the cultural features in the area.

Land use/land cover change over time

A comparison was made of the 1972 Landsat MSS data and the 1988 SPOT land cover layers in the GIS. This technique gives us quantified data concerning land use/land cover change in the region. It will be possible to update this with additional remote sensing data periodically in the future.

Line-of-sight analysis

The GRASS system includes a line-of-sight analysis module which uses the digital elevation layer. This module allows one to locate a given point and it will compute everywhere in a given radius that will be within sight from a given elevation above that point. This analysis was run on each of the known Celtic hillforts in the research area. Several interesting patterns were noticed. One is that the ancient road networks tend to follow locations that remain in sight of the hillforts, and avoid those routes that are hidden. In the example of Mont Dardon for example, the Celtic road pattern follows the approaches that are visible from the summit (Plate XIII). Portions of the road network of which the exact location is unknown are also correlated with areas that are out of sight of the summit.

Corridor analysis

GRASS has a corridor analysis package that allows one to choose 'optimum' routing between two points, based on weighted data that are derived from existing layers in the database. This was done between Mt. Dardon and Mt. Dône and Mt. Dardon and Mt. Beuvray to determine if the system could be made to approximate the known Celtic road patterns between these hillforts. If this could be done, then sections of the road network that are not known could then be projected, and this new information used to search for site locations that might be located near these and other road segments. It also could provide us with a quantification of the factors that determine road placement of the various cultural periods. Any layer can be used to 'weight' the model, but elevation, slope and soils, or combinations of these types of data, are most commonly used.

Attempts to replicate known Celtic road patterns between hillforts were made using single layers, such as slope, elevation, and the like, as the input. These attempts were generally unsuccessful, indicating that the placement of the roads systems between the hillforts was determined by other, more complex factors. Although it is known that the roads tend to follow the ridge line, and follow paths within the line-of-sight of the hillforts, it was not possible to replicate the known road patterns between the hillforts using a single variable. Work is continuing in this.

Three dimensional image display

Using the digital elevation data layer, GRASS can produce three dimensional projections of any data layer (or sub-area), with vector files overlaid, from any elevation or point of view. Vertical exaggeration of the elevation can be requested, as well as various fields of view. This is a particularly striking visual capability of the GRASS GIS, and while not an analytical function *per se*, it allows the analyst to view the research area, or any portion of it, as if from the air, which can be quite useful. Plate XIV is a 3-D perspective view of the entire research area from an elevation of 30,000 feet. The colour-coded elevation layer is overlaid on the elevation data. The ability to provide a synoptic view of the entire research area can be useful.

Future research directions

It is our intention to use the development of this GIS database as a common basis for future research in the area. We will continue to build new layers in the system as new information about the region becomes available. We will freely offer the information compiled in the database to other researchers who may wish to use it.

Much of the site specific data derived from the research activities is only now being entered into the GIS at the time of writing, and this is the next activity. Detailed site location data, along with numbers of artifacts, cultural affiliation, and other data, will be entered into the database and statistical analysis conducted to develop projective models of site patterning in the Iron Age, Gallo-Roman and Medieval periods. Similar activities will model the road networks to extrapolate the locations of undiscovered road sections and, hopefully, new sites located along them.

New maps showing the areas with the highest probability of site locations will be created, and new layers in the GIS containing these locations will be produced. These will be the subject of future field reconnaissance to test and refine the model. It is hoped that if this model of site occurrence can be fully developed, it might be used as a model for the location, analysis and eventual protection of archaeological resources in the Arroux valley. These data will continue to be provided to the proper authorities of Burgundy for management purposes. The geology map that covers the southern third of the research area will be available in 1991, and will be entered into the system to complete the geological and fault location layers of the GIS.

Digitizing aerial photographic coverage of the area is another top priority. A freedom of information act request has been submitted to the US Defense Intelligence Agency for the purchase of 158 9″×9″ aerial photographs dating from August/September 1945. The photographs cover 100 per cent of the research area at approximately 1:40,000 scale. These military reconnaissance photographs will be digitized and entered into the GIS as a single data layer. They will be classified to produce a land use/land cover map of the area if the photographs are of acceptable quality. Additional single photographs will be manually, photographically analyzed and also digitized with the Eikonix digitizer for individual site location work using digital image processing.

Under a National Geographic Society grant, we are currently working to integrate SPOT satellite 10 m panchromatic data that have been acquired over the study area with the SPOT 20 m data already in the database. This will provide us with additional spatial resolution remote sensing coverage, which may in turn provide us with actual site location capability from space. It will certainly improve our ability to locate accurately specific field-by-field land use/land cover changes in the research area. Additional work funded by the National Geographic Society provided for the acquisition and analysis of ARIES airborne thermal scanner data over several portions of the research area that have higher potential for site locations. These data will be integrated into the GIS database.

Regional climate modelling has become an area of intense interest in our project. Working with regional hydrological data and average global temperature data, we have been able to relate discharge rates of a small stream in the research area with changes in global temperature (Gunn and Crumley, 1989). We hope to utilize the modelling capabilities of GRASS to assemble information about the stream basin (surface cover, vegetation cover, land use patterns, geology, slope, etc.) to model future changes in regional climate due to global warming. Our previous historic work in the area has allowed us to analyze past climate-driven changes in human settlement and land use effectively, especially the Roman Climate Optimum and the Little Climate Optimum, which researchers think parallel anticipated global warming.

Collaboration with our French colleagues, including the Director of the Archaeological Circumscription in Dijon and local researchers, will be continued. We will allow them access to the information, and we hope to have access to their site location data to inform the model better. Demonstrations of the system and its capabilities will be made to French scholars and governmental agencies. It is our sincere hope that such a system will be adopted by the French authorities to manage the enormous responsibility of protecting their vanishing national heritage better. The adoption of geographic information systems, including relational database, remote sensing and statistical capabilities, is a powerful new tool for research and management that should begin to be considered seriously.

Summary and conclusions

This continuation of our long-term research activities in the Arroux River valley of France provided the research team with the first quantification of the natural environment on a regional scale. The SPOT data provided very high spatial resolution modern land use/land cover data (at 20 m spatial resolution) over the area which is nearly an order of magnitude improvement over the previously used Landsat MSS data (with 80 m spatial resolution). This is one of the first applications of SPOT imagery to archaeological analysis to have been attempted worldwide, and to our knowledge the first conducted in Western Europe.

The application of the multi-concept, where multiple types of information at multiple scales, from multiple sources, etc., are integrated and analyzed, is made far more capable by integration of these data in a GIS context. For example, the multiple remote sensing sources used previously, such as aerial photography, airborne scanners and satellite systems such as Landsat and SPOT, provide multiple scales of analysis and site location potential. By integrating these various data sources in a GIS, and comparing the individual spectral values with site locations, soils, etc., the user can build on the synergy of diverse information and capabilities. For example, general environmental data derived from satellite data can be improved in accuracy by using aerial photographs, and soils or geological maps. In the other direction, the products of GIS analysis, such as maps of areas of greatest potential for sites of a given period, can be produced, and the areas can then be flown in light aircraft, or analysis can be conducted using archival aerial photographs for actual site discovery.

The integration of existing and new data in a GIS for our Arroux River valley research area has also provided us with the foundation of a generic tool with which we will begin quantitatively to model the interaction between the environment and culture on a regional scale, and to attempt improved projective modelling of archaeological site and settlement patterns in the study area. We are only now beginning this task, and it will be a long and iterative process. The implementation of this new technology should allow the researchers on the team to continue to model changing patterns of settlement and land use from our research facilities in the United States, and to continue to refine and test our models in the field. It will also allow new data to be integrated with existing data layers as they become available from whatever sources in the future, including new remote sensing data, other researchers' archaeological surveys, and other cultural and environmental data. Models can be developed and tested in ways that were not possible when we began this research in the early 1970s. It is hoped that future continuations of our research activities in the area based on this work will provide us with new insights. It may also assist us in increasing the cost effectiveness of our field research, by providing us with maps of areas of the highest potential for new site discovery, which will hopefully allow us to maximize the efficiency of our field work.

The GRASS system has shown itself to be particularly appropriate to this type of analysis. Because it has several specific modules that relate directly to archaeological site analysis, corridor analysis, the integration of remote sensing imagery, weighted modelling, expert systems capability, and the generation of hardcopy at any scale, the GRASS system is superbly suited to such regional archaeological projects. The facts that it is public domain software (available with source code and documentation for US $ 950), very easy to learn and use, and runs on a variety of computer systems, (including the low cost Apple Mac II, Sun, Compaq 386, Masscomp, and AT&T systems), make it realistic that such systems can be implemented in other research projects such as ours, or in international academic, governmental and management contexts.

The wider availability and operational maturity of cost effective hardware/software environments for GIS technology could have an enormous impact on how we conduct regional analysis and cultural resource management. We must ensure that our methodological and theoretical capabilities at least attempt to keep pace with the technical tools now in our hands. With enhanced theoretical insights into regional modelling (Crumley and Marquardt, Chapter 7 of this volume), and the rapidly maturing technology of GIS, we can provide better information for making decisions concerning the conservation and protection of our rapidly dwindling cultural resources. It is now up to us to learn how to use these conceptual and technical tools properly.

Acknowledgements

Grateful acknowledgement is given to the people of the Arroux River valley for their kind assistance and continuing support of this research project. Special mention is due to Jean-Claude Jacquet, Claude and Laure Dejours and the members of Les Amis du Dardon. M. Pierre Beson, President of SPOT Image Corporation provided us with cloud-free coverage of the area with a rapid delivery. Dr. Fred Limp, Associate Director of the Arkansas Archaeological Survey, provided much assistance in the completion of this project. Special thanks are due to Henri Gaillard de Semainville, Director of the Antiquities Historiques de la Bourgogne. Field work for this project was conducted by Madry and Crumley, assisted by Bill Marquardt. The Institute for Technology Development's Space Remote Sensing Center and Rutgers University Cook College Remote Sensing Center provided in-kind assistance in the use of facilities and equipment. Jim Gasparich drafted the illustrations for this chapter.

References

Agache, R., 1970, Detection aerienne de vestiges protohistoriques gallo-romains et medievaux dans le bassin de la Somme et ses abords. *Bulletin de la Societe de Prehistoire*, special issue, **7**

CNES, 1984, *SPOT Satellite-based Remote Sensing System*. Toulouse, France

Cook, J. P. and Stringer, W., 1975, Feasibility study for locating archaeological village sites by satellite remote sensing. Final report for Contract NAS5–21833 ERTS project 100-N. Prepared for NASA Goddard Space Flight Center, Greenbelt. MD Copy on file at Geophysical Institute, University of Alaska, Fairbanks

Courtois, M. and Weill, G., 1985, The SPOT Satellite. In *Monitoring Earth's Ocean, Land, and Atmosphere from Space—Sensors, Systems, and Applications*, edited by A. Schnapf, Volume 97 of Progress in Astronautics and Aeronautics series (New York: American Institute of Aeronautics and Astronautics)

Crumley, C. L. and Marquardt, W., 1987, *Regional Dynamics: Burgundian Landscape in Historical Perspective* (San Diego: Academic Press)

Dassie, J., 1976, *Manuel d'archeologie aerienne* (Paris: Editions Technip)

Ebert, J., 1978, Remote Sensing and Large-Scale Cultural Resources Management. In *Remote Sensing and Nondestructive Archaeology*, edited by T.R. Lyons and J. Ebert (Washington, D.C.: National Park Service)

Ebert, J., 1984, Remote Sensing Applications in Archaeology. In *Advances in Archaeological Method and Theory*, edited by M. Schiffer, Volume 1, (New York: Academic Press) pp. 293–362

Farley, J., Limp, W. F. and Lockhart, J., 1988, The Archaeologist's Workbench: Integrating GIS, Remote Sensing, EDA, and Database Management. Invited paper presented at the Symposium on *Geographic Information Systems Applications in Archaeology*, 53rd Society of American Archaeology meetings, Phoenix, AZ

Goguey, R., 1968, *De l'avion à l'archeologie* (Paris: Editions Technip)

Goran, B., Westervelt, J., Schapero, M. and Johnson, M., 1988, *GRASS Users Manual*, Version 3.0. (Champaign, IL: US Army Corps of Engineers, Construction Engineering Research Laboratory, Environmental Division)

Gunn, J. and Crumley, C. L., 1989, Global energy balance and regional hydrology: A Burgundian case study. Discussion report on fluvial systems. *European Conference on Landscape-Ecological Impact of Climatic Change*, The Netherlands December 3–7, 1989

Inglis, M. T., Budge, T. and Ebert, J., 1984, Preliminary evaluation of Simulated SPOT Data for Cultural Resources Assessment and Management, New Mexico. In *SPOT Simulation Applications Handbook*. (Falls Church, VA: American Society of Photogrammetry and Remote Sensing) pp. 73–82

Johnson, J., Madry, S. and Sever, T., 1988, Remote Sensing and GIS Analysis in Large Scale Survey Design in North Mississippi. *Southeastern Archaeology* 7, pp. 124–131

Judge, W. J., and Sebastian, T., 1988, *Quantifying the Present and Predicting the Past: Theory, Method, and Application of Archaeological Predictive Modeling.* (Denver, CO: US Department of the Interior, Bureau of Land Management Service Center)

Limp, W. F., Parker, S. and Farley, J., 1988, The Use of Multispectal Digital Imagery in Archaeological Investigations. Final report submitted to the Southwestern Division of the US Army Corps of Engineers, Dallas, Texas

Madry, S. L. H., 1987, A Multiscalar Approach to Remote Sensing in a Temperate Regional Archaeological Survey. In *Regional Dynamics: Burgundian Landscapes in Historical Perspective*, edited by C. L. Crumley and W. H. Marquardt, (San Diego: Academic Press)

Madry, S. L. H., 1989, Geographic Resources Analysis Support System (GRASS): An Integrated, UNIX based, Public Domain GIS and Image Processing System for Resource Analysis and Management. In *Proceedings of GIS/LIS '89*, Orlando, FL, November 1989

Madry, S. L. H. and Crumley, C., 1985, La Grenouillere Perdu, Echos du Pass. Revue Periodique de l'association Les Amis du Dardon, **54**

Sever, T. and Wiseman, J., 1984, Remote Sensing and Archaeology, Potential for the Future. NASA—Earth Resources Laboratory, Stennis Space Center

Tabbagh, A., 1984, Evaluation of the Use of Thermal Airborne Prospection for Archaeology During the Last Ten Years. *Proceedings of IGARSS '84 Symposium*, Strasbourg, France. August 27–30, 1984, pp. 205–208

PART V

Conclusions

29

Interpreting space

Kathleen M.S. Allen, Stanton W. Green and Ezra B.W. Zubrow

Interpreting space

The utility of GIS for archaeology rests upon a simple concept, a spatially referenced database. Despite the complex mathematical algorithms, and computer software and hardware GIS is built upon, its basic contribution to archaeology is that it offers a methodology for representing the three-dimensional world within which archaeologists think and practise. Spatial modelling, sampling, surveying, excavating and data management all require three-dimensional control. This is the reason that archaeologists are jumping onto the GIS bandwagon along with their fellow geographers and geologists, so that the *International Journal of Geographical Information Systems* and *GIS World* are the fastest growing journals in the academic world. We hope that this book, through its general discussion and case studies, has clearly detailed the basis for this widespread enthusiasm. However, as in any bandwagon effect, we must maintain our critical edge.

The most dangerous potential pitfall, we believe, is the inclination to allow a powerful methodology with its accompanying techniques to drive the research and practice of a discipline. As in the case of many statistical methods, the power of grinding out the data and producing output can take on a life of its own. With GIS this might be particularly tempting because of the flexible and pleasing output it can render. Camera-ready three-dimensional, multi-colour maps can make one feel that one has done one's job—and done it well. For academics this might include writing a research report, publication or grant proposal; for someone working in a management context, this could involve a site or regional evaluation. In either case, we caution against the use of GIS as an end in itself. Good research and management is based on asking good questions—something GIS does not do for us. Plain and simple, archaeological significance derives from the questions one asks. The importance of GIS is that it provides ways of asking sophisticated questions—it is, of course, up to archaeologists to make good use of this methodology.

A second critical area involves GIS methodology itself. As we discuss in the chapter on method and theory, and as is demonstrated throughout the case studies, there are at least three different kinds of GIS and within these a variety of software and hardware options. In essence, the method is driven by its algorithms, software basis (raster, vector, object) and hardware (including the type of display device used, its resolution and colour spectrum). The theoretical implications that derive from this diversity suggest that certain systems may be more appropriate for some problems than others. One could imagine, for example, that raster systems may work best for excavations that rely primarily on grid provenance. Vector systems could be required for excavations that point-provenance most objects. The newly developing object-oriented GIS hold potential for the mapping and analysis of historical structures and regions. In this case, a house can actually be read in along with its characteristics, rather than having to outline it with points and lines (vectors) or fitting it into a grid. Quite simply, one must choose an appropriate GIS for one's problem. This, however, is not quite as simple as it might seem.

First, despite their elegance and relative 'user friendliness', GIS do require significant study. Their potential rests to a great extent on the user's experience. Moreover, expert users can often create uses and even functions as they work with a familiar GIS and data set. Second, data sets are, in many cases, created for specific GIS. This can lead to the problem of asking a question that one cannot answer with the chosen GIS database. Finally comes the practical aspect of being able to purchase and maintain multiple GIS systems. As we note in the Appendix, GIS and their related equipment are considerable investments—and this is especially true given the limited resources of most archaeologists. Even in the case of federal agencies such as the National Park Service, one finds the perceived need to commit to a particular system.

Much of the solution to these problems would seem to require making GIS and their databases accessible and compatible. Methods for converting raster-based data to vector, and vice versa, are currently available and being developed further. ERDAS, for example, now provides LIVELINK as a means of integrating raster-based satellite imagery and vector-based aerial photographic data. In addition, some GIS are being made more flexible with regard to the types of data matrices they can 'hook' onto. Many GIS are now providing hooks onto a variety of databases (Zubrow, Chapter 16). Compatibility, however, does not solve all of the problems outlined above. All conversions involve error and cause rectification and accuracy problems (gridding a road is a very good example). At the same time, with conversion possibilities and more GIS compatibility, cooperation between different researchers or agencies becomes possible.

A second broad set of problems involves the accessibility of the databases themselves. As Stine and Stine report (Chapter 5), basic issues of freedom of information and data ownership are just beginning to come forth. Geographic knowledge has historically been one of the most powerful forms of information, and this is especially true as we continue to connect the world economically and electronically. For this reason Abler (1989) states that we may be entering a period in which geographers and other geographic thinkers (among whom we would include archaeologists) may become the most powerful of the technological and intellectual elite. Similar to Ptolemy and the geographers of his time, and the explorers of the 14th and 15th centuries, controllers of geographic knowledge will hold much sought-after information. The US State Department relies heavily on its department of geography for its interpretation of world political events; environmental disasters, such as the oil spill in Prince William, Alaska, are monitored by GIS (illustrated on the cover of the November 1989 issue of *Photogrammetric Engineering & Remote Sensing*); and our everyday travel is conditioned by meteorological maps derived from GIS. As the value of geographic information soars, so does the political and economic competition for it. The Information Industry Association reports that revenues for on line databases will rise to $11.5 billion by 1993. At present, four hundred service organizations now repackage raw governmental data (Archer and Croswell, 1989). Who owns the data? Should a private company be allowed to buy the digital line graphs of a state and charge for their use? Should a national government restrict access to GIS databases? The answers to these questions rest on political values and surely will be developed through the world's judicial system. However, we would guess that most, if not all, current GIS practitioners have run into problems of data access, and there are few things more frustrating than not having access to or having to create an already existing digital map overlay.

The future of GIS and archaeology

We have no doubt that GIS will become increasingly popular in archaeology. Moreover, we expect that it will quickly become a regular member of the archaeologist's toolkit and that before long all archaeological academic and management units will incorporate or have access to GIS and related databases. The question then is how to make the most of this new aspect of archaeology.

Several areas would seem to hold particular promise for the use of GIS in archaeology. Primary to the archaeological endeavour would be the incorporation of time into our spatial models. Map overlays provide an analogue for stratigraphy and, by inference, time. Three-dimensional control of archaeological sites allows for the potential of tracing the pattern of artifacts and the human behaviour associated with this pattern. This might involve the possibilities of computerized cross-mending of ceramics and refitting of lithics; the tracing and interpretation of features; the inference of particular activities as they change (or stay the same) through time; the cross-dating and comparison of sites of a region and their components.

Several of these GIS fantasies involve material culture, which marks a second area of particular importance to archaeology. In addition to the more typical categories of lithic and ceramic artifacts, the recent burgeoning of historic archaeology holds great promise with regard to spatial analysis. Hasenstab and Resnick's farmstead research (Chapter 23) is a fine example of the broad-ranging land use analysis available to the archaeologist when s/he can control the spatial extent of the database. We can imagine archaeology enhancing its contribution to land use and settlement studies through the application of GIS to diachronic databases.

Finally, we would like to make a few remarks on the promise of GIS for increasing the accessibility of archaeological data. Leaving aside for the moment the compatibility and freedom of information problems we discussed above, the development of regional and site level GIS should improve our ability to communicate with one another through the sharing of our data. In such sharing lies tremendous opportunity for the regional syntheses we all yearn for. Regional databases become attainable, maintainable and updatable. Site excavation data can be integrated with regional site survey data, improving comprehensiveness and accessibility. In addition to the obvious research advantages of these possibilities, the development of such cooperation would make archaeology much more accessible to the environmental assessment process.

The development of GIS by state and federal agencies should profoundly improve the position of archaeology in governmental land use planning and environmental monitoring. At a regional level, archaeological databases will be encouraged and be increasingly incorporated (or at least made compatible with) statewide and federal GIS databases. This will diminish the expense and increase the value of connecting archaeological data with other environmental and cultural data. All this will benefit the archaeologist's cause.

We close our discussion by again highlighting the central point of our critique of archaeology and GIS. The enthusiasm exuded by us and the contributors to this volume is sparked by the potential of GIS for solving archaeological problems. The critical warning is that the problems are indeed archaeological and the method—as powerful as it is—is for us to use. We believe that its promise is manifold and hope that this book encourages the reader to evaluate GIS critically for use in their pursuit of archaeology.

References

Abler, R., 1989, Plenary address on GIS. *GIS conference*, University of South Carolina, Columbia, SC

Archer, H. and Croswell, P.L.G., 1989, Public access to Geographic Information Systems: an emerging legal issue. *Photogrammetric Engineering & Remote Sensing*, **55**, 1575–1581

Appendix

GIS acronyms

The following lists acronyms pertaining to GIS including related computer hardware and software, and federal agencies active in supporting GIS and/or GIS databases. References include chapters in this volume (noted by author and chapter number) and several other recent works. For a comprehensive list of GIS software systems we refer the reader to Table 15.1.

ACSM	American Congress on Surveying and Mapping	Parker (1989)
ASCS	Agricultural stabilization and conservation service	FICCDC (1988)
AM/FM	Automated mapping/facilities management	Parker (1989)
ANSI	American National Standards Committee	Date (1986)
APA	American Planning Association	Parker (1989)
ARC/INFO	Trademark for GIS from ESRI	Parker (1989)
ASPRS	American Society for Photogrammetry and Remote Sensing	Parker (1989)
AVHRR	Advanced Very High Resolution Radiometer	Parker (1989)
BIA	Bureau of Indian Affairs	FCCDIC (1988)
BLM	Bureau of Land Management	FCCDIC (1988)
BM	Bureau of Mines	FCCDIC (1988)
BOR	Bureau of Reclamation	FCCDIC (1988)
BPI	Bits per inch	Madry, Chapter 15
CAD	Computer aided/assisted drafting, design, drawing	Madry, Chapter 15
CAE	Computer aided/assisted engineering	Parker (1989)
CAM	Computer aided/assisted manufacturing	Parker (1989)
CAMA	Computer assisted mass appraisal	Archer and Croswell (1989)
CCD	Charged couple device	Madry, Chapter 15
CCT	Computer compatible tape	Madry, Chapter 15
CD-ROM	Compact disk-read only memory	Madry, Chapter 15
CGA	Colour graphics adapter	Welch (1989)
CGIS	Canadian GIS	Marble, Chapter 2
COE	Corps of Engineers	FICCDC (1988)
COGO	Coordinate geometry	Stine and Decker, Chapter 12
CPU	Central processing unit	Madry, Chapter 15

CRT	Cathode ray tube	Madry, Chapter 15
DBMS	Database management system	Archer and Croswell (1989)
DBTG	Database task group	Date (1986)
DC	Data communications	Date (1986)
DDL	Data definition language	Date (1986)
DEM	Digital elevation model	Parker (1989)
DIME	Dual Incoded Mapping Entry	
DLG	Digital line graph	Savage, Chapter 3
DMAP	Defense Mapping Agency	FICCDC (1988)
DML	Data manipulation language	Date (1986)
DMS	Desktop mapping system	Welch (1989)
DOE	Department of Energy	FICCDC (1988)
DOS	Disk operating system	Parker (1989)
DPI	Dots per inch	Madry, Chapter 15
DRAM	Dynamic RAM	Parker (1989)
DTM	Digital terrain model	Parker (1989)
EDA	Exploratory data analysis	Farley et al., Chapter 13
EGA	Enhanced graphic adapter	Welch (1989)
EOS	Earth Observation System	Ehlers et al., (1989)
EOSAT	Earth Observing Satellite	Madry, Chapter 15
EPA	Environmental Protection Agency	FICCDC (1988)
ESRI	Environmental Systems Research Institute	Zubrow, Chapter 15
ETL	(US Army) Engineer Topographic Laboratories	FICCDC (1988)
FICCDC	Federal Integrating Coordinating Commission on Digital Cartography	Stine and Stine, Chapter 5
FM	Facilities management	Parker (1989)
FOIA	Freedom of Information Act	Stine and Stine, Chapter 5
FS	Forest Service	FICCDC (1988)
FWS	(US) Fish and Wildlife Service	FICCDC (1988)
GB	Gigabyte	Parker (1989)
GBF	Geographic base file	Parker (1989)
GCDB	Geographic coordinate data base	Parker (1989)
GCP	Ground control points	Welch (1989)
GDMP	Geographic data management procedures	Dangermond (ms.)
GEMS	Global Environment Monitoring System	Parker (1989)
GPS	Global Positioning Satellite	Madry, Chapter 15
GRASS	Geographic Resources Analysis Support System	Madry, Chapter 15
GRID	Global Resource Information Database	Parker (1989)
HBMC	Historical Buildings and Monuments Commission (UK)	Harris and Lock, Chapter 4
HRV	High resolution visible (see SPOT)	Ehlers et al., (1989)
ICA	International Cartographic Association	Parker (1989)
IMS	Information management system	Date (1986)
INGRES	Integrated graphic and retrieval system	Date (1986)
I/O	Input/output	

IP	Image processing	Parker (1989)
IRM	Information resources management	Parker (1989)
KB	Kilobyte	Parker (1989)
LAN	Local area network	Date (1986)
LIDES	Local interactive digitizing and editing system	Madry, Chapter 15
LIS	Land information system	Ehlers *et al.*, (1989)
MAP	Map analysis package (other versions: MAPCGI, PMAP, PCMAP)	Savage, Chapter 26
MB	Megabyte	Parker (1989)
MC&G	Mapping, charting and geodesy	Parker (1989)
MSS	Mass spectral scanner	Stine and Decker, Chapter 12
NAR	National Archaeology Record	Harris and Lock, Chapter 4
NASS	National Agricultural Statistics Service	FICCDC (1988)
NCGA	National Computer Graphics Association	Parker (1989)
NCGIA	National Center for Geographic Information and Analysis	Ehlers *et al.*, (1989)
NCIC	National Cartographic Information Center	Parker (1989)
NDCD	National Digital Cartographic Database	Stine and Decker, Chapter 12
NEDIS	National Environmental Satellite, Data and Information Service	FICCDC (1988)
NGDC	National Geophysical Data Center	FICCDC (1988)
NMAS	National Map Accuracy Standard	Marozas and Zack, Chapter 14
NMFS	National Marine Fisheries Service	FICCDC (1988)
NMR	National Monuments Record (UK)	Harris and Lock, Chapter 4
NOAA	National Oceanic and Atmospheric Agency	FICCDC (1988)
NORDA	(US Navy) Naval Ocean Research and Development Activity	FICCDC (1988)
NOS	National Ocean Service	FICCDC (1988)
NPS	National Park Service	FICCDC (1988)
NTIS	National Technical Information Service	Stine and Stine, Chapter 5
OS	Ordnance Survey (UK)	Harris and Lock, Chapter 4
OSL	Operation specification language	Date (1986)
OSMIRE	Office of Surface Mining Reclamation and Enforcement	FICCDC (1988)
PMAP	(see MAP)	Parker (1989)
QBE	Query by example	Date (1986)
QBF	Query by form	Date (1986)
QC	Quality control	Campbell and Mortenson (1989)

RAM	Random access memory	Madry, Chapter 15
RCHME	Royal Commisson on the Historical Monuments of England	Harris and Lock, Chapter 4
RGB	Red/green/blue	Madry, Chapter 15
RID	Record identification	Date (1986)
RS	Remote sensing	Farley *et al*., Chapter 13
S	Statistical software	Williams *et al*., Chapter 21
SAS	Statistical Analysis System	Savage, Chapter 3
SCS	Soil Conservation Service	FICCDC (1988)
SEM	Scanning electron microscope	Welch (1989)
SMR	Sites Monuments Records (UK, Ireland)	Harris and Lock, Chapter 4
SOC	Spatial object classes	Menon and Smith (1989)
SOL	Spatial object language	Menon and Smith (1989)
SOS	Spatial object structure	Menon and Smith (1989)
SPOT	Satellite Pour l'Observation de la Terre	Stine and Decker, Chapter 12
SPSS	Statistical Package for the Social Sciences	Savage, Chapter 3
SQL	Standard query language	Ehlers *et al*., (1989)
SVF	Single variable files	Marozas and Zack, Chapter 14
TB	Terrabyte (1012)	Parker (1989)
TIGER	Topological Integrated Geographic Encoding and Referencing	Stine and Decker, Chapter 12
TIN	Triangular irregular network	Marozas and Zack, Chapter 14
TM	Thematic Mapper	Ehlers *et al*., (1989)
TVA	Tennessee Valley Authority	FICCDC (1988)
UDL	Unified data language	Date (1986)
URA	User requirement analysis	Guptill (1989)
USAF	US Air Force	FICCDC (1988)
USGS	US Geological Survey	FICCDC (1988)
UTM	Univeral Transverse Mercator	Welch (1989)
VDU	Video display unit	Madry, Chapter 15
VGA	Video graphics adapter	Welch (1989)
VMS	Virtual memory system	Madry, Chapter 15
WORM	Write once, read many	Parker (1989)

References

Archer, H. and Croswell, P.L., 1989, Public access to Geographic Information Systems. *Photogrammetric Engineering & Remote Sensing*, **55**, 1575–1581

Campbell, W.G. and Mortenson, D.C., 1989, Ensuring the quality of geographic information: a practical application of quality control, *Photogrammetric*

Engineering & Remote Sensing, **55,** 1613–1618

Dangermond, J., ms., GIS vs. CAD technology: Which is better for automated mapping? Environmental Systems Research Institute, Redlands, CA

Date, C. J., 1986, *An introduction to database systems: volume 1* (Don Mills, Ontario: Addison-Wesley)

Ehlers, M., Edwards, G. and Bedard, Y., 1989, Integration of remote sensing with geographic information systems: a necessary evolution. *Photogrammetric Engineering & Remote Sensing,* **55,** 1619–1627

Federal Interagency Coordinating Committee on Digital Cartography (FICCDC), 1988, *A summary of GIS activities in the federal government*

Guptill, S., 1989, Evaluating geographic information systems technology. *Photogrammetric Engineering & Remote Sensing,* **55,** 1583–1587

Menon, S. and Smith, T. R., 1989, A declarative spatial query processor for geographic information systems. *Photogrammetric Engineering & Remote Sensing,* **55,** 1593–1600

Parker, H.D., 1989, *The Parker: Geographic Information System Technology in 1989,* (Ft. Collins, Colorado: GIS World, Inc.)

Welch, S., 1989, Desktop mapping with personal computers, *Photogrammetric Engineering & Remote Sensing,* **55,** 1651–1662

Index